HATUR

Springer Series on
Atoms+Plasmas

1

Springer Series on
Atoms+Plasmas

Editors: G. Ecker   P. Lambropoulos   H. Walther

Joachim Kessler

Polarized Electrons

Second Edition

With 157 Figures

Springer-Verlag
Berlin Heidelberg New York Tokyo

Professor Dr. Joachim Kessler

Westfälische Wilhelms-Universität Münster, Physikalisches Institut, Domagkstraße 75
D-4400 Münster, Fed. Rep. of Germany

52885549

PHYS
Sep lac

Series Editors:

Professor Dr. Günter Ecker

Ruhr-Universität Bochum, Institut für Theoretische Physik, Lehrstuhl I
Universitätsstraße 150
D-4630 Bochum-Querenburg, Fed. Rep. of Germany

Professor Peter Lambropoulos, Ph. D.

University of Crete, P. O. Box 470, Iraklion, Crete, Greece, and

Department of Physics, University of Southern California, University Park
Los Angeles, CA 90089-0484, USA

Professor Dr. Herbert Walther

Sektion Physik der Universität München, Am Coulombwall 1
D-8046 Garching/München, Fed. Rep. of Germany

The first edition was published in 1976 in Texts and Monographs in Physics

Polarized Electrons

ISBN 3-540-07678-6 Springer-Verlag Berlin Heidelberg New York
ISBN 0-387-07678-6 Springer-Verlag New York Heidelberg Berlin

ISBN 3-540-15736-0 2. Auflage Springer-Verlag Berlin Heidelberg New York Tokyo
ISBN 0-387-15736-0 2nd edition Springer-Verlag New York Heidelberg Berlin Tokyo

Library of Congress Cataloging in Publication Data. Kessler, Joachim, 1930–. Polarized electrons. (Springer series on atoms and plasmas; v. 1). Bibliography: p. Includes index. 1. Electrons–Polarization. I. Title. II. Series. QC793.5.E628K47 1985 539.7′2112 85-14804

© Springer-Verlag Berlin Heidelberg 1976 und 1985
Printed in Germany

The use of registered names, trademarks, etc. in this publication does not imply, even in the absence of a specific statement, that such names are exempt from the relevant protective laws and regulations and therefore free for general use.

Monophoto typesetting, offset printing, and bookbinding: Brühlsche Universitätsdruckerei, Giessen

Preface to the Second Edition

The rapid growth of the subject since the first edition ten years ago has made it necessary to rewrite the greater part of the book. Except for the introductory portion and the section on Mott scattering, the book has been completely revised. In Chap. 3, sections on polarization violating reflection symmetry, on resonance scattering, and on inelastic processes have been added. Chapter 4 has been rewritten, taking account of the numerous novel results obtained in exchange scattering. Chapter 5 includes the recent discoveries on photoelectron polarization produced by unpolarized radiation with unpolarized targets and on Auger-electron polarization. In Chap. 6, a further discussion of relativistic polarization phenomena has been added to the book. The immense growth of polarization studies with solids and surfaces required an extension and new presentation of Chap. 7. All but one section of Chap. 8 has been rewritten and a detailed treatment of polarization analysis has been included.

Again, a nearly comprehensive treatment has been attempted. Even so, substantial selectivity among the wide range of available material has been essential in order to accomplish a compact presentation. The reference list, selected along the same lines as in the first edition, is meant to lead the reader through the literature giving a guide for finding further references.

I want to express my indebtedness to a number of people whose help has been invaluable. In addition to the persons referred to in the first edition I mention the coworkers who have joined my group in the meantime. Their enthusiasm and competence has been crucial to the success of our work. Among the results added to this edition are those obtained by Drs. O. Berger, K. Franz, R. Möllenkamp, G. Schönhense, A. Wolcke, and W. Wübker in their theses. The close cooperation on the theoretical aspects of polarization phenomena with Dr. K. Blum and Dr. K. Bartschat was an invaluable asset for our polarized-electron studies. The time-consuming task of updating my lectures and transforming them into a book was achieved during a stay at the National Bureau of Standards (NBS) in Gaithersburg, Maryland. I greatly enjoyed the warm hospitality of the Electron Physics Group which has played a fundamental role in the rapid expansion of polarized-electron physics in the past decade. I gratefully acknowledge helpful comments and many stimulating discussions, both in the past and present, with members of this group, in particular with the group leader Dr. R. J. Celotta and with Dr. D. T. Pierce. The manuscript has benefitted again from constructive suggestions by Dr. M. Reading, who critically, and with great understanding, read the lectures underlying this book. Warm thanks are due to my secretary H. Nicolai, who typed the manuscript with unique reliability and, despite innumerable changes, never lost her patience. I very much appreciate the accuracy of the illustrations made by W. David and the photographical work by K. Brinkmann. I am also grateful to the secretaries at NBS for their assistance in completing the typescript. The help of M. Chirazi in proofreading is gratefully acknowledged. Last but not least, I am pleased to acknowledge the constructive cooperation with Dr. H. Lotsch and R. Michels, Springer-Verlag. The project has been supported by Sonderforschungsbereich 216 of the Deutsche Forschungsgemeinschaft.

Washington D.C., January 1985   *Joachim Kessler*

From the Preface to the First Edition

This book deals with the physics of spin-polarized free electrons. Many aspects of this rapidly expanding field have been treated in review articles, but to date a self-contained monograph has not been available.

In writing this book, I have tried to oppose the current trend in science that sees specialists writing primarily for like-minded specialists, and even physicists in closely related fields understanding each other less than they are inclined to admit. I have attempted to treat a modern field of physics in a style similar to that of a textbook.

The presentation should be intelligible to readers at the graduate level, and while it may demand concentration, I hope it will not require deciphering. If the reader feels that it occasionally dwells upon rather elementary topics, he should remember that this pedestrian excursion is meant to be reasonably self-contained. It was, for example, necessary to give a simple introduction to the Dirac theory in order to have a basis for the discussion of Mott scattering – one of the most important techniques in polarized-electron studies.

This monograph is intended to be an introduction to the field of polarized electrons and not a replacement for review articles on the individual topics discussed. It does not include electron polarization in β decay, a field which has been covered in other books. Areas such as electron spin resonance, in which it is not the spins of *free* electrons that are oriented, are beyond the scope of this book. Well-established areas, like Mott scattering, have naturally been treated in more detail than areas that are just starting to develop, such as high-energy electron scattering. Ideas or general results that have not been quantitatively established, theoretically or experimentally, have not been considered, since physical results must be put on a quantitative basis.

In keeping with the introductory character of the book, the main purpose of the reference lists is to aid the reader in completing or supplementing the information in certain sections. The newcomer to the field should refer to the review articles wherever they exist. Primary sources have been cited if they are directly referred to in the text or if they have not yet been listed in review papers or other references.

It is a pleasure to express my gratitude to the many people who have contributed to the completion of this project. Several sections have been considerably influenced by the ideas and achievements of my coworkers – particularly Drs. G. F. Hanne, U. Heinzmann, and K. Jost – with whom, over the years, I have studied many of the topics discussed. The generous hospitality of the Joint Institute for Laboratory Astrophysics gave me the chance to write this book. The stimulating atmosphere of JILA which I enjoyed during my stay as a Visiting Fellow provided the ideal setting for this project. I gratefully acknowledge the excellent work of the JILA editorial office; thanks to the numerous helpful suggestions from L. Volsky and the typing skill of G. Romey, the transformation (in record time) of my stacks of messy, marked-up sheets into a beautiful manuscript was a joy to behold. I am particularly grateful to Dr. M. Lambropoulos who was kind enough to read the entire manuscript; her constructive criticisms have improved it considerably. Discussions with Prof. H. Merz in Münster and with many colleagues in Boulder have helped to clarify several passages. I appreciate the application and conscientiousness of H. Gerberon and B. Göhlsdorf who prepared most of the illustrations. I am also grateful for the assistance of E. Russel and Dr. C. B. Lucas in translating a number of my lectures which I used in preparing parts of the manuscript. Finally, I wish to thank those listeners, at home and abroad, who, by their reactions to my lectures, have helped to clarify this presentation.

Boulder, Colorado, August, 1975   *Joachim Kessler*

Contents

1. Introduction

1.1 The Concept of Polarized Electrons

An ensemble of electrons is said to be polarized if the electron spins have a preferential orientation so that there exists a direction for which the two possible spin states are not equally populated. Reasons are given for the interest in polarized electrons.

In early experiments with free electrons the direction of their spins was seldom considered. The spins in electron beams that were produced by conventional methods, such as thermal emission, had arbitrary directions. Whenever the spin direction played a role, one had to average over all spin orientations in order to describe the experiments properly.

Only in recent years has it been found possible to produce electron beams in which the spins have a preferential orientation. They are called polarized electron beams in analogy to polarized light in which it is the field vectors that have a preferred orientation. To put it more precisely: An electron beam (or any other electron ensemble) is said to be polarized if there exists a direction for which the two possible spin states are not equally populated.

If all spins have the same direction one has the extreme case of a totally polarized ensemble of electrons (Fig. 1.1). If not all, but only a majority of the spins has the same direction, the ensemble is called partially polarized.

There are many reasons for the interest in polarized electrons. One essential reason is that in physical investigations one endeavors to define as exactly as possible the initial and/or final states of the systems being considered. Let us illustrate this statement with two examples. It is important in many electron-scattering experiments to be able to select electrons of as uniform energy as possible. Otherwise one would have to carry out complicated averaging in order to understand the results, and many experiments (e.g., observing the excitation of particular energy states of atoms) could not even be performed. This also

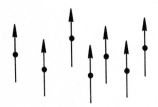

Fig. 1.1. Ensemble of totally polarized electrons

applies to momentum: Often one endeavors to have electrons in the form of a well-defined beam, that is, a beam in which the directions of the momenta of the individual electrons are as uniform as possible. A swarm of electrons with arbitrary momentum directions would, for example, be unsuitable for bombarding a target. For quite analogous reasons, it is important in the investigation of the large number of spin-dependent processes that occur in physics to have electrons available in well-defined spin states. Thus one is not obliged to average over all possibilities that may arise from different spin directions, thereby losing valuable information. One can rather investigate the individual possibilities separately.

This somewhat general statement will be substantiated in later chapters. Numerous other reasons for investigations with polarized electrons will then become clear, such as the possibility of obtaining a better understanding of the structure of magnetic substances or of atomic interactions, or the goal of determining precisely the magnetic moment of the electron.

1.2 Why Conventional Polarization Filters Do Not Work with Electrons

Conventional spin filters, the prototype of which is the Stern-Gerlach magnet, do not work with free electrons. This is because a Lorentz force which does not appear with neutral atoms arises in the Stern-Gerlach magnet. This, combined with the uncertainty principle, prevents the separation of spin-up and spin-down electrons.

When Malus in 1808 looked through a calcite crystal at the light reflected from a windowpane of the Palais Luxembourg, he detected the polarization of light. When Stern and Gerlach in 1921 sent an atomic beam through an inhomogeneous magnetic field they detected the polarization of atoms. Numerous exciting experiments with polarized light or polarized atoms have been made since these early discoveries. However, experiments of comparable quality with polarized electrons have been possible only in the past two decades.

This is not accidental; the reason can be easily given. Polarized light can be produced from unpolarized light by sending it through a polarizer which eliminates one of the two basic directions of polarization. One therefore loses a factor of 2 in intensity. Similarly, a polarized atomic beam can be produced by sending an unpolarized atomic beam through a spin filter. If, for example, an alkali atomic beam passes through a Stern-Gerlach magnet, it splits into two beams with opposite spin directions of the valence electrons. One can eliminate one of these beams and thus again have a polarized beam with an intensity loss of a factor of 2.

This procedure does not work with electrons. It is fundamentally impossible to polarize free electrons with the use of a Stern-Gerlach experiment as can be seen in the following [1.1].

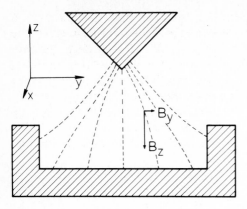

Fig. 1.2. Stern-Gerlach experiment with free electrons

In Fig. 1.2 the electron beam passes through the middle of the magnetic field in a direction perpendicular to the plane of the diagram (velocity $v = v_x$). The spins align parallel or antiparallel to the magnetic field and the electrons experience a deflecting force in the inhomogeneous field. In the plane of symmetry the force that tends to split the beam is

$$F = \pm \mu \frac{\partial B_z}{\partial z}, \tag{1.1}$$

where μ is the magnetic moment of the electrons. In addition, the electrons experience a Lorentz force due to their electric charge. Its component in the y direction, caused by the magnetic field component B_z, produces a right-hand shift of the image that could be detected by a photographic plate. As the electron beam has a certain width, it is also affected by the field component B_y which exists outside the symmetry plane. The component of the Lorentz force $F_L = (e/c) v_x B_y$, caused by B_y, deflects the electrons upwards if they are to the right of the symmetry plane and downwards if they are to the left of it. This causes a tilting of the traces on the photographic plate as is shown schematically in Fig. 1.3.

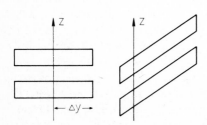

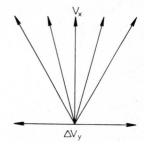

Fig. 1.3. Deflection of uncharged (*left-hand side*) and charged (*right-hand side*) particles with spin 1/2 in Stern-Gerlach field

Fig. 1.4. Transverse beam spread

Even in "thought" (Gedanken) experiments we must not consider an infinitely narrow beam, since the uncertainty principle must be taken into account, i.e., $\Delta y \cdot m \Delta v_y \approx h$. Because we want to work with a reasonable beam, the uncertainty of the velocity in the y direction Δv_y must be small compared to v_x (see Fig. 1.4). From this, together with the uncertainty relation given above, it follows that $h/m\Delta y \ll v_x$, or with $\lambda = h/mv_x$ (de Broglie wavelength)

$$\lambda \ll \Delta y; \tag{1.2}$$

correspondingly one has $\lambda \ll \Delta z$.

Nevertheless, to be able to draw Fig. 1.5 clearly, we assume for now that we can have a beam whose spread in the z direction, in which we hope to obtain the splitting, is smaller than the de Broglie wavelength. Let us consider two points A′ and B for which the y coordinate differs by λ. This is always possible since the beam width Δy is much greater than λ. As λ is small compared to the macroscopic dimensions of the field, the Taylor expansion

$$B_y(y+\lambda) = B_y(y) + \lambda \frac{\partial B_y}{\partial y}(y) \tag{1.3}$$

is, to a good approximation, valid. This means that the Lorentz force experienced by the electrons arriving at A′ has always been larger by about $\Delta F_L = (e/c)v_x\lambda(\partial B_y/\partial y)$ than that experienced by the electrons arriving at B.

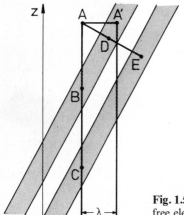

Fig. 1.5. Impossibility of the Stern-Gerlach experiment with free electrons

Thus A′ is higher than B by an amount AB shown in Fig. 1.5. We can easily compare this distance with the splitting BC caused by the force F from (1.1). Since AB and BC are proportional to the respective forces applied, one obtains

$$\frac{AB}{BC} = \frac{\Delta F_L}{2F} = \frac{(e/c)v_x\lambda(\partial B_z/\partial z)}{2(eh/2mc)(\partial B_z/\partial z)} = \frac{2\pi\lambda}{\lambda} = 2\pi, \tag{1.4}$$

where use has been made of div $\boldsymbol{B} = 0$, or $\partial B_y/\partial y = -\partial B_z/\partial z$. This means that the tilting of the traces is very large: AB is much larger than the splitting BC, although A'A is as small as λ. This has the following consequences:

If AE is the perpendicular from A to the traces, then, because AB > BC, AD is greater than DE. On the other hand, AD is smaller than the hypotenuse AA' = λ of the right triangle ADA'; hence DE, the distance between the centers of the traces, is such that DE < AD < λ. This means that this distance is smaller than the width of either of the traces, which we have shown is considerably larger than λ in every direction. Consequently, no splitting into traces with opposing spin directions can be observed. The uncertainty principle, together with the Lorentz force, prevents spin-up and spin-down electrons from being separated by a macroscopic field of the Stern-Gerlach type. The most one could expect would be a slight imbalance of the spin directions at the edges of the beam.

Attempts have frequently been made to disprove the above argument, originating from Bohr and Pauli, that a Stern-Gerlach type experiment is impossible with electrons (see [1.2]). Such attempts have the same challenge as "thought" experiments for constructing perpetual-motion machines. However, all suggestions for modifying the experiment so that it would work have failed. We shall, however, see at the end of the book that it is not, in principle, impossible to obtain different populations of spin-up and spin-down states of free electrons with the aid of macroscopic fields. Selection of spin states may, for instance, be performed by trapping electrons in suitable inhomogeneous magnetic fields (cf. end of Sect. 8.3).

Since the most direct method, the Stern-Gerlach filter, fails, one had to find other ways of producing polarized free electrons. Scattering of unpolarized electrons by heavy atoms, for example, yields highly polarized electrons. In this way, however, one does not lose only a factor of 2, as with a conventional polarization filter, but a factor of 10^4 to 10^7, depending on how high a polarization one wants. As we shall see later, there are methods other than scattering, but they have in common the fact that they yield only moderate intensities. Nobody has yet found a spin filter for electrons that reduces the intensity by just a factor of 2.

For a polarization experiment one also needs an analyzer for the polarization. Here we have the same situation. If the transmission axis of an optical analyzer is parallel to the polarization, a totally polarized light beam passes through the analyzer without loss of intensity. Similarly, if one uses a spin filter of the Stern-Gerlach type as an analyzer, a totally polarized atomic beam passes through without appreciable loss of intensity, if the direction of its polarization is parallel to the analyzing direction. With electrons, however, one cannot use such a spin filter as an analyzer for the same reason one cannot use it as a polarizer. One must use some spin-dependent collision process, usually electron scattering, where one again loses several orders of magnitude in intensity.

Since one needs a polarizer as well as an analyzer for a polarization experiment, the two factors together easily make an intensity reduction of a factor of 10^6 or more in an electron-polarization experiment. If we compare this to the factor of 2 for a light- or atom-polarization experiment (under ideal conditions), we see why electron-polarization studies became feasible only in recent years: Sufficiently advanced experimental techniques had to be developed before this field was accessible.

The fact that conventional polarization filters do not work with electrons does not mean that it is absolutely impossible to find effective electron polarization filters. As will be discussed in Sect. 7.1.2 there are interesting developments which show that it is worthwhile to search for "unconventional" electron polarization filters of high efficiency.

Before we can discuss quantitatively the processes in which electron polarization plays a role, we must look at the possibilities of describing polarized electrons mathematically.

2. Description of Polarized Electrons

2.1 A Few Results from Elementary Quantum Mechanics

The formal description of the spin of free electrons is summarized.

The following facts can be drawn from textbooks on quantum mechanics: The observable "spin" is represented by the operator s which satisfies the commutation relations characteristic of angular momenta:

$$s_x s_y - s_y s_x = i\hbar s_z \quad \text{(etc. by cyclic permutation)}. \tag{2.1}$$

If one separates out the factor $\hbar/2$ by the definition $s = (\hbar/2)\boldsymbol{\sigma}$, one obtains from the above commutation relations, with the additional condition that σ_z is diagonal, the Pauli matrices

$$\sigma_x = \begin{pmatrix} 0 & 1 \\ 1 & 0 \end{pmatrix}, \quad \sigma_y = \begin{pmatrix} 0 & -i \\ i & 0 \end{pmatrix}, \quad \sigma_z = \begin{pmatrix} 1 & 0 \\ 0 & -1 \end{pmatrix}. \tag{2.2}$$

These operators receive their meaning from their application to the two-component wave functions $\begin{pmatrix} a_1 \\ a_2 \end{pmatrix}$ with whose help the two possible orientations of the electron spin can be described. For example, one has the eigenvalue equations

$$\sigma_z \begin{pmatrix} 1 \\ 0 \end{pmatrix} = \begin{pmatrix} 1 & 0 \\ 0 & -1 \end{pmatrix} \begin{pmatrix} 1 \\ 0 \end{pmatrix} = 1 \cdot \begin{pmatrix} 1 \\ 0 \end{pmatrix}, \quad \sigma_z \begin{pmatrix} 0 \\ 1 \end{pmatrix} = \begin{pmatrix} 1 & 0 \\ 0 & -1 \end{pmatrix} \begin{pmatrix} 0 \\ 1 \end{pmatrix} = -1 \cdot \begin{pmatrix} 0 \\ 1 \end{pmatrix} \tag{2.3}$$

which mean that $\begin{pmatrix} 1 \\ 0 \end{pmatrix}$ is an eigenfunction of σ_z with the eigenvalue $+1$ (or $+\hbar/2$ of s_z) and $\begin{pmatrix} 0 \\ 1 \end{pmatrix}$ belongs to the eigenvalue -1.

We can use these two states as a basis for representing the general state

$$\chi = \begin{pmatrix} a_1 \\ a_2 \end{pmatrix}$$

as a linear superposition

$$a_1 \begin{pmatrix} 1 \\ 0 \end{pmatrix} + a_2 \begin{pmatrix} 0 \\ 1 \end{pmatrix} = \begin{pmatrix} a_1 \\ a_2 \end{pmatrix}. \tag{2.4}$$

When χ is assumed to be normalized one has

$$\langle \chi | \chi \rangle = (a_1^*, a_2^*) \begin{pmatrix} a_1 \\ a_2 \end{pmatrix} = |a_1|^2 + |a_2|^2 = 1. \tag{2.5}$$

Remembering the quantum mechanical interpretation of the expansion of a wave function, we see from the left side of (2.4) that $|a_1|^2$ is the probability of finding the value $+\hbar/2$ when measuring the spin component in the z direction; that is, $|a_1|^2$ is the probability of finding the electron in the state $\begin{pmatrix} 1 \\ 0 \end{pmatrix}$. $|a_2|^2$ is the probability of finding the eigenvalue $-\hbar/2$, that is, of finding the state $\begin{pmatrix} 0 \\ 1 \end{pmatrix}$. A measurement of the spin direction forces the electron with the probability $|a_i|^2$ $(i = 1, 2)$ into one of the two eigenstates. Even if the spins have been oriented in the x direction, by measuring the spin components in the z direction the values $+\hbar/2$ or $-\hbar/2$ are obtained, each with the probability $1/2$ (Sect. 2.2). Thus the spins are seen to be affected by the measurement.

This disturbance of the spin state by measurement makes it impossible to measure all the spin components simultaneously. This follows mathematically from the noncommutativity of the angular momentum components, (2.1). The operator s^2, however, which has the eigenvalue $s(s+1)\hbar^2 = 3\hbar^2/4$, commutes with the components s_x, s_y, and s_z. One can measure its eigenvalue simultaneously with those of any of the components of s. For these reasons, the statement that "the spin is in the z direction" means more precisely the following: The spin vector lies somewhere on a conical shell in such a way that its component in the z direction is $\hbar/2$; the two other components are not known, it is merely known that $s_x^2 + s_y^2 + s_z^2 = 3\hbar^2/4$ (Fig. 2.1).

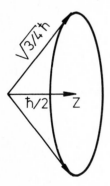

Fig. 2.1. Spin "in the z direction"

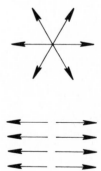

Fig. 2.2. Unpolarized electrons

It follows from the above statements that one cannot distinguish between electron beams in which all directions are equally likely and those in which half of the spins are parallel and half are antiparallel to some arbitrary reference direction (Fig. 2.2). This is because in both cases a measurement results in half of the spins being parallel and half being antiparallel to the direction specified by the observation. Every conceivable experiment with such electron beams yields the same result; the beams therefore must be regarded as identical.

2.2 Pure Spin States

The spin function describing the spin in an arbitrary direction is determined. The polarization is defined as the expectation value of the spin operator. Its magnitude for a pure state is 1.

In this section we consider electrons which are all in the same spin state. In such cases the system of electrons is said to be in a pure spin state.

The spin direction of a state which is described by $\chi = \begin{pmatrix} a_1 \\ a_2 \end{pmatrix}$ is specified by a_1 and a_2, as will now be shown. Let $\hat{e} = (e_x, e_y, e_z)$ be the unit vector in the direction ϑ, φ, i.e. (Fig. 2.3),

$$e_x = \sin\vartheta\cos\varphi, \qquad e_y = \sin\vartheta\sin\varphi, \qquad e_z = \cos\vartheta.$$

We now ask what the spin function that describes a spin in the direction ϑ, φ would look like. For this we must solve the eigenvalue equation $(\boldsymbol{\sigma} \cdot \hat{e})\chi = \lambda\chi$, since $\boldsymbol{\sigma} \cdot \hat{e}$ is the projection of the spin operator in the specified direction. Since

$$\sigma_x \begin{pmatrix} a_1 \\ a_2 \end{pmatrix} = \begin{pmatrix} a_2 \\ a_1 \end{pmatrix}, \quad \sigma_y \begin{pmatrix} a_1 \\ a_2 \end{pmatrix} = \begin{pmatrix} -ia_2 \\ ia_1 \end{pmatrix}, \quad \sigma_z \begin{pmatrix} a_1 \\ a_2 \end{pmatrix} = \begin{pmatrix} a_1 \\ -a_2 \end{pmatrix} \tag{2.6}$$

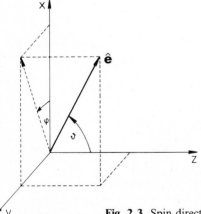

Fig. 2.3. Spin direction

one obtains

$$
(\boldsymbol{\sigma}\cdot\hat{e})\chi=\begin{pmatrix} a_2\sin\vartheta\cos\varphi-ia_2\sin\vartheta\sin\varphi+a_1\cos\vartheta \\ a_1\sin\vartheta\cos\varphi+ia_1\sin\vartheta\sin\varphi-a_2\cos\vartheta \end{pmatrix}
$$

$$
=\begin{pmatrix} a_1\cos\vartheta+a_2\sin\vartheta\,e^{-i\varphi} \\ a_1\sin\vartheta\,e^{i\varphi}-a_2\cos\vartheta \end{pmatrix}.
$$

The attempt to find the eigenfunction of the component of the spin operator in the $\hat{e}$ direction, i.e., to solve the equation $(\boldsymbol{\sigma}\cdot\hat{e})\chi=\lambda\chi$, thus gives

$$
a_1(\cos\vartheta-\lambda)+a_2\sin\vartheta\,e^{-i\varphi}=0
$$
$$
a_1\sin\vartheta\,e^{i\varphi}+a_2(-\cos\vartheta-\lambda)=0. \tag{2.7}
$$

The condition for a nontrivial solution (disappearance of the determinant) is

$$
-\cos^2\vartheta+\lambda^2-\sin^2\vartheta=0 \quad\text{or}\quad \lambda=\pm1.
$$

For $\lambda=+1$, one obtains from (2.7)

$$
\frac{a_2}{a_1}=\frac{\cos\vartheta-1}{-\sin\vartheta\,e^{-i\varphi}}=\tan\frac{\vartheta}{2}\,e^{i\varphi}, \tag{2.8}
$$

and for $\lambda=-1$

$$
\frac{a_2}{a_1}=\frac{\cos\vartheta+1}{-\sin\vartheta\,e^{-i\varphi}}=-\cot\frac{\vartheta}{2}\,e^{i\varphi}. \tag{2.9}
$$

Since a_1 and a_2 are solutions of a homogeneous system of equations, they are determined except for a constant. This constant can be specified by normalization according to (2.5). Then

$$
a_1=\cos\frac{\vartheta}{2},\quad a_2=\sin\frac{\vartheta}{2}\,e^{i\varphi}\quad\text{for}\quad \lambda=+1, \tag{2.10}
$$

$$
a_1=\sin\frac{\vartheta}{2},\quad a_2=-\cos\frac{\vartheta}{2}\,e^{i\varphi}\quad\text{for}\quad \lambda=-1. \tag{2.11}
$$

A common phase factor of a_1 and a_2 which remains undetermined has been chosen arbitrarily.

The spin functions with the a_1 and a_2 just calculated are eigenfunctions of the component of the spin operator $\boldsymbol{\sigma}$ in the $\hat{e}$ direction with the eigenvalues $+1$ and -1, that is, they represent the states where the spin in the direction ϑ, φ has the value $+\hbar/2$ or $-\hbar/2$. One can immediately see that the solutions (2.9) or (2.11) are none other than the solutions (2.8) or (2.10), respectively, for the direction $-\hat{e}$ which is described by the angles $\pi-\vartheta$, $\varphi+\pi$. Thus it is sufficient from now on to use only the solutions (2.8) or (2.10).

In the special cases $\vartheta=0$ or π, $\varphi=0$ (spin parallel or antiparallel to the z direction), (2.10) yields the expected eigenfunctions of σ_z, $\begin{pmatrix} 1 \\ 0 \end{pmatrix}$ or $\begin{pmatrix} 0 \\ 1 \end{pmatrix}$. For $\vartheta=\pi/2$, $\varphi=0$ (spin in the x direction) one obtains the spin function $\chi=\begin{pmatrix} 1/\sqrt{2} \\ 1/\sqrt{2} \end{pmatrix}$. The latter example should forewarn us of a false conclusion: A superposition of spin states in opposing directions with equal amplitudes and with a fixed phase relation, such as

$$\frac{1}{\sqrt{2}}\begin{pmatrix} 1 \\ 0 \end{pmatrix}+\frac{1}{\sqrt{2}}\begin{pmatrix} 0 \\ 1 \end{pmatrix}=\begin{pmatrix} 1/\sqrt{2} \\ 1/\sqrt{2} \end{pmatrix}$$

does not produce a cancellation of spins, but a spin in another direction. This is analogous to a coherent superposition of right- and left-circularly polarized light waves with a fixed phase relation, which also does not produce an unpolarized wave but a linearly polarized wave.

Let us now consider the polarization of the electrons described by $\begin{pmatrix} a_1 \\ a_2 \end{pmatrix}$. Whereas the eigenvalue represents the result of a single measurement, the polarization tells us something about the average spin direction of the ensemble. It is therefore an expectation value.

We recall that the expectation value is the average over all values which the considered property of a particle can assume in a given state ψ. A single measurement of a property, described by the operator Q, yields an eigenvalue. If such measurements are made on a large number of identical systems one generally observes all possible eigenvalues. The average of the eigenvalues thus obtained is the expectation value

$$\langle Q \rangle = \langle \psi | Q | \psi \rangle. \tag{2.12}$$

For example, for the hydrogen atom in the ground state ψ_0, the expectation value of the momentum $\boldsymbol{p}$ is clearly zero as the momentum of the orbiting electron constantly changes its direction. In fact, one has

$$\langle \boldsymbol{p} \rangle = \langle \psi_0 | \boldsymbol{p} | \psi_0 \rangle = -i\hbar \int \psi_0^* \, \text{grad} \, \psi_0 \, d\tau = 0,$$

as can easily be seen by using the simple wave function of the ground state.

If a system is in the eigenstate of an operator each measurement of the corresponding observable will definitely yield the eigenvalue, so that the expectation value coincides with the eigenvalue. For example, for a system in a normalized energy state ψ_n with the eigenvalue E_n, one obtains the expectation value

$$\langle H \rangle = \langle \psi_n | H | \psi_n \rangle = \langle \psi_n | E_n | \psi_n \rangle = E_n \langle \psi_n | \psi_n \rangle = E_n.$$

After this brief look at elementary quantum mechanics, we return to the polarization and define it as the expectation value of the Pauli spin operator

$$\boldsymbol{P} = \langle \boldsymbol{\sigma} \rangle = \langle \chi | \boldsymbol{\sigma} | \chi \rangle = (a_1^*, a_2^*) \boldsymbol{\sigma} \begin{pmatrix} a_1 \\ a_2 \end{pmatrix}. \tag{2.13}$$

With this definition one finds from (2.6) and (2.10) that the components of the polarization vector are

$$P_x = (a_1^*, a_2^*) \begin{pmatrix} a_2 \\ a_1 \end{pmatrix} = a_1^* a_2 + a_2^* a_1$$

$$= \cos\frac{\vartheta}{2}\sin\frac{\vartheta}{2}e^{i\varphi} + \cos\frac{\vartheta}{2}\sin\frac{\vartheta}{2}e^{-i\varphi} = \sin\vartheta\cos\varphi$$

$$P_y = (a_1^*, a_2^*) \begin{pmatrix} -ia_2 \\ ia_1 \end{pmatrix} = i(a_2^* a_1 - a_1^* a_2) \tag{2.14}$$

$$= \cos\frac{\vartheta}{2}\sin\frac{\vartheta}{2}\frac{1}{i}e^{i\varphi} - \cos\frac{\vartheta}{2}\sin\frac{\vartheta}{2}\frac{1}{i}e^{-i\varphi} = \sin\vartheta\sin\varphi$$

$$P_z = (a_1^*, a_2^*) \begin{pmatrix} a_1 \\ -a_2 \end{pmatrix} = |a_1|^2 - |a_2|^2$$

$$= \cos^2\frac{\vartheta}{2} - \sin^2\frac{\vartheta}{2} = \cos\vartheta.$$

It can be seen from these equations that the polarization has the direction ϑ, φ and that the degree of polarization which is defined by

$$P = \sqrt{P_x^2 + P_y^2 + P_z^2}$$

is 1 in the case discussed here. This is reasonable, as we have assumed that the electron spins can be described by a single spin function $\begin{pmatrix} a_1 \\ a_2 \end{pmatrix}$ (pure state) so that there is only one spin direction in the beam, namely that in the direction ϑ, φ specified in (2.10).

If the state $\chi = \begin{pmatrix} a_1 \\ a_2 \end{pmatrix}$ is not normalized, a sensible extension of the definition (2.13) is

$$P = \frac{\langle \chi | \sigma | \chi \rangle}{\langle \chi | \chi \rangle}. \tag{2.15}$$

Thus the magnitude of the components of P remains between 0 and 1, e.g.,

$$P_z = \frac{|a_1|^2 - |a_2|^2}{|a_1|^2 + |a_2|^2}.$$

The polarization vector can be completely determined. One can measure all its components, but one must take care not to use the same particles for this, since the states of the particles are affected by a measurement, so that a subsequent measurement could give a wrong value. One might, for example, with a beam that is polarized in the x direction [polarization $P=(1,0,0)$], proceed as follows. First, one could measure the spin components in the z direction. Half of the measurements would yield $+\hbar/2$, the other half $-\hbar/2$, which would imply $P_z=0$. If one then carried out a measurement in the y or x direction on the same electrons which would then have an equal number of spins parallel and antiparallel to the z direction, one would again obtain $+\hbar/2$ and $-\hbar/2$ with equal probability, implying $P_x=0$, $P_y=0$. If the measurement had been carried out properly, however, in the x direction one should have found only the value $+\hbar/2$, since we initially assumed the beam to be totally polarized in this direction.

The objection that one should have started with the measurement in the x direction is not valid, as one does not know before making the measurement that the beam is polarized in this direction (otherwise the measurement would be superfluous).

In order to conduct the experiment properly, one must make the measurements on different subsystems of the electron ensemble. In doing so, one must of course be sure that the subsystems are in the same polarization state as the total system. For example, one can make the measurements sequentially on a beam of constant polarization. One then always uses electrons which have not been affected by a previous measurement and obtains, in the above example, $P_x=1$, $P_y=0$, and $P_z=0$.

Problem 2.1.   Two plane-wave electron beams, one with spins parallel, the other with spins antiparallel to the momentum, propagate with different momenta along the same direction. Find the spin direction of the electrons resulting from (coherent) superposition of the two beams.

Solution.   The superposition yields

$$\psi = \begin{pmatrix} 1 \\ 0 \end{pmatrix} e^{i(kz-\omega t)} + \begin{pmatrix} 0 \\ 1 \end{pmatrix} e^{i(k'z-\omega' t)} \quad \text{or}$$

$$\psi = \begin{pmatrix} a_1 \\ a_2 \end{pmatrix} = e^{i(kz-\omega t)} \begin{pmatrix} 1 \\ e^{i[(k'-k)z-(\omega'-\omega)t]} \end{pmatrix}.$$

From (2.8)

$$\frac{a_2}{a_1} = \tan\frac{\vartheta}{2}\, e^{i\varphi} = e^{i[(k'-k)z-(\omega'-\omega)t]}$$

one obtains $\vartheta = 90°$, $\varphi = (k'-k)z - (\omega'-\omega)t$, describing rotation of the spins around the z axis. A snapshot at a fixed time shows φ varying with z.

2.3 Statistical Mixtures of Spin States.
Description of Electron Polarization by Density Matrices

Partially polarized beams represent a statistical mixture of different spin states. To describe them, one can suitably apply density matrices. The connection between polarization and density matrices is given.

Until now, only totally polarized electron beams have been considered, that is, ensembles in which all particles are in the same spin state. Now we will consider partially polarized beams, which are statistical mixtures of spin states. In this case, the polarization of the total system is the average of the polarization vectors $P^{(n)}$ of the individual systems which are in pure spin states $\chi^{(n)}$:

$$P = \sum_n w^{(n)} P^{(n)} = \sum_n w^{(n)} \langle \chi^{(n)} | \sigma | \chi^{(n)} \rangle, \tag{2.16}$$

where the weighting factors $w^{(n)}$ take into account the relative proportion of the states $\chi^{(n)}$:

$$w^{(n)} = \frac{N^{(n)}}{\sum_n N^{(n)}},$$

where $N^{(n)}$ is the number of electrons in the state $\chi^{(n)}$. The $\chi^{(n)}$ have been assumed to be normalized here.

As an expedient means of describing the polarization in this case, one can use the density matrix ϱ which is defined as [2.1–3]

$$\varrho = \sum_n w^{(n)} \begin{pmatrix} |a_1^{(n)}|^2 & a_1^{(n)} a_2^{(n)*} \\ a_1^{(n)*} a_2^{(n)} & |a_2^{(n)}|^2 \end{pmatrix} = \sum_n w^{(n)} \begin{pmatrix} a_1^{(n)} \\ a_2^{(n)} \end{pmatrix} (a_1^{(n)*}, a_2^{(n)*})$$

$$= \sum_n w^{(n)} \chi^{(n)} \chi^{(n)\dagger} = \sum_n w^{(n)} |\chi^{(n)}\rangle\langle\chi^{(n)}|. \tag{2.17}$$

The individual matrices of this sum are the density matrices of the pure states.
The density matrix is connected to the polarization by the relation

$$P = \mathrm{tr}\{\varrho\sigma\} \quad \text{or} \quad P_i = \mathrm{tr}\{\varrho\sigma_i\}. \tag{2.18}$$

By using (2.2) one sees immediately that these equations are correct. For example, we have

$$\varrho\sigma_x = \sum_n w^{(n)} \begin{pmatrix} a_1^{(n)} a_2^{(n)*} & |a_1^{(n)}|^2 \\ |a_2^{(n)}|^2 & a_1^{(n)*} a_2^{(n)} \end{pmatrix},$$

and thus

$$\mathrm{tr}\{\varrho\sigma_x\} = \sum_n w^{(n)} (a_1^{(n)} a_2^{(n)*} + a_1^{(n)*} a_2^{(n)}) = \sum_n w^{(n)} P_x^{(n)} = P_x,$$

where use has been made of part of (2.14). Similarly

$$\text{tr}\{\varrho\sigma_z\} = \sum_n w^{(n)}(|a_1^{(n)}|^2 - |a_2^{(n)}|^2) = \sum_n w^{(n)} P_z^{(n)} = P_z.$$

One can express the elements of the density matrix in terms of the components of the polarization. Using (2.2, 14, 17) one then obtains

$$\varrho = \frac{1}{2}\begin{pmatrix} 1+P_z & P_x-iP_y \\ P_x+iP_y & 1-P_z \end{pmatrix} = \tfrac{1}{2}(1+\boldsymbol{P}\cdot\boldsymbol{\sigma}), \tag{2.19}$$

where **1** is the unit matrix.

In making the definition (2.16) we assumed that the states $\chi^{(n)}$ were normalized; the relative proportions of the single states were taken into account by using weighting factors. One can also start with unnormalized $\chi^{(n)}$. Then weighting factors are unnecessary, since the relative proportion of the nth state is already expressed by the unnormalized amplitude of $\chi^{(n)}$; it is given by

$$\frac{\langle\chi^{(n)}|\chi^{(n)}\rangle}{\sum_n \langle\chi^{(n)}|\chi^{(n)}\rangle}.$$

In this case the polarization is

$$\boldsymbol{P} = \frac{\sum_n \langle\chi^{(n)}|\boldsymbol{\sigma}|\chi^{(n)}\rangle}{\sum_n \langle\chi^{(n)}|\chi^{(n)}\rangle} \tag{2.20}$$

[which can also be written as

$$\boldsymbol{P} = \frac{\sum_n \langle\chi^{(n)}|\chi^{(n)}\rangle[\langle\chi^{(n)}|\boldsymbol{\sigma}|\chi^{(n)}\rangle/\langle\chi^{(n)}|\chi^{(n)}\rangle]}{\sum_n \langle\chi^{(n)}|\chi^{(n)}\rangle}$$

thus leading back to the form (2.16) with normalized functions]. Instead of (2.18) we then have

$$\boldsymbol{P} = \text{tr}\{\varrho\boldsymbol{\sigma}\} / \text{tr}\{\varrho\}, \tag{2.21}$$

where the density matrix has the form

$$\varrho = \sum_n \begin{pmatrix} |a_1^{(n)}|^2 & a_1^{(n)}a_2^{(n)*} \\ a_1^{(n)*}a_2^{(n)} & |a_2^{(n)}|^2 \end{pmatrix}. \tag{2.22}$$

The denominator

$$\sum_n \langle\chi^{(n)}|\chi^{(n)}\rangle = \sum_n (|a_1^{(n)}|^2 + |a_2^{(n)}|^2) = \text{tr}\{\varrho\}$$

now appearing in the polarization formulae (2.20 and 21) must also be taken into account on the left-hand side of (2.19), so that the corresponding relation is

$$\frac{\varrho}{\text{tr}\{\varrho\}} = \frac{1}{2}\begin{pmatrix} 1+P_z & P_x-iP_y \\ P_x+iP_y & 1-P_z \end{pmatrix} = \tfrac{1}{2}(1+\boldsymbol{P}\cdot\boldsymbol{\sigma}). \tag{2.23}$$

The density matrix assumes its simplest form if one takes the direction of the resultant polarization as the z axis of the coordinate system shown in Fig. 2.3, i.e., chooses $P_x = P_y = 0$, $P = P_z$. Then from (2.19) (if we return to normalized states) one has

$$\varrho = \frac{1}{2}\begin{pmatrix} 1+P & 0 \\ 0 & 1-P \end{pmatrix}. \tag{2.24}$$

The density matrix thus is transformed to a diagonal form.

This form of the density matrix illustrates again the meaning of P: Since $|a_1^{(n)}|^2$ is the probability that the eigenvalue $+\hbar/2$ will be obtained from a spin measurement in the z direction on the nth subsystem, the probability is $\sum_n w^{(n)}|a_1^{(n)}|^2$ that this measurement on the total beam will give the value $+\hbar/2$. This probability can also be expressed as $N_\uparrow/(N_\uparrow+N_\downarrow)$, where $N_\uparrow$ is the number of measurements that yield the value $+\hbar/2$ and $N_\uparrow+N_\downarrow$ is the total number of measurements. (Correspondingly, $\sum_n w^{(n)}|a_2^{(n)}|^2 = N_\downarrow/(N_\uparrow+N_\downarrow)$ is the probability that the value $-\hbar/2$ will be obtained.) Thus one has

$$N_\uparrow/(N_\uparrow+N_\downarrow) = \sum_n w^{(n)}|a_1^{(n)}|^2 = \tfrac{1}{2}(1+P),$$

where the last part of the equation comes from a comparison of (2.17) and (2.24). Consequently, one obtains for the polarization

$$P = \frac{N_\uparrow - N_\downarrow}{N_\uparrow + N_\downarrow}. \tag{2.25}$$

For a beam totally polarized in the $+z$ direction ($N_\downarrow = 0$), the diagonal form of the density matrix (2.24) becomes

$$\varrho = \begin{pmatrix} 1 & 0 \\ 0 & 0 \end{pmatrix}. \tag{2.26}$$

For an unpolarized beam ($N_\uparrow = N_\downarrow$) one obtains

$$\varrho = \begin{pmatrix} \tfrac{1}{2} & 0 \\ 0 & \tfrac{1}{2} \end{pmatrix}. \tag{2.27}$$

From the identity

$$\varrho = \frac{1}{2}\begin{pmatrix} 1+P & 0 \\ 0 & 1-P \end{pmatrix} = (1-P)\begin{pmatrix} \tfrac{1}{2} & 0 \\ 0 & \tfrac{1}{2} \end{pmatrix} + P\begin{pmatrix} 1 & 0 \\ 0 & 0 \end{pmatrix} \tag{2.28}$$

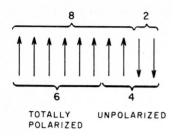

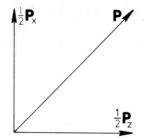

Fig. 2.4. Partially polarized beam   **Fig. 2.5.** Superposition of polarization vectors

one sees by comparison with (2.26, 27) that an electron beam with an arbitrary polarization P can be considered to be made up of a totally polarized fraction and an unpolarized fraction which are mixed in the ratio $P/(1-P)$.

We will illustrate the general definitions introduced here by two simple examples.

Example 2.1. In an ensemble of 100 electrons, one finds 80 electrons with spin $+\hbar/2$ in the z direction and 20 with $-\hbar/2$. With $N_\uparrow = 80$, $N_\downarrow = 20$, one has, from (2.25)

$$P = 0.6.$$

From (2.28), this can also be expressed by saying that 60% of the beam is totally polarized and 40% is unpolarized (Fig. 2.4).

Example 2.2. An electron beam, which is totally polarized in the z direction, is mixed with another beam of the same intensity, which is totally polarized in the x direction. According to (2.16) the resulting polarization is then (Fig. 2.5)

$$P = \tfrac{1}{2} P_x + \tfrac{1}{2} P_z,$$

forming a $45°$ angle with both the z and the x direction. Its magnitude is $1/\sqrt{2} = 0.7071$. One has $P < 1$ because the two component beams are independent of each other so that they are incoherently superimposed. (With opposing spin directions for the two beams one would have $P = 0$.)

This result can also be obtained by using the density matrix: From (2.10) the two component beams have the normalized eigenfunctions $\begin{pmatrix} 1 \\ 0 \end{pmatrix}$ and $\begin{pmatrix} \cos 45° \\ \sin 45° \end{pmatrix} = \begin{pmatrix} 1/\sqrt{2} \\ 1/\sqrt{2} \end{pmatrix}$ so that from (2.17) the density matrix is

$$\varrho = \frac{1}{2} \begin{pmatrix} 1 & 0 \\ 0 & 0 \end{pmatrix} + \frac{1}{2} \begin{pmatrix} \frac{1}{2} & \frac{1}{2} \\ \frac{1}{2} & \frac{1}{2} \end{pmatrix} = \frac{1}{2} \begin{pmatrix} \frac{3}{2} & \frac{1}{2} \\ \frac{1}{2} & \frac{1}{2} \end{pmatrix}.$$

From this and (2.19) it follows that

$$P_z = \tfrac{3}{2} - 1 = \tfrac{1}{2}, \qquad P_y = 0, \qquad P_x = \tfrac{1}{2} \quad \text{and}$$

$$P = \sqrt{P_x^2 + P_z^2} = 1/\sqrt{2}.$$

By proper choice of the coordinate system the density matrix could have been transformed to the diagonal form (see Problem 2.2 below).

In such simple cases one can, of course, obtain the result more quickly without density matrices. Their real use becomes obvious in more complicated cases (Sects. 3.3 and 5.2).

We emphasize once more the difference between coherent and incoherent superposition of spin states.[1] The summation of products of amplitudes when forming the density matrix of a mixed state implies a loss of phase relations. The simplest example of such an incoherent superposition is the superposition of two opposing spin states with equal weighting factors, which yields $P = 0$. In contrast to this, with the coherent superposition of amplitudes, as mentioned in Sect. 2.2, the phase relations between the amplitudes are retained and one obtains a completely polarized state. This is because every new state

$$\begin{pmatrix} \sum_n a_1^{(n)} \\ \sum_n a_2^{(n)} \end{pmatrix} = \begin{pmatrix} A_1 \\ A_2 \end{pmatrix}$$

formed in this way is an eigenfunction of the component $(\boldsymbol{\sigma} \cdot \hat{\boldsymbol{e}})$ of the spin operator in the direction ϑ, φ which is specified by $A_2/A_1 = \tan(\vartheta/2) \exp(i\varphi)$ (cf. Sect. 2.2). (Every complex number $A_2/A_1 = |R| \exp(i\varphi)$ can be represented in this way as the tangent passes through all real numbers.) Thus the polarization in this direction is

$$P = \frac{(A_1^*, A_2^*) \, (\boldsymbol{\sigma} \cdot \hat{\boldsymbol{e}}) \begin{pmatrix} A_1 \\ A_2 \end{pmatrix}}{(A_1^*, A_2^*) \begin{pmatrix} A_1 \\ A_2 \end{pmatrix}} = 1, \quad \text{since} \quad (\boldsymbol{\sigma} \cdot \hat{\boldsymbol{e}}) \begin{pmatrix} A_1 \\ A_2 \end{pmatrix} = 1 \cdot \begin{pmatrix} A_1 \\ A_2 \end{pmatrix}.$$

Let us conclude this chapter with the remark that, in the jargon of physics, the term "polarization of the beam" may stand for the degree of polarization or for a polarization component of the beam, the latter particularly when the other components disappear. Usually the danger of confusion is small, and if the "polarization" is negative it is certainly a polarization component which is referred to.

---

[1] For a detailed discussion of coherent and incoherent superposition in quantum mechanics see [2.3].

Problem 2.2. For Example 2.2, give the density matrix in diagonal form using the degree of polarization calculated there. Check the result by using the spin functions.

Solution. As the magnitude of the polarization was found above to be $P=0.7071$ the density matrix in this case must be, according to (2.24),

$$\varrho=\frac{1}{2}\begin{pmatrix}1.7071 & 0 \\ 0 & 0.2929\end{pmatrix}.$$

We will check to see if this is correct: In the coordinate system, where the z axis is in the direction of the resultant polarization vector, the spin directions of the two constituent beams have the angles $\vartheta=45°$, and $\varphi=0°$ and $180°$, respectively. Their spin functions, from (2.10), are

$$\begin{pmatrix}\cos\dfrac{\vartheta}{2} \\ \sin\dfrac{\vartheta}{2}\end{pmatrix} \quad\text{and}\quad \begin{pmatrix}\cos\dfrac{\vartheta}{2} \\ \sin\dfrac{\vartheta}{2}\,e^{i\pi}\end{pmatrix} \quad\text{with}\quad \vartheta=45°.$$

Thus the density matrix of the total system has the expected diagonal form

$$\varrho=\frac{1}{2}\,\varrho^{(1)}+\frac{1}{2}\,\varrho^{(2)}=\frac{1}{2}\begin{pmatrix}2\cos^2\dfrac{\vartheta}{2} & \cos\dfrac{\vartheta}{2}\sin\dfrac{\vartheta}{2}(1+e^{-i\pi}) \\ \cos\dfrac{\vartheta}{2}\sin\dfrac{\vartheta}{2}(1+e^{i\pi}) & 2\sin^2\dfrac{\vartheta}{2}\end{pmatrix}$$

$$=\frac{1}{2}\begin{pmatrix}1.7071 & 0 \\ 0 & 0.2929\end{pmatrix}\quad\text{for}\quad \vartheta=45°.$$

3. Polarization Effects in Electron Scattering Caused by Spin-Orbit Interaction

3.1 The Dirac Equation and Its Interpretation

By linearizing the relativistic generalization of the Schrödinger equation, one obtains the Dirac equation. It is Lorentz invariant and describes the electron spin and spin-orbit coupling without the need to introduce further assumptions. The definition of the polarization as the expectation value of the spin operator is not Lorentz invariant and will therefore be referred to the rest system of the electrons.

In analogy to the case of light beams, electron beams can be polarized by scattering, and the angular distribution of scattered electrons depends on the state of polarization of the incident beam. These effects can be treated by the Dirac equation, which is the basic equation for describing the electron, including its spin and its relativistic behavior.

Dirac discovered this equation in 1928 when he tried to find a relativistic generalization of the Schrödinger equation. We can best see how the relativistic generalization can be made by recalling the path which formally leads to the Schrödinger equation.

One starts from the Hamiltonian function for a free particle

$$H = p^2/2m \tag{3.1}$$

and substitutes for p and H the operators

$$p = -i\hbar\nabla, \quad H = i\hbar \frac{\partial}{\partial t}. \tag{3.2}$$

By applying these to a wave function $\psi(r, t)$, one obtains the Schrödinger equation

$$i\hbar\dot{\psi} = -\frac{\hbar^2}{2m} \nabla^2 \psi. \tag{3.3}$$

Does this method also lead to a useful result in the relativistic case? To examine this question we start with the relativistic energy law

$$H^2 = c^2 p^2 + m^2 c^4 \tag{3.4}$$

(m = rest mass). By substituting the operators (3.2) for p and H, we obtain

$$\left(\nabla^2 - \frac{1}{c^2} \frac{\partial^2}{\partial t^2} - \frac{1}{\lambda^2} \right) \psi = 0, \quad \text{where} \tag{3.5}$$

$$\lambda = \hbar/mc \tag{3.6}$$

is the Compton wavelength.

When we include electric and magnetic potentials ϕ and A in the consideration, the Hamiltonian function for the nonrelativistic case is

$$H = \frac{[p - (\varepsilon/c) A]^2}{2m} + \varepsilon\phi, \tag{3.7}$$

and thus

$$H - \varepsilon\phi = \frac{[p - (\varepsilon/c) A]^2}{2m},$$

where $\varepsilon = $ electron charge $= -e$. This follows from (3.1), if one substitutes $p - (\varepsilon/c) A$ for p ($p = $ canonical momentum) and $H - \varepsilon\phi$ for H. Correspondingly, it follows from (3.4) for the relativistic case

$$(H - \varepsilon\phi)^2 = (cp - \varepsilon A)^2 + m^2 c^4. \tag{3.8}$$

By interpreting H and p as operators, one obtains a wave equation for an electron in an external electromagnetic field (Klein-Gordon equation).

Serious difficulties are encountered in the use of this equation. For example, it predicts far too large a fine-structure splitting of the hydrogen spectrum. It is also problematic in its mathematical structure as a second-order differential equation in t, since one requires the initial values of ψ and $\dot{\psi}$ to solve it. The Schrödinger equation requires only the initial value of ψ, and it is difficult to see why the consideration of relativistic effects should lead to such radical differences in the initial information required to describe the behavior of the electron.

Dirac had the idea of splitting up the equation into a product of two linear expressions and of considering these individually. The equation

$$(H^2 - c^2 \sum_{\mu} p_{\mu}^2 - m^2 c^4) \psi = 0 \tag{3.9}$$

($p_{\mu} = p_x, p_y, p_z$, components of the momentum operator), which follows from the force-free form of (3.8), can be expressed in the form

$$(H - c \sum_{\mu} \alpha_{\mu} p_{\mu} - \beta mc^2) \, (H + c \sum_{\mu} \alpha_{\mu} p_{\mu} + \beta mc^2) \psi = 0 \tag{3.10}$$

if the constant coefficients α_{μ} and β satisfy the relations

$$\alpha_\mu \alpha_{\mu'} + \alpha_{\mu'} \alpha_\mu = 2\delta_{\mu\mu'}$$
$$\alpha_\mu \beta + \beta \alpha_\mu = 0 \tag{3.11}$$
$$\beta^2 = 1.$$

This can easily be seen by multiplication. If one can solve the equation

$$(H - c \sum_\mu \alpha_\mu p_\mu - \beta mc^2)\psi = 0 \tag{3.12}$$

(or the corresponding equation from (3.10) with the plus sign) then (3.10) is also solved. The linearized equation (3.12) has the advantage that it is of the first order in $\partial/\partial t$ just as the Schrödinger equation is. The derivatives with respect to the space coordinates are of the same order, which is necessary for relativistic covariance.

The relations (3.11) cannot be satisfied with ordinary numbers. One can, however, solve (3.11) with matrices (at least 4×4), for example with

$$
\alpha_x = \begin{pmatrix} 0 & 0 & 0 & 1 \\ 0 & 0 & 1 & 0 \\ 0 & 1 & 0 & 0 \\ 1 & 0 & 0 & 0 \end{pmatrix}, \quad
\alpha_y = \begin{pmatrix} 0 & 0 & 0 & -i \\ 0 & 0 & i & 0 \\ 0 & -i & 0 & 0 \\ i & 0 & 0 & 0 \end{pmatrix},
$$

$$\tag{3.13}$$

$$
\alpha_z = \begin{pmatrix} 0 & 0 & 1 & 0 \\ 0 & 0 & 0 & -1 \\ 1 & 0 & 0 & 0 \\ 0 & -1 & 0 & 0 \end{pmatrix}, \quad
\beta = \begin{pmatrix} 1 & 0 & 0 & 0 \\ 0 & 1 & 0 & 0 \\ 0 & 0 & -1 & 0 \\ 0 & 0 & 0 & -1 \end{pmatrix}.
$$

Thus the Dirac equation for a free particle is, if one arbitrarily chooses the left factor of (3.10),[1]

$$\left[i\hbar \frac{\partial}{\partial t} + i\hbar c \left(\alpha_x \frac{\partial}{\partial x} + \alpha_y \frac{\partial}{\partial y} + \alpha_z \frac{\partial}{\partial z} \right) - \beta mc^2 \right]\psi = 0. \tag{3.14}$$

As α_μ and β are 4×4 matrices, the formula makes sense only when ψ has the form

$$\psi = \begin{pmatrix} \psi_1 \\ \psi_2 \\ \psi_3 \\ \psi_4 \end{pmatrix}.$$

---

[1] If one starts from the right factor as was usual in earlier literature, the two pairs of components ψ_1, ψ_2 and ψ_3, ψ_4 are interchanged. This choice is made by *Mott* and *Massey* [3.1].

Then (3.14) represents a system of four simultaneous first-order partial differential equations:

$$i\hbar \frac{\partial}{\partial t}\psi_1 + i\hbar c\left(\frac{\partial}{\partial x}\psi_4 - i\frac{\partial}{\partial y}\psi_4 + \frac{\partial}{\partial z}\psi_3\right) - mc^2\psi_1 = 0$$

$$i\hbar \frac{\partial}{\partial t}\psi_2 + i\hbar c\left(\frac{\partial}{\partial x}\psi_3 + i\frac{\partial}{\partial y}\psi_3 - \frac{\partial}{\partial z}\psi_4\right) - mc^2\psi_2 = 0 \tag{3.15}$$

$$i\hbar \frac{\partial}{\partial t}\psi_3 + i\hbar c\left(\frac{\partial}{\partial x}\psi_2 - i\frac{\partial}{\partial y}\psi_2 + \frac{\partial}{\partial z}\psi_1\right) + mc^2\psi_3 = 0$$

$$i\hbar \frac{\partial}{\partial t}\psi_4 + i\hbar c\left(\frac{\partial}{\partial x}\psi_1 + i\frac{\partial}{\partial y}\psi_1 - \frac{\partial}{\partial z}\psi_2\right) + mc^2\psi_4 = 0.$$

Let us now turn to the description of free electrons by the Dirac equation. We set the z axis in the direction of propagation and take the wave function to be

$$\psi_j = a_j e^{i(kz - \omega t)}, \tag{3.16}$$

where the a_j must be determined such that equations (3.15) are solved. Substituting (3.16) into (3.15) and using the abbreviations $E = \hbar\omega$ and $p_z = \hbar k$, one obtains

$$
\begin{aligned}
(E - mc^2)a_1 && -cp_z a_3 && &= 0 \\
& (E - mc^2)a_2 && +cp_z a_4 &= 0 \\
-cp_z a_1 && +(E + mc^2)a_3 && &= 0 \\
& cp_z a_2 && +(E + mc^2)a_4 &= 0.
\end{aligned}
\tag{3.17}
$$

As a homogeneous linear system of equations, (3.17) has only a nontrivial solution if the determinant

$$
\begin{vmatrix}
(E - mc^2) & 0 & -cp_z & 0 \\
0 & (E - mc^2) & 0 & cp_z \\
-cp_z & 0 & (E + mc^2) & 0 \\
0 & cp_z & 0 & (E + mc^2)
\end{vmatrix}
\tag{3.18}
$$

vanishes, that is, if $(E^2 - m^2 c^4 - c^2 p_z^2)^2 = 0$, or

$$E = \pm\sqrt{c^2 p_z^2 + m^2 c^4}. \quad [2] \tag{3.19}$$

We therefore obtain the reasonable result that the system of equations (3.17) can be solved only on the condition that the relativistic energy law holds.

---

[2] In this section E includes the rest energy, whereas we usually denote by E the energy without rest energy as is customary in low-energy physics.

We shall now consider only the positive root in (3.19) (electrons) and not the negative energy states (positrons). Under the condition (3.19), the determinant (3.18) is of rank 2 (i.e., all 3×3 minor determinants vanish). This means that we obtain two linearly independent solutions from (3.17). We can write these as

$$a_1 = 1, \qquad a_2 = 0, \qquad a_3 = \frac{cp_z}{E + mc^2}, \qquad a_4 = 0 \qquad \text{and} \tag{3.20}$$

$$a_1 = 0, \qquad a_2 = 1, \qquad a_3 = 0, \qquad a_4 = \frac{-cp_z}{E + mc^2}, \tag{3.21}$$

as can easily be verified by substituting into (3.17). By linearly combining these independent solutions, one obtains the general solution, so that we find for the general form of the plane wave

$$\left[A \begin{pmatrix} 1 \\ 0 \\ \dfrac{cp_z}{E + mc^2} \\ 0 \end{pmatrix} + B \begin{pmatrix} 0 \\ 1 \\ 0 \\ \dfrac{-cp_z}{E + mc^2} \end{pmatrix} \right] e^{i(kz - \omega t)}, \tag{3.22}$$

where A and B are constants.

Let us now investigate which spin states are described by the solutions obtained. For this we must first find the form of the spin operator in the Dirac theory. We start from the fact that the orbital angular momentum operator $l = r \times p$ does not commute with the Hamiltonian operator of the Dirac theory. This means that l is not a constant of the motion, since for every operator Q that is not explicitly time dependent one has the relation

$$\frac{dQ}{dt} = \frac{1}{i\hbar} [Q, H] = \frac{1}{i\hbar} (QH - HQ).$$

This is different from what one would expect from classical physics or nonrelativistic quantum mechanics where, for a central field, l is a constant of the motion.

We will show for the component l_x that l does not commute with the Hamiltonian

$$H = c\boldsymbol{\alpha} \cdot \boldsymbol{p} + \beta mc^2 + V(r). \tag{3.23}$$

One has

$$[l_x, H] = c \left[(yp_z - zp_y) \left(\sum_\mu \alpha_\mu p_\mu + \beta mc + \frac{V(r)}{c} \right) \right.$$
$$\left. - \left(\sum_\mu \alpha_\mu p_\mu + \beta mc + \frac{V(r)}{c} \right) (yp_z - zp_y) \right].$$

The terms containing $V(r)$ cancel each other since l commutes with a spherically symmetric potential as one can recall from elementary quantum mechanics or easily check.

As the matrices α_μ and β contain only constants, they commute with coordinates and derivatives with respect to coordinates; thus

$$[l_x, H] = c\left[\sum_\mu \alpha_\mu (yp_z - zp_y)p_\mu - \sum_\mu \alpha_\mu p_\mu (yp_z - zp_y)\right].$$

If one further observes that the space coordinates and likewise the momentum coordinates commute with each other and that one has

$$yp_y - p_y y = i\hbar, \qquad xp_y - p_y x = 0, \quad \text{etc.,}$$

then it follows that

$$[l_x, H] = -i\hbar c(\alpha_z p_y - \alpha_y p_z). \tag{3.24}$$

Corresponding results are obtained for the components l_y and l_z.

A theory which violates conservation of angular momentum for central forces is not satisfactory. Thus we must find an operator whose commutator with H is the negative of $[l, H]$; the sum of this operator and l would then commute with H (i.e., represent a constant of the motion) and could thus be conceived of as the total angular momentum operator. Such an operator can indeed be found; it is

$$s = \frac{\hbar}{2}\sigma \quad \text{with}$$

$$\sigma_x = \begin{pmatrix} 0 & 1 & 0 & 0 \\ 1 & 0 & 0 & 0 \\ 0 & 0 & 0 & 1 \\ 0 & 0 & 1 & 0 \end{pmatrix}, \quad \sigma_y = \begin{pmatrix} 0 & -i & 0 & 0 \\ i & 0 & 0 & 0 \\ 0 & 0 & 0 & -i \\ 0 & 0 & i & 0 \end{pmatrix},$$

$$\sigma_z = \begin{pmatrix} 1 & 0 & 0 & 0 \\ 0 & -1 & 0 & 0 \\ 0 & 0 & 1 & 0 \\ 0 & 0 & 0 & -1 \end{pmatrix}, \tag{3.25}$$

which is a generalization of the Pauli spin operator (2.2). One can easily see (cf. Problem 3.1) that

$$[s, H] = -[l, H] \tag{3.26}$$

and thus

$$[(l+s), H] = 0. \tag{3.27}$$

Since $l+(\hbar/2)\boldsymbol{\sigma}$ can be interpreted as the total angular momentum operator it appears obvious that $(\hbar/2)\boldsymbol{\sigma}$ is the spin operator. (This argument will later be strengthened.)

We now come back to the question of which spin states are represented by our solutions of the Dirac equation. It will be shown that (3.20) and (3.21) represent electron waves with spin parallel and antiparallel to the direction of propagation.

The first solution satisfies the equation[3]

$$\sigma_z \begin{pmatrix} 1 \\ 0 \\ \dfrac{cp_z}{E+mc^2} \\ 0 \end{pmatrix} e^{ikz} = 1 \cdot \begin{pmatrix} 1 \\ 0 \\ \dfrac{cp_z}{E+mc^2} \\ 0 \end{pmatrix} e^{ikz}, \qquad (3.28)$$

that is, the wave function given in (3.20) is an eigenfunction of $s_z = (\hbar/2)\sigma_z$ with the eigenvalue $+(\hbar/2)$. Similarly, because

$$\sigma_z \begin{pmatrix} 0 \\ 1 \\ 0 \\ \dfrac{-cp_z}{E+mc^2} \end{pmatrix} e^{ikz} = \begin{pmatrix} 0 \\ -1 \\ 0 \\ \dfrac{cp_z}{E+mc^2} \end{pmatrix} e^{ikz} = -1 \cdot \begin{pmatrix} 0 \\ 1 \\ 0 \\ \dfrac{-cp_z}{E+mc^2} \end{pmatrix} e^{ikz} \qquad (3.29)$$

the solution (3.21) is an eigenfunction of s_z with eigenvalue $-(\hbar/2)$.

Contrary to the nonrelativistic case, an eigenfunction of σ_x cannot be constructed now by the superposition of the eigenfunctions of σ_z with the eigenvalues ± 1. It can immediately be seen that the wave function formed by this superposition is not an eigenstate of σ_x:

$$\begin{pmatrix} 0 & 1 & 0 & 0 \\ 1 & 0 & 0 & 0 \\ 0 & 0 & 0 & 1 \\ 0 & 0 & 1 & 0 \end{pmatrix} \begin{pmatrix} 1 \\ 1 \\ \dfrac{cp_z}{E+mc^2} \\ \dfrac{-cp_z}{E+mc^2} \end{pmatrix} e^{ikz} = \begin{pmatrix} 1 \\ 1 \\ \dfrac{-cp_z}{E+mc^2} \\ \dfrac{cp_z}{E+mc^2} \end{pmatrix} e^{ikz}. \qquad (3.30)$$

This is not an eigenvalue equation, except for the special case $p_z = 0$.

This result was to be expected when one considers that it is not the spin but the total angular momentum that is a constant of the motion. Only if l or particular components l_μ vanish, are s or the corresponding s_μ constant. This is

---

[3] We neglect the irrelevant factor $\exp(-i\omega t)$.

why for our plane wave in the z direction, for which $l_z = 0$ (but $l_x, l_y \neq 0$), one can find eigenvalues only of s_z. The components of orbital angular momentum l_x and l_y vanish only if $p = 0$. In this limiting case, eigenvalues of s_x and s_y will exist, as (3.30) shows for s_x.

Hence it can be seen that in the relativistic case it is only in the rest frame of the electron that one can speak of a transverse spin direction of the plane wave (i.e., spins perpendicular to the direction of propagation). Only in the rest frame can one assign an eigenvalue to the spin operator in an arbitrary direction ϑ, φ. The spin part of the appropriate eigenfunction in the rest frame is, from (3.22),

$$\begin{pmatrix} A \\ B \\ 0 \\ 0 \end{pmatrix}, \tag{3.31}$$

so that in the limiting case where the momentum vanishes only two of the spinor components are different from zero. By referring to the results of Sect. 2.2 for two-component spinors, we see that the components of the spin function for the direction ϑ, φ which is defined in the rest frame are connected by the relation [cf. (2.8)]

$$\frac{B}{A} = \tan \frac{\vartheta}{2} \, e^{i\varphi}. \tag{3.32}$$

Since an arbitrary spin direction can be defined only in the rest frame, a definition of the polarization also makes sense only in the rest frame. Except for this restriction, we define the polarization as the expectation value of the spin operator, exactly as before.

To show what happens if one does calculate the polarization in the laboratory system instead of the rest system, we take, for example, the state

$$\chi^{(x)} = \begin{pmatrix} 1 \\ 1 \\ \dfrac{cp_z}{E+mc^2} \\ \dfrac{-cp_z}{E+mc^2} \end{pmatrix} e^{ikz}.$$

We have already seen from (3.30) that $\chi^{(x)}$ is not an eigensolution of σ_x in the laboratory system, but only in the rest frame. We calculate

$$P_x = \frac{\langle \chi^{(x)} | \sigma_x | \chi^{(x)} \rangle}{\langle \chi^{(x)} | \chi^{(x)} \rangle} \tag{3.33}$$

and find, using (3.30), that

$$P_x = \frac{2 \cdot \left(1 - \dfrac{c^2 p_z^2}{(E + mc^2)^2}\right)}{2 \cdot \left(1 + \dfrac{c^2 p_z^2}{(E + mc^2)^2}\right)} \cdot$$

As $E = m\gamma c^2$ (with $\gamma = 1/\sqrt{1 - \beta^2}$) and

$$c^2 p_z^2 = E^2 - m^2 c^4 = m^2 c^4 (\gamma^2 - 1),$$

we get

$$P_x = \frac{m^2 c^4 (\gamma + 1)^2 - m^2 c^4 (\gamma^2 - 1)}{m^2 c^4 (\gamma + 1)^2 + m^2 c^4 (\gamma^2 - 1)} = \frac{2\gamma + 2}{2\gamma^2 + 2\gamma}$$

$$= \frac{1}{\gamma} = \sqrt{1 - \beta^2}. \tag{3.34}$$

Thus the polarization depends on the reference system; for an unambiguous definition it is therefore practical to refer it to the rest frame.

It is also possible to make "Lorentz invariant" (more precisely: covariant) definitions of the polarization, where it is not necessary to refer to the rest frame [3.2]. For our purposes, however, the definition as the expectation value of the spin operator in the rest frame suffices.

To conclude, we summarize what the Dirac equation describes:

a) relativistic electrons (although it has not been shown here that the theory is Lorentz invariant, this appears plausible as it started from the relativistic energy equation),
b) the spin 1/2 of electrons,
c) the magnetic moment $\varepsilon\hbar/2\,mc$ of electrons,
d) spin-orbit coupling.

The last two points have not yet been shown here and their derivations will not be given in full. We will only outline the method of the somewhat tedious calculation.

If we consider electrons in external fields, we must again substitute $p - (\varepsilon/c)A$ and $H - \varepsilon\phi$ for p and H. The Dirac equation is then, see (3.12),

$$\left[H - \varepsilon\phi - c\boldsymbol{\alpha} \cdot \left(\boldsymbol{p} - \frac{\varepsilon}{c}\,\boldsymbol{A}\right) - \beta mc^2\right]\psi = 0. \tag{3.35}$$

In order to be able to compare it with the Schrödinger equation, we reduce (3.35) to the nonlinearized form of (3.10), writing

$$\left[H - \varepsilon\phi - c\boldsymbol{\alpha} \cdot \left(\boldsymbol{p} - \frac{\varepsilon}{c}\boldsymbol{A} \right) - \beta mc^2 \right]$$
$$\times \left[H - \varepsilon\phi + c\boldsymbol{\alpha} \cdot \left(\boldsymbol{p} - \frac{\varepsilon}{c}\boldsymbol{A} \right) + \beta mc^2 \right] \psi = 0. \tag{3.36}$$

By multiplying out and making the approximation that the kinetic and potential energies are small compared with the rest energy mc^2 so that two components of the spin function can be neglected, one obtains

$$\left[\frac{1}{2m}\left(\boldsymbol{p} - \frac{\varepsilon}{c}\boldsymbol{A} \right)^2 + \varepsilon\phi - \frac{\varepsilon\hbar}{2mc}\,\boldsymbol{\sigma}\cdot\boldsymbol{B} + \mathrm{i}\,\frac{\varepsilon\hbar}{4m^2c^2}\,\boldsymbol{E}\cdot\boldsymbol{p} \right.$$
$$\left. - \frac{\varepsilon\hbar}{4m^2c^2}\,\boldsymbol{\sigma}\cdot(\boldsymbol{E}\times\boldsymbol{p}) \right]\psi = W\psi \tag{3.37}$$

when $W + mc^2$ is the total energy.[4]

The first two terms on the left are identical with those of the Schrödinger equation for external fields. The third term corresponds to the interaction energy $-\boldsymbol{\mu}\cdot\boldsymbol{B}$ between a magnetic dipole, whose moment is represented by the operator $\boldsymbol{\mu} = (\varepsilon\hbar/2\,mc)\boldsymbol{\sigma} = (\varepsilon/mc)\boldsymbol{s}$, and the external magnetic field. The fact that $(\varepsilon\hbar/2\,mc)\boldsymbol{\sigma}$ appears here as the operator of the magnetic moment, is a further reason for taking $(\hbar/2)\boldsymbol{\sigma}$ as the operator of the spin which is connected with this moment.

The fourth term is a relativistic correction to the energy and has no classical analogue. The meaning of the last term can again be illustrated. It describes the spin-orbit coupling. Since according to Maxwell's electrodynamics the vectors of the electromagnetic field are dependent on the reference system, an observer on an electron moving with velocity $\boldsymbol{v}$ relative to an electric field $\boldsymbol{E}$ finds a magnetic field[5] $\boldsymbol{B} = -c^{-1}\boldsymbol{v}\times\boldsymbol{E} = (mc)^{-1}(\boldsymbol{E}\times\boldsymbol{p})$ [terms of the order $(v/c)^2$ are neglected]. Thus, in its rest frame, an electron moving relative to the electric field of an atomic nucleus experiences this magnetic field, which affects its spin. The energy of the electron, due to its magnetic moment $\boldsymbol{\mu}$, in this field is

$$-\boldsymbol{\mu}\cdot\boldsymbol{B} = -\frac{\varepsilon}{mc}\,\boldsymbol{s}\cdot\boldsymbol{B} = -\frac{\varepsilon}{m^2c^2}\,\boldsymbol{s}\cdot(\boldsymbol{E}\times\boldsymbol{p}). \tag{3.38}$$

Hence an additional energy term is obtained in the classical Hamiltonian function. If we substitute the spin operator $(\hbar/2)\boldsymbol{\sigma}$ for $\boldsymbol{s}$, we obtain the fifth term in the Hamiltonian operator (3.37) except for the factor 2. This factor is missing because our interpretation was too rough. We have not taken into account that

---

[4] For a more detailed discussion of this equation and the approximations made see [3.3, 4].
[5] For the derivation of the formula refer to a textbook on electrodynamics.

in changing the frame of reference, the time transformation changes the precession frequency of the electron spin in the magnetic field (Thomas precession).

The term

$$-\frac{\varepsilon\hbar}{4m^2c^2}\,\sigma\cdot(E\times p) \tag{3.39}$$

is called the spin-orbit energy, as it arises from the interaction of the spin with the magnetic field produced by the orbital motion of the electron, as we have illustrated. If the motion takes place in a central field of potential energy $V(r)$ where $E=-\varepsilon^{-1}(dV/dr)(r/r)$, we obtain from (3.39)

$$-\frac{\varepsilon}{2m^2c^2}\,s\cdot\left(-\frac{1}{\varepsilon}\frac{dV}{dr}\frac{r}{r}\times p\right)=\frac{1}{2m^2c^2}\frac{1}{r}\frac{dV}{dr}\,(s\cdot l). \tag{3.40}$$

Thus the spin-orbit energy, in the case of the Coulomb potential, decreases more quickly with increasing distance than does the Coulomb energy itself and can therefore be neglected at fairly large distances from the nucleus.

Before we turn to the treatment of the scattering problem by the Dirac equation, it should be emphasized that spin, magnetic moment, and spin-orbit coupling, which are very important in the following sections, were not introduced by making additional assumptions. They follow automatically from the first principles from which Dirac's derivation started.

Problem 3.1. Show the validity of the formula

$$\frac{\hbar}{2}\,[\sigma,H]=-[l,H]$$

taking the x components as an example.

Solution.   From (3.24) one has $[l_x,H]=-i\hbar c(\alpha_z p_y-\alpha_y p_z)$; thus only $[\sigma_x,H]$ remains to be calculated. As $\sigma_x V(r)=V(r)\sigma_x$ it follows that

$$[\sigma_x,H]=c\left[\sigma_x\left(\sum_\mu\alpha_\mu p_\mu+\beta mc\right)-\left(\sum_\mu\alpha_\mu p_\mu+\beta mc\right)\sigma_x\right].$$

One sees immediately from the matrices β and α_x that $\sigma_x\beta=\beta\sigma_x$ and $\sigma_x\alpha_x=\alpha_x\sigma_x$. Thus

$$[\sigma_x,H]=c[\sigma_x(\alpha_y p_y+\alpha_z p_z)-(\alpha_y p_y+\alpha_z p_z)\sigma_x].$$

Due to the relations

$$\sigma_x\alpha_y=\begin{pmatrix}0&0&i&0\\0&0&0&-i\\i&0&0&0\\0&-i&0&0\end{pmatrix}=i\alpha_z,\qquad \alpha_y\sigma_x=\begin{pmatrix}0&0&-i&0\\0&0&0&i\\-i&0&0&0\\0&i&0&0\end{pmatrix}=-i\alpha_z,$$

$$\sigma_x\alpha_z = \begin{pmatrix} 0 & 0 & 0 & -1 \\ 0 & 0 & 1 & 0 \\ 0 & -1 & 0 & 0 \\ 1 & 0 & 0 & 0 \end{pmatrix} = -i\alpha_y, \qquad \alpha_z\sigma_x = \begin{pmatrix} 0 & 0 & 0 & 1 \\ 0 & 0 & -1 & 0 \\ 0 & 1 & 0 & 0 \\ -1 & 0 & 0 & 0 \end{pmatrix} = i\alpha_y$$

and the fact that p commutes with σ one obtains

$$\frac{\hbar}{2}[\sigma_x, H] = i\hbar c(\alpha_z p_y - \alpha_y p_z) = -[l_x, H].$$

3.2 Calculation of the Differential Scattering Cross Section

The differential cross section for elastic scattering of an electron beam with arbitrary spin direction is calculated using the Dirac equation. The scattering cross section depends on the azimuthal angle ϕ; this means that there is generally no axial symmetry of the scattered intensity with respect to the incident direction. The asymmetry is described by the Sherman function.

We are now in a position to deal with the scattering of relativistic electrons with spin by a central field. The electrons will be taken as an incident plane wave in the z direction. In analogy to nonrelativistic scattering theory, we look for solutions to the Dirac equation with the asymptotic form

$$\psi_\lambda \xrightarrow[r\to\infty]{} a_\lambda e^{ikz} + a'_\lambda(\theta, \phi)\, \frac{e^{ikr}}{r} \tag{3.41}$$

for the four components of the wave function ($\lambda = 1, \ldots, 4$). Generalizing the differential elastic scattering cross section obtained from the Schrödinger theory one finds

$$\frac{d\sigma}{d\Omega}(\theta, \phi) \equiv \sigma(\theta, \phi) = \frac{\sum\limits_{\lambda=1}^{4} |a'_\lambda(\theta, \phi)|^2}{\sum\limits_{\lambda=1}^{4} |a_\lambda|^2}. \tag{3.42}$$

This follows from the general definition of the differential cross section and from the fact that the current density can be written as $\varrho v = \psi^\dagger\psi v$. If one uses normalized wave functions, the denominator in (3.42) is 1.

In order to simplify this expression, we use the fact that the a_λ are not independent of each other. This can be seen from the solution (3.22) for a plane wave with arbitrary spin direction which shows that

$$r \equiv \frac{|a_3|}{|a_1|} = \frac{|a_4|}{|a_2|} = \frac{cp_z}{E+mc^2}. \tag{3.43}$$

The same relation exists asymptotically between the a'_λ because at very large distances from the scattering center, the scattered spherical wave can be regarded as made up of plane waves proceeding in different directions from the center. Therefore we have

$$\sigma(\theta, \phi) = \frac{|a'_1|^2 + |a'_2|^2 + |a'_1|^2 r^2 + |a'_2|^2 r^2}{|a_1|^2 + |a_2|^2 + |a_1|^2 r^2 + |a_2|^2 r^2} = \frac{|a'_1|^2 + |a'_2|^2}{|a_1|^2 + |a_2|^2}. \tag{3.44}$$

In the following we consider the scattering of electron waves whose spins are oriented in the $+z$ or $-z$ direction (longitudinal polarization). If we have the solutions to the scattering problem for these two basic states we can construct all other cases from them. By coherent superposition we obtain, for example, scattering of a beam with arbitrary spin direction and $P=1$ (see Sect. 2.2); by incoherent superposition we obtain scattering of an unpolarized beam (see Sect. 2.3).

In the case in which the spin of the incident wave is in the $+z$ direction, its wave function, from (3.28), is

$$\begin{pmatrix} \psi_1 \\ \psi_2 \\ \vdots \end{pmatrix} = \begin{pmatrix} 1 \\ 0 \\ \vdots \end{pmatrix} e^{ikz}.$$

The "small" components ψ_3 and ψ_4, due to their dependence on the "large" components ψ_1 and ψ_2 [see (3.43)], yield no additional information for our scattering problem as is shown clearly by (3.44); they therefore need not be considered. To solve the scattering problem for this particular choice of incident wave one must look for solutions of the Dirac equation with the asymptotic form

$$\begin{pmatrix} \psi_1 \\ \psi_2 \end{pmatrix} \xrightarrow{r \to \infty} \begin{pmatrix} 1 \\ 0 \end{pmatrix} e^{ikz} + \begin{pmatrix} S_{11}(\theta, \phi) \\ S_{21}(\theta, \phi) \end{pmatrix} \frac{e^{ikr}}{r}. \tag{3.45}$$

This takes account of the fact that the second component of the wave function is no longer necessarily zero after scattering, since the spin may change its direction due to spin-orbit coupling as described by the scattering amplitude S_{21}. The electron "sees" in its rest frame the moving charge of the scattering center; that means it sees a current and thus a magnetic field, which acts upon its magnetic moment and may change its spin direction (see end of Sect. 3.1). This possibility is taken into account through the inclusion of S_{21}, which is therefore called the spin-flip amplitude.

Analogously, for the incident wave with the other spin direction, we expect a solution with the asymptotic form

$$\begin{pmatrix} \psi_1 \\ \psi_2 \end{pmatrix} \xrightarrow{r \to \infty} \begin{pmatrix} 0 \\ 1 \end{pmatrix} e^{ikz} + \begin{pmatrix} S_{12}(\theta, \phi) \\ S_{22}(\theta, \phi) \end{pmatrix} \frac{e^{ikr}}{r}. \tag{3.46}$$

As in the case of the Schrödinger equation, the scattering problem can be solved by using the method of partial waves in which one looks for particular solutions with specific angular momenta and constructs from them a general solution with the required boundary conditions. In our case, the procedure is the same, but the solution is considerably more complicated. This is because one does not (as in the Schrödinger theory) have only one differential equation for one function, but instead has a system of simultaneous differential equations. We shall not reproduce the calculation step by step – it can be found in [3.1]. We will only explain the essential ideas and emphasize the physical background.

Let us first consider the case in which the spin of the incident beam is in the $+z$ direction which we take as the quantization axis. By separating the variables in the Dirac equation for a central field, one obtains the particular solutions

$$\begin{pmatrix} \psi_1 \\ \psi_2 \end{pmatrix} = \begin{pmatrix} (l+1)G_l(r)P_l(\cos\theta) \\ -G_l(r)P_l^1(\cos\theta)\,e^{i\phi} \end{pmatrix} \quad \text{and}$$

$$\begin{pmatrix} \psi_1 \\ \psi_2 \end{pmatrix} = \begin{pmatrix} lG_{-l-1}(r)P_l(\cos\theta) \\ G_{-l-1}(r)P_l^1(\cos\theta)\,e^{i\phi} \end{pmatrix}, \qquad (3.47)$$

where $P_l^1(\theta)$ are associated Legendre functions; ϕ is the azimuthal angle. Thus one finds a *pair* of solutions just as in the case of the plane wave. There the two solutions differed by the spin directions which they described. What we have obtained for a central potential is quite analogous, as will now be illustrated.

The functions $G_l(r)$ and $G_{-l-1}(r)$ are solutions of the two r-dependent ordinary differential equations which arise from separating the variables in the Dirac equation. The fact that here, contrary to the Schrödinger theory, not one but two radial differential equations appear can be explained by the spin-orbit coupling term. For this we consider once again the case of small velocities as at the end of Sect. 3.1. Then the operator for spin-orbit coupling is proportional to $r^{-1}(dV/dr)(\mathbf{l}\cdot\mathbf{s})$ [see (3.40)], and since $j^2 = (\mathbf{l}+\mathbf{s})^2 = l^2 + s^2 + 2\mathbf{l}\cdot\mathbf{s}$, it is proportional to $r^{-1}(dV/dr)(j^2 - l^2 - s^2)$. When this r-dependent operator is applied to the wave function it produces a term in the radial differential equation which is proportional to $r^{-1}(dV/dr)[j(j+1) - l(l+1) - s(s+1)]$. Since

$$[j(j+1) - l(l+1) - s(s+1)] = \begin{cases} l, & \text{if } j = l+\tfrac{1}{2} \\ -l-1, & \text{if } j = l-\tfrac{1}{2} \end{cases} \qquad (3.48)$$

one obtains a different differential equation for each of the two spin orientations and thus two different solutions G_l and G_{-l-1}.

The physically important occurrence in (3.47) of a ϕ-dependent term which does not occur in the Schrödinger treatment of scattering can also be intuitively explained. Due to the conservation law for j, the angular momentum component m_j in the z direction must be $+(1/2)$ after the scattering, since before the scattering $m_s = +(1/2)$, $m_l = 0$, i.e., $m_j = +(1/2)$, according to the initial assumption. For a spin flip as described by ψ_2 the decrease of the z component of the spin must be compensated for by an increase of the orbital angular momentum component in this direction. Thus a nonzero expectation value for the z

component of the orbital angular momentum must exist. Since the operator of this component is

$$l_z = -i\hbar \frac{\partial}{\partial \phi},$$

this is possible only if the solution contains a ϕ-dependent term.

As with the treatment of scattering in the Schrödinger theory, the solutions of the radial differential equations for potentials which decrease faster than r^{-1} have the asymptotic form

$$G_l \xrightarrow[r \to \infty]{} \frac{\sin [kr - (l\pi/2) + \eta_l]}{kr}$$

$$(3.49)$$

$$G_{-l-1} \xrightarrow[r \to \infty]{} \frac{\sin [kr - (l\pi/2) + \eta_{-l-1}]}{kr}.$$

In the important case of the r^{-1} potential a logarithmic term must be included in the argument of the sine function; everything else remains valid.

Let us now construct from the particular solutions (3.47) (partial waves) a solution which has the required form (3.45). This can be done if they are combined as follows

$$\psi_1 = \sum_{l=0}^{\infty} [(l+1) e^{i\eta_l} G_l + l e^{i\eta_{-l-1}} G_{-l-1}] i^l P_l(\cos \theta)$$

$$(3.50)$$

$$\psi_2 = \sum_{l=1}^{\infty} [-e^{i\eta_l} G_l + e^{i\eta_{-l-1}} G_{-l-1}] i^l P_l^1 (\cos \theta) e^{i\phi}.$$

One can easily check (see Problem 3.2) that the condition (3.45) is then fulfilled; one needs only to choose

$$S_{11}(\theta, \phi) = \frac{1}{2ik} \sum_{l=0}^{\infty} [(l+1) (e^{2i\eta_l} - 1) + l(e^{2i\eta_{-l-1}} - 1)] P_l(\cos \theta) \equiv f(\theta)$$

and

$$(3.51)$$

$$S_{21}(\theta, \phi) = \frac{1}{2ik} \sum_{l=1}^{\infty} (-e^{2i\eta_l} + e^{2i\eta_{-l-1}}) P_l^1 (\cos \theta) e^{i\phi} \equiv g(\theta) e^{i\phi}.$$

The solution of the scattering problem is thus reduced to a calculation of the scattering phases η. These depend, as in the nonrelativistic case, on the energy of the incident electrons and on the scattering potential (scattering substance). Hence the scattering amplitudes $S_{\mu\nu}$, apart from depending on the scattering angle, also depend on these two variables.

An analogous treatment for the spin of the incident wave in the $-z$ direction yields for S_{22} and S_{12} in (3.46)

$$S_{22} = S_{11} = f(\theta)$$
$$S_{12} = -S_{21}\, e^{-2i\phi} = -g(\theta)\, e^{-i\phi}, \tag{3.52}$$

with S_{11} and S_{21} from (3.51).

By (coherent) superposition of the two basic solutions with spin directions parallel and antiparallel to z, we can easily treat the case of an incident wave with an arbitrary spin direction,

$$A \begin{pmatrix} 1 \\ 0 \end{pmatrix} e^{ikz} + B \begin{pmatrix} 0 \\ 1 \end{pmatrix} e^{ikz} = \begin{pmatrix} A \\ B \end{pmatrix} e^{ikz}, \tag{3.53}$$

where A and B, according to (3.32), specify the spin direction in the rest frame. Using (3.45, 46, 52, 53), one obtains by coherent superposition

$$A \begin{pmatrix} S_{11} \\ S_{21} \end{pmatrix} \frac{e^{ikr}}{r} + B \begin{pmatrix} S_{12} \\ S_{22} \end{pmatrix} \frac{e^{ikr}}{r} = \begin{pmatrix} Af - Bge^{-i\phi} \\ Bf + Age^{i\phi} \end{pmatrix} \frac{e^{ikr}}{r} = \begin{pmatrix} a_1' \\ a_2' \end{pmatrix} \frac{e^{ikr}}{r} \tag{3.54}$$

for the asymptotic form of the scattered wave. Thus from (3.44) the differential cross section is

$$\sigma(\theta, \phi) = \frac{|a_1'|^2 + |a_2'|^2}{|A|^2 + |B|^2}$$

$$= |f|^2 + |g|^2 + \frac{-AB^* e^{i\phi} + A^* B e^{-i\phi}}{|A|^2 + |B|^2}\, (fg^* - f^* g), \tag{3.55}$$

which shows that for a polarized primary beam the scattering intensity generally depends on ϕ. By substituting

$$S(\theta) = i\, \frac{fg^* - f^* g}{|f|^2 + |g|^2} \tag{3.56}$$

(Sherman function [3.5]; it is real as its numerator is the difference of two conjugate complex functions), it follows that

$$\sigma(\theta, \phi) = (|f|^2 + |g|^2) \left[1 + S(\theta)\, \frac{-AB^* e^{i\phi} + A^* B e^{-i\phi}}{i(|A|^2 + |B|^2)} \right]. \tag{3.57}$$

Example 3.1. $A = 1$ and $B = 1$, i.e., transverse polarization, since according to (3.32–34) the primary beam is totally polarized in the x direction. From (3.57) one has $\sigma(\theta, \phi) = (|f|^2 + |g|^2)\, [1 - S(\theta)\sin\phi]$. The ϕ dependence of the cross section is not surprising since the primary beam is not axially symmetric with respect to the direction of propagation. As shown in Figs. 3.1 and 3.2, the scattering asymmetry is maximum when the scattering plane (plane described by primary beam and direction of observation) is perpendicular to P, that is, when

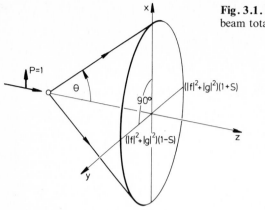

Fig. 3.1. Left-right scattering asymmetry of a beam totally polarized in the x direction

$\phi = 90°$ and $270°$. This "left-right" asymmetry of the scattering is used to measure the polarization of electron beams ("Mott[6] detector"). The scattering intensities "up" and "down" ($\phi = 0°$ and $180°$) are equal.

Example 3.2. $A = 1$, $B = 0$ or $A = 0$, $B = 1$, i.e., longitudinal polarization. These are our basic functions with spin parallel or antiparallel to the incident direction. In this case, the ϕ-dependent part of the scattering cross section disappears according to (3.57). This is to be expected, as here – unlike the first example – the incident beam is axially symmetric.

Problem 3.2. Prove that the solution (3.50) has the required asymptotic form (3.45), if (3.49) and (3.51) are fulfilled.

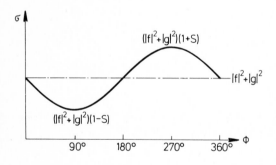

Fig. 3.2. Dependence of the differential cross section on the azimuthal angle ϕ (for $S > 0$; see Sect. 3.6)

---

[6] The relativistic theory of electron scattering dealt with in this section originated with *Mott* (see [3.1]; further references are given there).

Solution. Using the relations (3.49) one obtains from (3.50)

$$\psi_1 \xrightarrow[r\to\infty]{} \sum_{l=0}^{\infty} \left[(l+1)\exp(i\eta_l) \frac{\exp\left[i\left(kr-\frac{l\pi}{2}+\eta_l\right)\right]-\exp\left[-i\left(kr-\frac{l\pi}{2}+\eta_l\right)\right]}{2\,ikr} \right.$$

$$\left. + l\exp(i\eta_{-l-1}) \frac{\exp\left[i\left(kr-\frac{l\pi}{2}+\eta_{-l-1}\right)\right]-\exp\left[-i\left(kr-\frac{l\pi}{2}+\eta_{-l-1}\right)\right]}{2\,ikr} \right] i^l P_l$$

$$= \frac{1}{2\,ik}\sum_{l=0}^{\infty}\left\{ [(l+1)\exp(2\,i\eta_l)+l\exp(2\,i\eta_{-l-1})]\frac{e^{ikr}}{r} \right.$$

$$\left. - \left[(l+1)\frac{\exp\left[-i\left(kr-\frac{l\pi}{2}\right)\right]}{r}+l\frac{\exp\left[-i\left(kr-\frac{l\pi}{2}\right)\right]}{r}\right] i^l \right\} P_l.$$

Adding

$$\left\{ -[(l+1)+l]\frac{e^{ikr}}{r}+i^l(2l+1)\frac{\exp\left[i\left(kr-\frac{l\pi}{2}\right)\right]}{r} \right\} P_l = 0$$

yields

$$\psi_1 \xrightarrow[r\to\infty]{} \frac{1}{2\,ik}\sum_{l=0}^{\infty}\left(\{(l+1)[\exp(2\,i\eta_l)-1]+l[\exp(2\,i\eta_{-l-1})-1]\}P_l\frac{e^{ikr}}{r} \right.$$

$$\left. +(2l+1)i^l\cdot\frac{\exp\left[i\left(kr-\frac{l\pi}{2}\right)\right]-\exp\left[-i\left(kr-\frac{l\pi}{2}\right)\right]}{r}P_l \right)$$

$$= S_{11}(\theta,\phi)\frac{e^{ikr}}{r}+\sum_{l=0}^{\infty}(2l+1)i^l\frac{\sin\left(kr-\frac{l\pi}{2}\right)}{kr}P_l;$$

hence

$$\psi_1 \xrightarrow[r\to\infty]{} S_{11}(\theta,\phi)\frac{e^{ikr}}{r}+e^{ikz},$$

where we have used the asymptotic expansion for exp(ikz) which can be found in the treatment of scattering with the Schrödinger equation in textbooks on quantum mechanics.

Similarly one obtains

$$\psi_2 \xrightarrow[r\to\infty]{} \frac{1}{2\,ik}\sum_{l=1}^{\infty}\left[-\exp(i\eta_l)\frac{\exp\left[i\left(kr-\frac{l\pi}{2}+\eta_l\right)\right]-\exp\left[-i\left(kr-\frac{l\pi}{2}+\eta_l\right)\right]}{r} \right.$$

$$\left. +\exp(i\eta_{-l-1})\frac{\exp\left[i\left(kr-\frac{l\pi}{2}+\eta_{-l-1}\right)\right]-\exp\left[-i\left(kr-\frac{l\pi}{2}+\eta_{-l-1}\right)\right]}{r} \right] i^l P_l^1 e^{i\phi}$$

$$= \frac{1}{2ik} \sum_{l=1}^{\infty} \frac{e^{ikr}}{r} [-\exp(2i\eta_l) + \exp(2i\eta_{-l-1})] P_l^1 e^{i\phi}$$

$$= S_{21}(\theta, \phi) \frac{e^{ikr}}{r}.$$

3.3 The Role of Spin Polarization in Elastic Scattering

The following will be shown with the use of density matrix formalism: Only the component of the polarization vector which is perpendicular to the scattering plane contributes to the scattering asymmetry. An originally unpolarized electron beam has, after scattering, a polarization of magnitude $S(\theta)$ [$S(\theta)$=Sherman function] perpendicular to the scattering plane. The direction and amplitude of the polarization vector of an arbitrarily polarized primary beam are usually changed by scattering. However, a totally polarized beam remains totally polarized, and a beam polarized perpendicular to the scattering plane retains its direction of polarization. Double scattering experiments are suitable for determining the Sherman function (except for the sign). "Perfect" experiments can be made which yield the maximum information on the scattering process.

3.3.1 Polarization Dependence of the Cross Section

We will now write the differential scattering cross section in a form which shows the influence of the polarization more clearly. For this, we recall the equations (3.53) and (3.54), which show that the spinor $\chi = \begin{pmatrix} A \\ B \end{pmatrix}$ of a pure initial state is transformed by the scattering process to the spinor

$$\chi' = \begin{pmatrix} a_1' \\ a_2' \end{pmatrix} = \begin{pmatrix} AS_{11} + BS_{12} \\ AS_{21} + BS_{22} \end{pmatrix} = \begin{pmatrix} Af - Bge^{-i\phi} \\ Bf + Age^{i\phi} \end{pmatrix} \tag{3.58}$$

of the final state. This can be mathematically represented as transformation

$$\chi' = \begin{pmatrix} a_1' \\ a_2' \end{pmatrix} = \begin{pmatrix} S_{11} & S_{12} \\ S_{21} & S_{22} \end{pmatrix} \begin{pmatrix} A \\ B \end{pmatrix} = \begin{pmatrix} f & -ge^{-i\phi} \\ ge^{i\phi} & f \end{pmatrix} \begin{pmatrix} A \\ B \end{pmatrix} = S\chi \tag{3.59}$$

by means of a matrix, the scattering matrix S for the spin.

The density matrix ϱ' for the scattered state[7] is [cf. (2.22)]

$$\varrho' = \begin{pmatrix} |a_1'|^2 & a_1'a_2'^* \\ a_1'^*a_2' & |a_2'|^2 \end{pmatrix} = \begin{pmatrix} a_1' \\ a_2' \end{pmatrix} (a_1'^*, a_2'^*) = \chi'\chi'^\dagger = S\chi\chi^\dagger S^\dagger. \tag{3.60}$$

---

[7] Since the scattering problem is completely described by two amplitudes (cf. Sect. 3.2), we need only two spinor components, just as in the calculation of the scattering cross section. We can therefore use the density matrix formalism which was developed in Sect. 2.3 for two-component spinors.

Since $\chi\chi^\dagger = \varrho$ (density matrix of the unscattered state), it follows that

$$\varrho' = S\varrho S^\dagger = \tfrac{1}{2} S(1 + \boldsymbol{P} \cdot \boldsymbol{\sigma}) S^\dagger \mathrm{tr}\{\varrho\}. \tag{3.61}$$

Equation (2.23) for unnormalized wave functions has been used here because we did not consider normalization in our treatment of scattering.

If we don't have a pure initial state but a statistical mixture of spin states, i.e., a partially polarized primary beam, (3.61) is valid as it stands; ϱ and ϱ' are then the respective density matrices of the mixed initial and final states (see Problem 3.3).

Equation (3.61) can be used to directly write the dependence of the differential cross section on the polarization $\boldsymbol{P}$ of the incident beam. According to (3.44 and 60), one has

$$\sigma(\theta, \phi) = \mathrm{tr}\{\varrho'\}/\mathrm{tr}\{\varrho\}. \tag{3.62}$$

Therefore with (3.61) the dependence of the differential cross section on the polarization of the primary beam is

$$\sigma(\theta, \phi) = \tfrac{1}{2}\mathrm{tr}\{S(1 + \boldsymbol{P} \cdot \boldsymbol{\sigma})S^\dagger. \tag{3.63}$$

To evaluate this one must form the trace of the product

$$\begin{pmatrix} f & -ge^{-i\phi} \\ ge^{i\phi} & f \end{pmatrix} \begin{pmatrix} 1 + P_z & P_x - iP_y \\ P_x + iP_y & 1 - P_z \end{pmatrix} \begin{pmatrix} f^* & g^*e^{-i\phi} \\ -g^*e^{i\phi} & f^* \end{pmatrix}.$$

The simple calculation yields

$$\sigma(\theta, \phi) = (|f|^2 + |g|^2) \left\{ 1 - \frac{S(\theta)}{2i} \left[e^{i\phi}(P_x - iP_y) - e^{-i\phi}(P_x + iP_y) \right] \right\}. \tag{3.64}$$

[$S(\theta)$ is the Sherman function (3.56) and should not be confused with the scattering matrix S]. The differential cross section is thus independent of the longitudinal polarization component P_z.

With $P_t \exp(\pm i\varphi) = P_x \pm iP_y$, where P_t is the magnitude of the transverse polarization component $\boldsymbol{P}_t$ (cf. Fig. 3.3), and

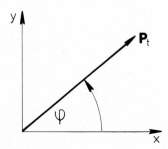

Fig. 3.3. Transverse polarization component

$$|f|^2 + |g|^2 = I(\theta) \tag{3.65}$$

one has

$$\sigma(\theta, \phi) = I(\theta) \left\{ 1 - \frac{S(\theta) P_t}{2 i} \left[e^{i(\phi - \varphi)} - e^{-i(\phi - \varphi)} \right] \right\}$$

$$= I(\theta) \left\{ 1 - P_t S(\theta) \sin (\phi - \varphi) \right\}. \tag{3.66}$$

This shows that for an electron beam which has no transverse polarization the differential cross section is independent of the azimuthal angle ϕ and simply has the value $I(\theta) = |f|^2 + |g|^2$.

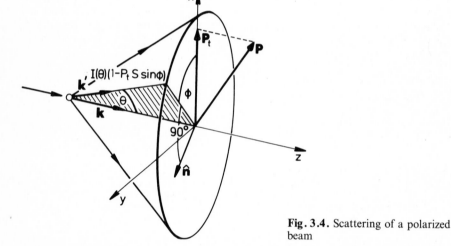

Fig. 3.4. Scattering of a polarized beam

Equation (3.66) can be further simplified if one defines the direction of the transverse polarization component as the x direction (see Fig. 3.4). Then it follows that the differential cross section for a primary beam with arbitrary polarization is

$$\sigma(\theta, \phi) = I(\theta) \left[1 - P_t S(\theta) \sin \phi \right]. \tag{3.67}$$

Frequently this formula is written using the unit vector perpendicular to the scattering plane

$$\hat{n} = \frac{k \times k'}{|k \times k'|} \tag{3.68}$$

(k and k' are, except for the factor $\hbar$, the electron momenta before and after scattering). Since in our suitably chosen coordinate system we have

$P = (P_x, 0, P_z)$, $P_x = P_t$, and

$$\hat{n} = (-\sin \phi, \cos \phi, 0) \tag{3.69}$$

(cf. Fig. 3.4), we obtain $-P_t \sin \phi = P \cdot \hat{n}$ and thus from (3.67)

$$\sigma(\theta, \phi) = I(\theta) [1 + S(\theta) P \cdot \hat{n}]. \tag{3.70}$$

This formula is independent of the choice of coordinate system as the scalar product is invariant under coordinate transformations.

This is the basic equation for the measurement of electron polarization by "Mott scattering", as illustrated by Figs. 3.1 and 3.2. An essential feature of this formula is that only the component of the polarization vector perpendicular to the scattering plane contributes to the scattering asymmetry; components parallel to the scattering plane make no contribution (see also Problem 3.6).

Problem 3.3. Relation (3.61) has been derived for the scattering of a pure spin state. Show that it is also valid for the scattering of a mixture of spin states, when the density matrices of the mixed states are defined by (2.22).

Solution. One has

$$\varrho' = \sum_n \varrho'^{(n)} = \sum_n S \varrho^{(n)} S^\dagger$$

$$= \tfrac{1}{2} S \left[\sum_n (1 + P^{(n)} \cdot \sigma) \operatorname{tr}\{\varrho^{(m)}\} \right] S^\dagger.$$

As $\sum_n \operatorname{tr}\{\varrho^{(m)}\} = \operatorname{tr}\{\varrho\}$ and $\sum_n P^{(m)} \operatorname{tr}\{\varrho^{(m)}\} = P \operatorname{tr}\{\varrho\}$ [see (2.20)], it follows

$$\varrho' = \tfrac{1}{2} S(1 + P \cdot \sigma) S^\dagger \operatorname{tr}\{\varrho\}.$$

3.3.2 Polarization of an Electron Beam by Scattering

The density matrix formalism readily produces the remarkable fact that an initially unpolarized electron beam is polarized by the scattering process. From (2.21) one has

$$P' = \operatorname{tr}\{\varrho' \sigma\} / \operatorname{tr}\{\varrho'\}.$$

If the primary beam is unpolarized one obtains from (3.61), as $P = 0$,

$$\varrho' = \tfrac{1}{2} S(1 + P \cdot \sigma) S^\dagger \operatorname{tr}\{\varrho\} = \tfrac{1}{2} SS^\dagger \operatorname{tr}\{\varrho\} \tag{3.71}$$

so that the polarization of the scattered beam is

$$P' = \frac{1}{2} \frac{\operatorname{tr}\{SS^\dagger \sigma\}}{\operatorname{tr}\{\varrho'\}} \operatorname{tr}\{\varrho\}. \tag{3.72}$$

From (3.62) one has $\mathrm{tr}\{\varrho'\}/\mathrm{tr}\{\varrho\}=\sigma(\theta,\phi)=|f|^2+|g|^2$, where (3.66) has been applied in the special case of an unpolarized primary beam. Furthermore, with $\boldsymbol{\sigma}=\sum_\mu \sigma_\mu \hat{\boldsymbol{e}}_\mu$ [where the $\hat{\boldsymbol{e}}_\mu$ are unit vectors along the coordinate axes and the σ_μ are defined by (2.2)], one has

$$
\begin{aligned}
\tfrac{1}{2}\mathrm{tr}\{SS^\dagger\boldsymbol{\sigma}\} = \tfrac{1}{2}\mathrm{tr}\Bigg\{ &\begin{pmatrix} |f|^2+|g|^2 & fg^*e^{-i\phi}-f^*ge^{-i\phi} \\ f^*ge^{i\phi}-fg^*e^{i\phi} & |f|^2+|g|^2 \end{pmatrix} \\
&\times \begin{pmatrix} \hat{\boldsymbol{e}}_z & \hat{\boldsymbol{e}}_x-i\hat{\boldsymbol{e}}_y \\ \hat{\boldsymbol{e}}_x+i\hat{\boldsymbol{e}}_y & -\hat{\boldsymbol{e}}_z \end{pmatrix} \Bigg\} \\
= \tfrac{1}{2}[(|f|^2+|g|^2)\hat{\boldsymbol{e}}_z &+ (fg^*-f^*g)(\hat{\boldsymbol{e}}_x+i\hat{\boldsymbol{e}}_y)e^{-i\phi} \\
-(fg^*-f^*g)(\hat{\boldsymbol{e}}_x &-i\hat{\boldsymbol{e}}_y)e^{i\phi}-(|f|^2+|g|^2)\hat{\boldsymbol{e}}_z] \\
= i(fg^*-f^*g)(&-\sin\phi\cdot\hat{\boldsymbol{e}}_x+\cos\phi\cdot\hat{\boldsymbol{e}}_y).
\end{aligned}
$$

Since from (3.69) $\hat{\boldsymbol{n}}=(-\sin\phi,\cos\phi,0)$, one finally obtains, by using the Sherman function (3.56),

$$
\boldsymbol{P}'=i\frac{fg^*-f^*g}{|f|^2+|g|^2}\,\hat{\boldsymbol{n}}=S(\theta)\hat{\boldsymbol{n}}. \tag{3.73}
$$

Thus, through scattering, an initially unpolarized beam acquires a polarization of magnitude $S(\theta)$ perpendicular to the scattering plane. This, together with the result of the last section, shows that the Sherman function describes two important features: the extent of the asymmetry in the scattering of a polarized beam and the amount of polarization produced by scattering an unpolarized beam.

3.3.3 Behavior of the Polarization in Scattering

The problem just treated is a special case of the following: Given an incident electron beam with arbitrary polarization $\boldsymbol{P}$, how is this polarization changed by scattering?

The polarization $\boldsymbol{P}'$ after scattering is

$$
\boldsymbol{P}'=\frac{\mathrm{tr}\{\varrho'\boldsymbol{\sigma}\}}{\mathrm{tr}\{\varrho'\}}=\frac{\tfrac{1}{2}\mathrm{tr}\{S(1+\boldsymbol{P}\cdot\boldsymbol{\sigma})S^\dagger\boldsymbol{\sigma}\}}{\tfrac{1}{2}\mathrm{tr}\{S(1+\boldsymbol{P}\cdot\boldsymbol{\sigma})S^\dagger\}}. \tag{3.74}
$$

The denominator has already been evaluated [cf. (3.63–70)] with the result $(|f|^2+|g|^2)[1+S(\theta)\boldsymbol{P}\cdot\hat{\boldsymbol{n}}]$. The numerator is calculated in Problem 3.4. As a result, one obtains for the polarization after scattering

$$
\boldsymbol{P}'=\frac{[\boldsymbol{P}\cdot\hat{\boldsymbol{n}}+S(\theta)]\hat{\boldsymbol{n}}+T(\theta)\,[\boldsymbol{P}-(\boldsymbol{P}\cdot\hat{\boldsymbol{n}})\hat{\boldsymbol{n}}]+U(\theta)\,(\hat{\boldsymbol{n}}\times\boldsymbol{P})}{1+\boldsymbol{P}\cdot\hat{\boldsymbol{n}}S(\theta)} \tag{3.75}
$$

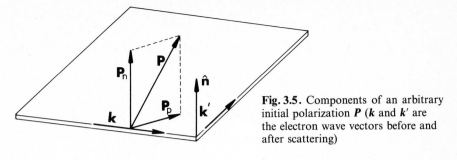

Fig. 3.5. Components of an arbitrary initial polarization P (k and k' are the electron wave vectors before and after scattering)

with

$$S(\theta) = i \frac{fg^* - f^*g}{|f|^2 + |g|^2}, \quad T(\theta) = \frac{|f|^2 - |g|^2}{|f|^2 + |g|^2}, \quad U(\theta) = \frac{fg^* + f^*g}{|f|^2 + |g|^2}. \tag{3.76}$$

If one resolves the polarization vector into components P_p parallel to the scattering plane and P_n perpendicular to it (i.e., parallel to $\hat{n}$, see Fig. 3.5), one has $P = P_n + P_p$. As $(P \cdot \hat{n})\hat{n} = P_n$ and $\hat{n} \times P_n = 0$ or $\hat{n} \times P = \hat{n} \times P_p$, (3.75) can also be written in the form

$$P' = \frac{[P_n + S(\theta)]\hat{n} + T(\theta)P_p + U(\theta)(\hat{n} \times P_p)}{1 + P_n S(\theta)}. \tag{3.77}$$

In this formula only the essential components of the initial polarization appear. By use of the vector relation

$$a \times (b \times c) = (a \cdot c)b - (a \cdot b)c,$$

which in our case yields

$$\hat{n} \times (P \times \hat{n}) = P - (P \cdot \hat{n})\hat{n} = P_p,$$

(3.77) can be rearranged. Then one obtains the frequently used formula in which the complete vector P appears:

$$P' = \frac{[P \cdot \hat{n} + S(\theta)]\hat{n} + T(\theta)\hat{n} \times (P \times \hat{n}) + U(\theta)(\hat{n} \times P)}{1 + P \cdot \hat{n}S(\theta)}. \tag{3.78}$$

From (3.77) we easily see that the scattering process affects the polarization as follows: The (positive or negative) vector $S(\theta)\hat{n}$ is added to the component $P_n\hat{n}$ perpendicular to the scattering plane. The component parallel to the scattering plane is reduced from P_p to TP_p($|T| \leq 1$ by definition). The polarization vector is rotated out of its original plane (P_n, P_p) [identical to the plane $(\hat{n}, P_p)$] as there is an additional component, determined by $U(\theta)$, that is perpendicular to this plane. Last but not least, all components are modified by

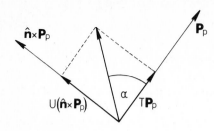

Fig. 3.6. Angle of rotation of the polarization component P_p

the common factor $1 + \mathbf{P} \cdot \hat{\mathbf{n}} S(\theta)$. The change of the polarization vector due to the scattering is determined by the spin-flip amplitude g. If $g = 0$, then $T = 1$, $S = U = 0$, and according to (3.77) the polarization does not change.

The angle α, through which the polarization component $\mathbf{P}_p$ in the scattering plane is rotated, is given by (see Fig. 3.6)

$$\tan \alpha = \frac{U(\theta)|\hat{\mathbf{n}}| |\mathbf{P}_p|}{T(\theta)|\mathbf{P}_p|} = \frac{U}{T}(\theta). \qquad (3.79)$$

In general, scattering changes not only the direction but also the magnitude of the polarization vector. For example, when $\mathbf{P} = 0$, it follows from (3.77) that $\mathbf{P}' = S(\theta)\hat{\mathbf{n}}$, as already shown in Sect. 3.3.2. If, however, $|\mathbf{P}| = 1$, that is with total initial polarization, the degree of polarization of the electrons scattered into a certain direction remains unchanged. This can be seen from (3.77) if one uses the relation $S^2 + T^2 + U^2 = 1$, which follows immediately from the definition (3.76):

$$|\mathbf{P}'|^2 = \frac{(P_n + S)^2 + T^2 P_p^2 + U^2 P_p^2}{(1 + P_n S)^2} ;$$

with $|\mathbf{P}|^2 = 1$ or $P_p^2 = 1 - P_n^2$ one has

$$|\mathbf{P}'|^2 = \frac{P_n^2 + 2 P_n S + S^2 + T^2 + U^2 - (T^2 + U^2) P_n^2}{(1 + P_n S)^2}$$

$$= \frac{P_n^2 + 2 P_n S + 1 + (S^2 - 1) P_n^2}{(1 + P_n S)^2} = 1.$$

Thus, from $|\mathbf{P}|^2 = 1$ it follows that $|\mathbf{P}'|^2 = 1$. Another special case where the direction of the polarization changes while the degree of polarization remains constant is found for $S = 0$. From the formulas just given, one then has

$$|\mathbf{P}'|^2 = P_n^2 + (T^2 + U^2) P_p^2 = |\mathbf{P}|^2.$$

In one special case, only the magnitude and not the direction of $\mathbf{P}$ is altered by the scattering; this is when $P_p = 0$, or the polarization is perpendicular to the scattering plane. Then from (3.77) it follows that

$$\boldsymbol{P'} = \frac{[P_n + S(\theta)]\hat{\boldsymbol{n}}}{1 + P_n S(\theta)}, \tag{3.80}$$

which shows that the polarization retains its direction perpendicular to the scattering plane.

Using the relations given in this section, the quantities S, T, U can be determined experimentally, as will be discussed in Sect. 3.7. In conjunction with experimental data of the cross section $d\sigma/d\Omega$ for scattering of an unpolarized beam (3.65), such measurements enable one to determine the complex scattering amplitudes $f = |f| \exp(i\gamma_1)$ and $g = |g| \exp(i\gamma_2)$. One can, for example, from $I = |f|^2 + |g|^2$ and $T = (|f|^2 - |g|^2)/(|f|^2 + |g|^2)$ determine $|f|$ and $|g|$. Then from (3.76) the measurement of S and U yields $-2|f||g|\sin(\gamma_1 - \gamma_2)$ and $2|f||g|\cos(\gamma_1 - \gamma_2)$, so that one also obtains the difference in the phases of the scattering amplitudes. The four observables $d\sigma/d\Omega$, S, T, and U yield only the three parameters $|f|$, $|g|$, and $\gamma_1 - \gamma_2$ since the observables are not independent of each other as the relation $S^2 + T^2 + U^2 = 1$ shows. Still, all of the observables have to be measured, since two measurements (S and U) are required for an unambiguous determination of the parameter $\gamma_1 - \gamma_2$. This is because the measurements do not yield the phase difference directly; instead, one obtains $\sin(\gamma_1 - \gamma_2)$ and $\cos(\gamma_1 - \gamma_2)$.

Since an absolute determination of the phases from an analysis of the scattered wave is impossible, the measurements discussed yield the maximum possible information on the scattering process, thus representing a complete or "perfect" experiment.

Problem 3.4. Calculate $\boldsymbol{P'}$ from (3.74).

Solution. The denominator has already been evaluated in Sect. 3.3.1. The numerator is

$$\frac{1}{2}\,\mathrm{tr}\left\{\begin{pmatrix} f & -ge^{-i\phi} \\ ge^{i\phi} & f \end{pmatrix}\begin{pmatrix} 1+P_z & P_x-iP_y \\ P_x+iP_y & 1-P_z \end{pmatrix}\begin{pmatrix} f^* & g^*e^{-i\phi} \\ -g^*e^{i\phi} & f^* \end{pmatrix}\begin{pmatrix} \hat{\boldsymbol{e}}_z & \hat{\boldsymbol{e}}_x-i\hat{\boldsymbol{e}}_y \\ \hat{\boldsymbol{e}}_x+i\hat{\boldsymbol{e}}_y & -\hat{\boldsymbol{e}}_z \end{pmatrix}\right\}$$

$$= \frac{1}{2}\,\mathrm{tr}\left\{\begin{pmatrix} a_{11} & a_{12} \\ a_{21} & a_{22} \end{pmatrix}\begin{pmatrix} \hat{\boldsymbol{e}}_z & \hat{\boldsymbol{e}}_x-i\hat{\boldsymbol{e}}_y \\ \hat{\boldsymbol{e}}_x+i\hat{\boldsymbol{e}}_y & -\hat{\boldsymbol{e}}_z \end{pmatrix}\right\}$$

where

$$a_{11} = (1+P_z)|f|^2 - (P_x-iP_y)fg^*e^{i\phi} - (P_x+iP_y)f^*ge^{-i\phi} + (1-P_z)|g|^2$$

$$a_{12} = (1+P_z)fg^*e^{-i\phi} + (P_x-iP_y)|f|^2 - (1-P_z)f^*ge^{-i\phi} - (P_x+iP_y)|g|^2e^{-2i\phi}$$

$$a_{21} = (1+P_z)f^*ge^{i\phi} - (P_x-iP_y)|g|^2e^{2i\phi} + (P_x+iP_y)|f|^2 - (1-P_z)fg^*e^{i\phi}$$

$$a_{22} = (1+P_z)|g|^2 + (P_x-iP_y)f^*ge^{i\phi} + (1-P_z)|f|^2 + (P_x+iP_y)fg^*e^{-i\phi}.$$

Hence $\frac{1}{2}\,\mathrm{tr}\{S(1+\boldsymbol{P}\cdot\boldsymbol{\sigma})S^\dagger\boldsymbol{\sigma}\}$ equals

$$\frac{1}{2}\{[2P_z(|f|^2-|g|^2) - (P_x-iP_y)e^{i\phi}(fg^*+f^*g) - (P_x+iP_y)e^{-i\phi}(fg^*+f^*g)]\hat{\boldsymbol{e}}_z$$

$$+ [(1+P_z)fg^*e^{-i\phi} + (P_x-iP_y)(|f|^2-|g|^2e^{2i\phi}) - (1-P_z)f^*ge^{-i\phi}$$

$$- (P_x+iP_y)(|g|^2e^{-2i\phi}-|f|^2) + (1+P_z)f^*ge^{i\phi} - (1-P_z)fg^*e^{i\phi}]\hat{\boldsymbol{e}}_x$$

$$+[(1+P_z)fg^*e^{-i\phi}+(P_x-iP_y)(|f|^2+|g|^2e^{2i\phi})-(1-P_z)f^*ge^{-i\phi}$$
$$-(P_x+iP_y)(|g|^2e^{-2i\phi}+|f|^2)-(1+P_z)f^*ge^{i\phi}+(1-P_z)fg^*e^{i\phi}]i\hat{e}_y\}$$
$$=\tfrac{1}{2}(\{(fg^*-f^*g)(e^{-i\phi}-e^{i\phi})+P_z(fg^*+f^*g)(e^{-i\phi}+e^{i\phi})$$
$$+P_x[2|f|^2-|g|^2(e^{2i\phi}+e^{-2i\phi})]+iP_y|g|^2(e^{2i\phi}-e^{-2i\phi})\}\hat{e}_x$$
$$+\{(fg^*-f^*g)(e^{i\phi}+e^{-i\phi})-P_z(fg^*+f^*g)(e^{i\phi}-e^{-i\phi})$$
$$+P_x|g|^2(e^{2i\phi}-e^{-2i\phi})-iP_y[2|f|^2+|g|^2(e^{2i\phi}+e^{-2i\phi})]\}i\hat{e}_y$$
$$+\{2P_z(|f|^2-|g|^2)-(fg^*+f^*g)[P_x(e^{i\phi}+e^{-i\phi})-iP_y(e^{i\phi}-e^{-i\phi})]\}\hat{e}_z)$$
$$=[-i(fg^*-f^*g)\sin\phi+P_x(|f|^2-|g|^2\cos2\phi)-P_y|g|^2\sin2\phi$$
$$+P_z(fg^*+f^*g)\cos\phi]\hat{e}_x$$
$$+[i(fg^*-f^*g)\cos\phi-P_x|g|^2\sin2\phi+P_y(|f|^2+|g|^2\cos2\phi)$$
$$+P_z(fg^*+f^*g)\sin\phi]\hat{e}_y$$
$$+[-(fg^*+f^*g)(P_x\cos\phi+P_y\sin\phi)+P_z(|f|^2-|g|^2)]\hat{e}_z.$$

After rearranging and using $\cos2\phi=1-2\sin^2\phi$, i.e.,

$$P_x(|f|^2-|g|^2\cos2\phi)=P_x(|g|^2\sin^2\phi+|g|^2\sin^2\phi+|f|^2-|g|^2)$$
$$P_y(|f|^2+|g|^2\cos2\phi)=P_y(|f|^2-|g|^2+|g|^2\cos^2\phi+|g|^2\cos^2\phi)$$

and $\sin2\phi=2\sin\phi\cos\phi$, we get

$$\tfrac{1}{2}\mathrm{tr}\{S(1+\boldsymbol{P}\cdot\boldsymbol{\sigma})S^\dagger\boldsymbol{\sigma}\}=(|f|^2+|g|^2)(-P_x\sin\phi+P_y\cos\phi)(-\sin\phi\hat{e}_x+\cos\phi\hat{e}_y)$$
$$+i(fg^*-f^*g)(-\sin\phi\hat{e}_x+\cos\phi\hat{e}_y)+(|f|^2-|g|^2)[(P_x\hat{e}_x+P_y\hat{e}_y+P_z\hat{e}_z)$$
$$-(-P_x\sin\phi+P_y\cos\phi)(-\sin\phi\hat{e}_x+\cos\phi\hat{e}_y)]$$
$$+(fg^*+f^*g)[P_z\cos\phi\hat{e}_x+P_z\sin\phi\hat{e}_y-(P_y\sin\phi+P_x\cos\phi)\hat{e}_z].$$

With the definitions (3.76) and the relations (3.63) and (3.70) it follows that

$$\boldsymbol{P}'=\frac{[\boldsymbol{P}\cdot\hat{\boldsymbol{n}}-S(\theta)]\hat{\boldsymbol{n}}-T(\theta)[\boldsymbol{P}-(\boldsymbol{P}\cdot\hat{\boldsymbol{n}})\hat{\boldsymbol{n}}]+U(\theta)(\hat{\boldsymbol{n}}\times\boldsymbol{P})}{1+\boldsymbol{P}\cdot\hat{\boldsymbol{n}}S(\theta)}.$$

The possibility of evaluating $\boldsymbol{P}'$ by use of vector algebra is demonstrated in Problem 3.9.

3.3.4 Double Scattering Experiments

Double scattering experiments are important for determining the Sherman function $S(\theta)$. An unpolarized beam is first scattered by the scattering center 1 (see Fig. 3.7). The electrons scattered through the angles θ_1, $\phi_1=0$ [8] are polarized by the scattering process and undergo a second scattering (scattering angles θ_2, $\phi_2=\phi$, where $\phi=0$ if the two scattering planes coincide).

---

[8] As the incident beam is unpolarized, scattering into all azimuthal angles is equally probable, so that we can assign the azimuthal angle $\phi_1=0$ to an arbitrary scattering direction.

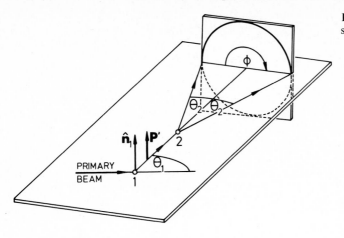

Fig. 3.7. Double scattering experiment

PRIMARY BEAM

From (3.62) the cross section for the scattering by the second target is

$$\sigma_2(\theta_2, \phi_2) = \mathrm{tr}\{\varrho''\}/\mathrm{tr}\{\varrho'\},$$

where ϱ' and ϱ'' are the respective density matrices of the single and double scattered states. One has from (3.71) that $\varrho' = \frac{1}{2} SS^\dagger \mathrm{tr}\{\varrho\}$, since the incident beam is unpolarized, and from (3.61) that

$$\varrho'' = S'\varrho'S'^\dagger = \frac{1}{2} S'SS^\dagger S'^\dagger \mathrm{tr}\{\varrho\}, \tag{3.81}$$

where S and S' are the respective scattering matrices of the first and second scattering processes. Thus we have

$$\sigma_2(\theta_2, \phi_2) = \frac{1}{2} \mathrm{tr}\{S'SS^\dagger S'^\dagger\} \frac{\mathrm{tr}\{\varrho\}}{\mathrm{tr}\{\varrho'\}}. \tag{3.82}$$

According to Sect. 3.3.1, the cross section for the scattering by the first target where the primary beam is unpolarized is

$$\mathrm{tr}\{\varrho'\}/\mathrm{tr}\{\varrho\} = |f(\theta_1)|^2 + |g(\theta_1)|^2 = I(\theta_1).$$

Since $\phi_1 = 0$, one has

$$SS^\dagger = \begin{pmatrix} f(\theta_1) & -g(\theta_1) \\ g(\theta_1) & f(\theta_1) \end{pmatrix} \begin{pmatrix} f^*(\theta_1) & g^*(\theta_1) \\ -g^*(\theta_1) & f^*(\theta_1) \end{pmatrix} = I(\theta_1) \begin{pmatrix} 1 & \dfrac{S(\theta_1)}{i} \\ -\dfrac{S(\theta_1)}{i} & 1 \end{pmatrix},$$

where $S(\theta_1)$ is the Sherman function as defined previously. Hence with $\phi_2 = \phi$ one has

$$\sigma_2(\theta_2, \phi) = \frac{1}{2} \mathrm{tr} \left\{ \begin{pmatrix} f(\theta_2) & -g(\theta_2)e^{-i\phi} \\ g(\theta_2)e^{i\phi} & f(\theta_2) \end{pmatrix} \right.$$

$$\times \left. \begin{pmatrix} f^*(\theta_2) - g^*(\theta_2)\dfrac{S(\theta_1)}{i}e^{i\phi} & g^*(\theta_2)e^{-i\phi} + f^*(\theta_2)\dfrac{S(\theta_1)}{i} \\ -f^*(\theta_2)\dfrac{S(\theta_1)}{i} - g^*(\theta_2)e^{i\phi} & -g^*(\theta_2)\dfrac{S(\theta_1)}{i}e^{-i\phi} + f^*(\theta_2) \end{pmatrix} \right\}$$

$$= \frac{1}{2} \left[|f(\theta_2)|^2 - f(\theta_2) g^*(\theta_2) \frac{S(\theta_1)}{i} e^{i\phi} + f^*(\theta_2) g(\theta_2) \frac{S(\theta_1)}{i} e^{-i\phi} \right.$$

$$+ |g(\theta_2)|^2 + |g(\theta_2)|^2 + f^*(\theta_2) g(\theta_2) \frac{S(\theta_1)}{i} e^{i\phi}$$

$$\left. - f(\theta_2) g^*(\theta_2) \frac{S(\theta_1)}{i} e^{-i\phi} + |f(\theta_2)|^2 \right]$$

$$= I(\theta_2) + \tfrac{1}{2} S(\theta_1) I(\theta_2) S(\theta_2) (e^{i\phi} + e^{-i\phi})$$

$$= I(\theta_2) [1 + S(\theta_1) S(\theta_2) \cos \phi]. \tag{3.83}$$

If the angle between the first and the second scattering plane is zero, we obtain

$$\sigma_2(\theta_2, \phi = 0°) = I(\theta_2) [1 + S(\theta_1) S(\theta_2)].$$

If $\phi = 180°$, then

$$\sigma_2(\theta_2, \phi = 180°) = I(\theta_2) [1 - S(\theta_1) S(\theta_2)].$$

Hence observation of the left-right asymmetry of the intensity in a double scattering experiment yields

$$\frac{\sigma_2(\phi = 0°) - \sigma_2(\phi = 180°)}{\sigma_2(\phi = 0°) + \sigma_2(\phi = 180°)} = S(\theta_1) S(\theta_2). \tag{3.84}$$

If the targets are the same and the scattering angles are equal in the first and second scattering processes ($\theta_1 = \theta_2 = \theta$), one can measure $S^2(\theta)$ for the target in question.

By using the results derived in Sects. 3.3.1 and 3.3.2, the double scattering experiment could have been understood without making further calculations. According to (3.73), the first scattering process produces the polarization $P' = S(\theta_1) \hat{n}_1$. Hence from (3.70) the cross section for the second scattering is

$$\sigma_2 = I(\theta_2) [1 + S(\theta_2) P' \cdot \hat{n}_2] = I(\theta_2) [1 + S(\theta_1) S(\theta_2) \hat{n}_1 \cdot \hat{n}_2],$$

where $\hat{n}_1 \cdot \hat{n}_2 = \cos \phi$, if ϕ is the angle between the normals of the scattering planes (= angle between the scattering planes).

3.4 Simple Physical Description of the Polarization Phenomena

> The main object of physical science is not the provision of pictures, but is the formulation of laws governing phenomena and the application of these laws to the discovery of new phenomena. If a picture exists, so much the better....
>
> P. A. M. DIRAC
> (*The Principles of Quantum Mechanics*, Chapter I,4)

With the use of simple pictures and basic principles, the results which have been mathematically derived in the previous sections will be illustrated. A qualitative explanation will be given for the change in the direction and magnitude of the polarization vector, the asymmetry in the scattering of a polarized electron beam, the fact that the polarization arising from the scattering of an unpolarized beam is perpendicular to the scattering plane, and the fact that the degree of polarization in this case can be described by the same function (Sherman function) which describes the scattering asymmetry of a polarized beam.

3.4.1 Illustration of the Rotation of the Polarization Vector

We shall now illustrate by simple models the results obtained in the last section. First – why does the polarization vector retain its direction only if it is perpendicular to the scattering plane?

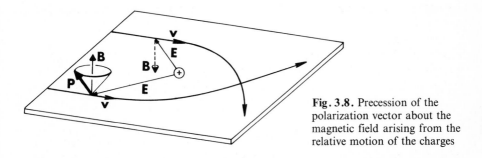

Fig. 3.8. Precession of the polarization vector about the magnetic field arising from the relative motion of the charges

The polarization effects in scattering are caused by spin-orbit coupling, in other words, by the magnetic field which the electrons experience in their rest frame (cf. Sect. 3.1). The charged scattering center moves in the rest frame of the electrons; the current that is represented by this moving charge produces a magnetic field $B = E \times v/c$ which acts upon the magnetic moments of the electrons. As E and v lie in the scattering plane, B is perpendicular to this plane (see Fig. 3.8). If the polarization P does not lie parallel or antiparallel to B, the magnetic moment which is connected with P experiences a torque that induces P to change its direction and to precess. Only if the polarization is parallel or antiparallel to B does it retain its direction.

3.4.2 Illustration of the Change in the Magnitude of the Polarization Vector

The picture just used does not answer the question of how it is possible for the magnitude of P to change during the scattering. We first explain this for the conspicuous case in which the degree of polarization changes from zero to a finite value (scattering of an unpolarized beam).

For an unpolarized beam, if we take 300 eV incident energy and scattering by Hg as an example, the scattering cross section has the shape shown in Fig. 3.9. The typical interference structure of the cross section arises because the electron wavelength is of the same order of magnitude as the atomic radius ($\lambda = 0.07$ nm for 300 eV).

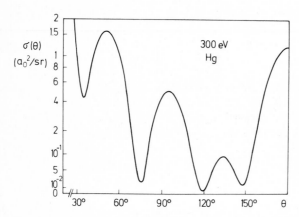

Fig. 3.9. Differential cross section for elastic scattering of an unpolarized electron beam ($a_0 =$ Bohr radius). Ordinate pseudologarithmic according to $\log[1 + 10\sigma(\theta)/(a_0^2/\text{sr})]$

It follows from Chap. 2 that one can consider the unpolarized beam as a mixture of two equal fractions with opposing spin directions. It is expedient to choose the arbitrary spin directions of these two constituent beams to be perpendicular to the scattering plane because they then remain unchanged in the scattering process.

The cross sections of the two beams with opposite polarization differ slightly from each other. This is because the scattering potential essentially consists of the electrostatic and the spin-orbit potential: $V = V_0 + V_{ls}$. Since V_{ls} contains the scalar product $l \cdot s$, it has different signs for electrons of the same orbit but different spin directions. As Fig. 3.10 shows, the resulting scattering potential will therefore be higher or lower for spin-up electrons ($e\uparrow$) than for spin-down electrons ($e\downarrow$), depending on which side of the atom they pass. If we consider, for example, electrons which pass by the atom on the right, i.e., that are scattered to the left, it can be seen from Fig. 3.10 that the effective radius R of the atom for scattering (defined as the radius where the potential has dropped to a certain value) will be smaller for $e\uparrow$ than for $e\downarrow$. Since the positions of the extrema in interference patterns like those shown in Fig. 3.9 are determined by λ/R, their abscissas depend on the effective atomic radius, so that different cross-section

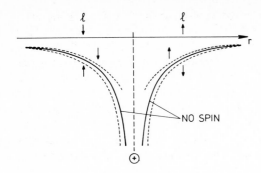

Fig. 3.10. Potential curves with (−−−) and without (——) spin-orbit coupling for electrons with spins up ↑ and down ↓

curves result for e↑ and e↓.[9] Since a change in the scattering potential also affects the ordinates of the cross sections, one obtains the curves shown in the upper half of Fig. 3.11 for scattering to the left.

The numbers of e↑ and e↓ scattered in a particular direction θ are therefore usually different from each other; in other words, the scattered beam is polarized. Since the number of scattered electrons is proportional to the corresponding cross section, the polarization is, from (2.25),

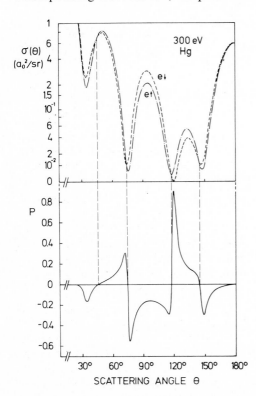

Fig. 3.11. Construction of the polarization from the cross sections for e↑ and e↓. Ordinate of cross sections pseudologarithmic according to $\log[1 + 50\sigma(\theta)/(a_0^2/\mathrm{sr})]$

---

[9] Needless to say, quantitative results cannot be derived from these qualitative arguments.

$$P = \frac{N_\uparrow - N_\downarrow}{N_\uparrow + N_\downarrow} = \frac{\sigma_\uparrow - \sigma_\downarrow}{\sigma_\uparrow + \sigma_\downarrow}, \tag{3.85}$$

so that one can construct P directly from the cross-section curves as indicated in Fig. 3.11.

What has been shown here for a mixture of 50 % e↑ and 50 % e↓ (unpolarized beam) is also valid for every other mixture (partially polarized beam, $|P| < 1$). Since the cross sections for e↑ and e↓ are different, the proportions of the mixture change with scattering, which means, according to (3.85), that the polarization changes.

From the construction just described it follows that a particularly high polarization arises at those angles where one of the two cross sections has a deep minimum so that its value is very small compared to that of the other cross section at the same angle. Electrons of a single spin direction then predominate in the scattered beam, so that one approaches the ideal case of a totally polarized electron beam. Due to the fact that the spin-orbit interaction is relatively weak, the mutual shift of $\sigma_\uparrow$ and $\sigma_\downarrow$ is not significant (see Fig. 3.11), so that the complete differential cross section $\sigma(\theta) = \sigma_\uparrow(\theta) + \sigma_\downarrow(\theta)$ likewise has its minima near the extreme values of P. This would be different if the spin-orbit coupling were so large that $\sigma_\uparrow$ and $\sigma_\downarrow$ were strongly shifted with respect to each other. Such cases occur in nucleon scattering. For electron scattering we can, however, note that high degrees of polarization occur only near cross-section minima.

Problem 3.5. Show that (3.85) is compatible with (3.73).

Solution. According to (3.73) scattering of an unpolarized beam yields a polarization

$$P = i \frac{fg^* - f^*g}{|f|^2 + |g|^2}$$

perpendicular to the scattering plane. This is easily seen to be equal to

$$P = \frac{|f - ig|^2 - |f + ig|^2}{|f - ig|^2 + |f + ig|^2}.$$

From (3.58) and (3.44), $|f - ig|^2$ is seen to be the cross section $\sigma_\uparrow$ for scattering of e↑: For incident spins directed along the x direction in Fig. 3.4 ($A = B = 1$) and scattering to the left ($\phi = \frac{3}{2}\pi$) one obtains $\chi' = \begin{pmatrix} f - ig \\ f - ig \end{pmatrix}$ for the spinor of the final state, which yields $\sigma_\uparrow = |f - ig|^2$. Similarly one finds $\sigma_\downarrow = |f + ig|^2$, so that

$$P = i \frac{fg^* - f^*g}{|f|^2 + |g|^2} = \frac{\sigma_\uparrow - \sigma_\downarrow}{\sigma_\uparrow + \sigma_\downarrow}.$$

3.4.3 Illustration of the Asymmetry in the Scattering of a Polarized Beam

The left-right asymmetry which is observed in the scattering of a polarized electron beam can immediately be interpreted with the help of Fig. 3.10. We assume without loss of generality that the beam consists only of e↑. Then the electrons that pass the atom on the left, i.e., that are scattered to the right, experience a stronger potential than those scattered to the left. Different scattering potentials cause different scattering intensities, so that one observes a scattering asymmetry.

A quantitative example can be taken from the upper part of Fig. 3.11 which shows the cross sections for scattering of e↑ and e↓ to the left. Remembering the discussion in Sect. 3.4.2, we see that this graph also represents the case of scattering to the right, if one interchanges the labels ↑ and ↓. The two cross sections in Fig. 3.11 can therefore also be taken to represent the scattering of e↑ to the left and to the right, respectively. This shows that there is a left-right asymmetry of the scattering intensity – except for those angles where the cross sections happen to intersect.

3.4.4 Transversality of the Polarization as a Consequence of Parity Conservation. Counterexample: Longitudinal Polarization in β Decay

The polarization resulting from the scattering of an unpolarized electron beam is perpendicular to the scattering plane, as proved in Sect. 3.3.2. This can be seen directly from simple symmetry considerations. According to all physical experience, parity is conserved for the electromagnetic interaction which governs electron scattering. In other words, the result of an electron-scattering experiment must not depend on whether it is described in a right- or left-handed (e. g., reflected) coordinate system. This means that the mirror image of such an experiment must also be a process which can occur in nature.

From this principle it follows that the polarization P of the scattered beam cannot have a longitudinal component $P \cdot k'$. Otherwise this component would define a certain screw sense (helicity) in the laboratory system, for example, a right-handed screw, $P \cdot k' > 0$. With the reflection of the experiment one would obtain a left-handed screw, $P \cdot k' < 0$ because the sense of rotation connected with the polarization is inverted (see Fig. 3.12). Hence, the mirror image would yield a different final state though the initial state remains unchanged. As we must expect a well-defined final state when we have a well-defined initial state, this course of the experiment contradicts reality: the results obtained in the laboratory and in the mirror image are not compatible with one another.

From the same argument it follows that a component $P \cdot k$ in the incident direction cannot exist either. This means that P must be perpendicular to k and k', i.e., perpendicular to the scattering plane.

The mirror inversion being considered $(x, y, z) \rightarrow (x, y, -z)$ is not identical to the parity inversion $(x, y, z) \rightarrow (-x, -y, -z)$. It can, however, be transformed to

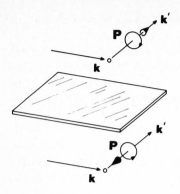

Fig. 3.12. Reflection of a scattering process with unpolarized initial state. To avoid confusion, the axial vector **P** is depicted by its direction of rotation

this by a rotation through 180° (see Fig. 3.13). As the screw sense is not changed by a rotation, use of the simple mirror inversion is justified for our considerations.

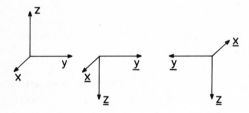

Fig. 3.13. Mirror inversion followed by a rotation through 180° yielding the parity inversion

To prevent misunderstanding, it must be emphasized that disappearance of the longitudinal polarization component can no longer be inferred from parity conservation, if one has a nonzero initial polarization. We explain this by an example. Let the initial polarization be transverse, the polarization vector lying in the scattering plane (see Fig. 3.14). For simplicity let us further assume that the spin-flip amplitude is small so that the polarization vector is virtually not

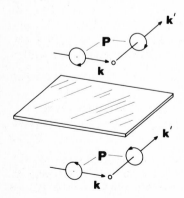

Fig. 3.14. Reflection of a scattering process with polarized initial state

changed by the scattering, as is actually often the case (cf. Sect. 3.6). Then **P** approximately retains its direction in space. In the example shown, the right-handed screw formed by **P** and **k'** in the laboratory is transformed into a left-handed screw in the mirror image. In this case, however, the mirror image also represents an experiment that is possible in nature, because it is not only the final state that has been changed by the reflection but also the initial state. It is, of course, no surprise to obtain a different result of an experiment when the initial situation is altered.

With interactions that violate parity conservation, arguments of the kind just used cannot be applied. Parity violation can then even be the reason for the spin polarization of the particles concerned. A famous example of this is β decay.

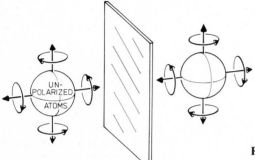

Fig. 3.15. Radioactive β source and its mirror image

A longitudinal polarization of the electrons emitted by unpolarized nuclei, as actually occurs in β decay, is incompatible with parity conservation. Fig. 3.15 shows that all left-handed screws are transformed to right-handed screws by reflection, so that the final states in the mirror image and in the laboratory differ from each other although the initial states are the same. Accordingly, the result of the experiments is not invariant under spatial inversion; in other words, the law of conservation of parity is violated.

The currents of polarized electrons which one can obtain from radioactive sources are very small. Electron polarization in β decay is, however, of fundamental importance as it led to the discovery of parity violation in weak interactions. The theoretical and experimental aspects of β decay are treated in detail in modern textbooks on nuclear physics and in review articles so that we will not consider them further [3.6].

Problem 3.6. It was shown in Sect. 3.3.1 that the polarization components which lie in the scattering plane do not contribute to the scattering asymmetry. Derive this result from symmetry considerations.

Solution. We resolve the polarization component P_p which lies parallel to the scattering plane (see Fig. 3.5) into a component $P_{\parallel}$ in the incident direction and a component $P_{\perp}$ perpendicular to it (see Fig. 3.16). The former, due to the rotational symmetry, cannot give rise

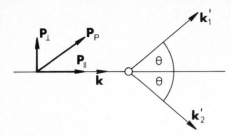

Fig. 3.16. Resolution of the polarization component P_p. (Scattering plane = plane of diagram)

to an asymmetrical intensity distribution (cf. Example 3.2). Let us assume that $P_\perp$ causes different intensities in the directions k_1' and k_2' which are symmetrical to the incident direction. The reflection through the incident direction (mirror perpendicular to scattering plane) would then invert the scattering asymmetry but leave $P_\perp$ (sense of rotation!) unchanged. Hence the final state represented by the mirror image would differ from the final state in the laboratory although the initial states are the same. Accordingly, the result of the experiment would depend on whether it is described in the original or the reflected coordinate system.

3.4.5 Equality of Polarizing and Analyzing Power

We now give the connection between the following two facts which were proved in Sect. 3.3 for elastic scattering:

a) Scattering as a polarizer: A beam that is originally unpolarized obtains the polarization $P = S(\theta)\hat{n}$ from scattering, i.e., the polarization is determined by the Sherman function.

b) Scattering as an analyzer: The left-right asymmetry A observed with the scattering of a beam that is polarized perpendicular to the scattering plane is also determined by $S(\theta)$. For a beam of polarization P one has from (3.70)

$$A = \frac{N_l - N_r}{N_l + N_r} = \frac{1 + PS(\theta) - (1 - PS(\theta))}{1 + PS(\theta) + 1 - PS(\theta)} = PS(\theta) \tag{3.86}$$

(N_l, N_r = number of electrons scattered to left and right, respectively). With total polarization $P = 1$ one has $A = S(\theta)$.

Polarizer and analyzer are thus characterized here by one and the same function.[10] It can easily be seen that this must be true because the first fact follows immediately from the second:

Once again we consider the incident unpolarized beam to be a mixture of equal numbers of electrons polarized in opposite directions. One half are totally polarized in the direction ↑ perpendicular to the scattering plane; the other half are totally polarized in the opposite direction (see Fig. 3.17). From (3.70) it follows for the e↑ beam that the scattering intensity to the left is proportional to $1 + SP \cdot \hat{n} = 1 + S$, whereas the intensity to the right is proportional to $1 - S$ due to the opposite vector $\hat{n}$. For the e↓ beam the corresponding values are $1 - S$ and

---

[10] That this is not so with all scattering processes is shown in Sect. 3.9 and in Problem 4.1.

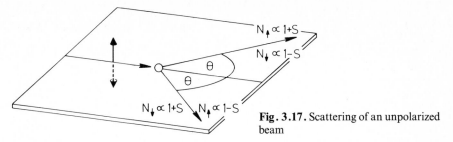

Fig. 3.17. Scattering of an unpolarized beam

$1+S$. In both cases the polarization vectors remain unchanged as they are perpendicular to the scattering plane and have the magnitude 1 (cf. Sects. 3.3.3 and 3.4.1).

The fact that there are different numbers of $e\uparrow$ and $e\downarrow$ in the scattered beam means that this beam is polarized. Figure 3.17 illustrates that for scattering to the left the polarization is

$$P = \frac{N_\uparrow - N_\downarrow}{N_\uparrow + N_\downarrow} = \frac{1+S-(1-S)}{1+S+1-S} = S, \tag{3.87}$$

whereas for scattering to the right one has $P = -S$ (as long as $\uparrow$ and $\downarrow$ always refer to the same reference direction). The polarization after the scattering of an unpolarized beam is thus seen to be a direct consequence of the scattering asymmetry of a polarized beam. Both quantities are described by the same function $S(\theta)$.

Measuring left-right asymmetry is not the only way to analyze the polarization of an electron beam. Instead one may observe the relative difference of the intensities scattered to the left (or right) for polarization P and $-P$ of the incident beam. Since the scattered intensities are proportional to $1+SP \cdot \hat{n}$, one has the asymmetry

$$A = \frac{1+PS-(1-PS)}{1+PS+1-PS} = PS,$$

the same value as obtained by observing the left-right asymmetry.

Problem 3.7. In the preceding considerations, the incident unpolarized beam was separated into two beams which were polarized perpendicular to the scattering plane. Show that the result does not depend on this particular separation, i.e., that the separation can be made in any arbitrary direction.

Solution. We choose an arbitrary separation of the incident beam. Let one half have the polarization $P = (P_n, P_p)$ and the other half $-P = (-P_n, -P_p)$. The number N_+ of electrons from the part of the beam with polarization P which are scattered at the angle θ is proportional to $1+S(\theta)P \cdot \hat{n} = 1+S(\theta)P_n$. The polarization of these electrons after scattering is, according to (3.77),

$$P'_+ = \frac{[P_n + S(\theta)]\hat{n} + T(\theta)P_p + U(\theta)(\hat{n} \times P_p)}{1 + P_n S(\theta)}.$$

The number N_- of electrons from the part of the beam with polarization $-\boldsymbol{P}$ which are scattered at the same angle is proportional to $1 - S(\theta)\boldsymbol{P} \cdot \hat{\boldsymbol{n}} = 1 - S(\theta)P_n$. The polarization vector of these electrons after scattering is

$$\boldsymbol{P}'_- = \frac{[-P_n + S(\theta)]\hat{\boldsymbol{n}} - T(\theta)\boldsymbol{P}_p - U(\theta)(\hat{\boldsymbol{n}} \times \boldsymbol{P}_p)}{1 - P_n S(\theta)}.$$

The resultant polarization $\boldsymbol{P}'$ of the scattered beam is according to (2.16)

$$\boldsymbol{P}' = \frac{N_+}{N_+ + N_-}\, \boldsymbol{P}'_+ + \frac{N_-}{N_+ + N_-}\, \boldsymbol{P}'_- = \frac{2\,S(\theta)\hat{\boldsymbol{n}}}{2} = S(\theta)\hat{\boldsymbol{n}}.$$

Again one has a beam polarized perpendicularly to the scattering plane with polarization $S(\theta)$.

3.5 Polarization Violating Reflection Symmetry

When the interaction or the target violates parity, scattering of an unpolarized beam yields polarization components parallel to the scattering plane. Such components also produce a scattering asymmetry and the connection between polarizing and analyzing power becomes more complicated.

Our attention has so far been focused on the common situation where reflection symmetry holds; let us now consider what happens when it is violated. This can occur for two reasons. First, the interaction may violate parity. Examples may be found in Sects. 3.4.4 and 8.4. Second, the target may be an object which is not in an eigenstate of the parity operator so that the experiment considered and its mirror image differ from each other. This occurs when the target is a right- or left-handed molecule or mineral, such as those which are responsible for optical activity (cf. Sect. 8.6) [3.7]. Reflection symmetry is also violated when the scattering plane is chosen such that it is not a mirror plane of the target crystal, a situation which may, for instance, be arranged in a LEED experiment (cf. Sect. 7.3) [3.8].

It is obvious that in the cases of parity violation the scattering matrix must be of a more general form than the matrix S given in (3.59) which is easily seen to be reflection invariant. From (2.2) and (3.69) one obtains ($\mathbf{1}$ = unit matrix)

$$S = \begin{pmatrix} f & -g(\cos\phi - i\sin\phi) \\ g(\cos\phi + i\sin\phi) & f \end{pmatrix} = f\mathbf{1} - ig\boldsymbol{\sigma} \cdot \hat{\boldsymbol{n}}. \tag{3.88}$$

Since under space reflection one has

$$\boldsymbol{\sigma} \to \boldsymbol{\sigma}, \quad \boldsymbol{k} \to -\boldsymbol{k} \tag{3.89}$$

one sees, recalling (3.68), that S is invariant under reflection.

Even if reflection invariance is dropped, the experiments must certainly be rotationally invariant. If b denotes a three-dimensional vector, the most general form of S that is allowed is therefore

$$S = f\mathbf{1} + \boldsymbol{\sigma} \cdot \boldsymbol{b},$$

since the scalar product is invariant under rotation and since the Pauli matrices together with the unit matrix form a complete set of 2×2 matrices, so that an arbitrary 2×2 matrix can be written in this fashion. To find the general form of the vector $\boldsymbol{b}$, we note that in elastic scattering the three vectors

$$\hat{\boldsymbol{n}}, \quad \hat{\boldsymbol{e}}_1 = \frac{\boldsymbol{k}' + \boldsymbol{k}}{|\boldsymbol{k}' + \boldsymbol{k}|}, \quad \text{and} \quad \hat{\boldsymbol{e}}_2 = \frac{\boldsymbol{k}' - \boldsymbol{k}}{|\boldsymbol{k}' - \boldsymbol{k}|}$$

form an orthonormal set and can therefore be used to expand any vector

$$\boldsymbol{b} = b_n \hat{\boldsymbol{n}} + b_1 \hat{\boldsymbol{e}}_1 + b_2 \hat{\boldsymbol{e}}_2.$$

Under time reversal, angular momentum (spin) and momentum change sign,

$$\boldsymbol{\sigma} \to -\boldsymbol{\sigma}, \quad \boldsymbol{k} \to -\boldsymbol{k}', \quad \boldsymbol{k}' \to -\boldsymbol{k},$$

so that the last term of $\boldsymbol{\sigma} \cdot \boldsymbol{b} = b_n \boldsymbol{\sigma} \cdot \hat{\boldsymbol{n}} + b_1 \boldsymbol{\sigma} \cdot \hat{\boldsymbol{e}}_1 + b_2 \boldsymbol{\sigma} \cdot \hat{\boldsymbol{e}}_2$ changes sign and is therefore ruled out, because time-reversal invariance is required. Adopting the notation of (3.88) one obtains thus the general form of the scattering matrix which allows for parity violation but is invariant under rotation and time reversal:

$$S = f\mathbf{1} - \mathrm{i}(g\boldsymbol{\sigma} \cdot \hat{\boldsymbol{n}} + h\boldsymbol{\sigma} \cdot \hat{\boldsymbol{e}}_1). \tag{3.90}$$

Let us now consider a few consequences of the additional parity-violating term. If a beam of polarization $\boldsymbol{P}$ is scattered, the scattering cross section is given according to (3.63) by $\frac{1}{2} \operatorname{tr}\{S(1 + \boldsymbol{P} \cdot \boldsymbol{\sigma})S^\dagger\}$ or, with (3.90),

$$|f|^2 + |g|^2 + |h|^2 + 2(\operatorname{Im}\{f^*g\}\hat{\boldsymbol{n}} + \operatorname{Im}\{f^*h\}\hat{\boldsymbol{e}}_1 - \operatorname{Im}\{g^*h\}\hat{\boldsymbol{e}}_2) \cdot \boldsymbol{P}$$

as derived in Problem 3.8. The scattering asymmetry which is the relative intensity difference obtained with polarizations $\boldsymbol{P}$ and $-\boldsymbol{P}$ of the incident beam is therefore

$$\boldsymbol{A} \cdot \boldsymbol{P} \quad \text{with} \quad \boldsymbol{A} = 2 \frac{\operatorname{Im}\{f^*g\}\hat{\boldsymbol{n}} + \operatorname{Im}\{f^*h\}\hat{\boldsymbol{e}}_1 - \operatorname{Im}\{g^*h\}\hat{\boldsymbol{e}}_2}{|f|^2 + |g|^2 + |h|^2}. \tag{3.91}$$

If the parity-violating term disappears ($h \to 0$), we obtain the asymmetry derived in the preceding sections, since the Sherman function is $2 \operatorname{Im}\{f^*g\}/(|f|^2 + |g|^2)$. Equation (3.91) shows that parity violation results in a scattering asymmetry which is caused not only by the polarization components normal to the scattering plane, but also by those lying in the scattering plane.

If an unpolarized beam is scattered, the polarization P' after scattering is given by the general expression (3.72) which in the present case yields

$$P' = \frac{\mathrm{tr}\{SS^\dagger\sigma\}}{\mathrm{tr}\{SS^\dagger\}} = 2\,\frac{\mathrm{Im}\{f^*g\}\hat{n} + \mathrm{Im}\{f^*h\}\hat{e}_1 + \mathrm{Im}\{g^*h\}\hat{e}_2}{|f|^2 + |g|^2 + |h|^2}, \tag{3.92}$$

where the evaluation has been made analogous to Problem 3.8. Apart from the normal polarization component one also has components parallel to the scattering plane. They are determined by interference of the parity-violating scattering amplitude h with the other amplitudes.

Comparison of (3.91 and 92) shows that the connection between the polarization P' and the analyzing power is somewhat more complicated than in Mott scattering. Since $|P'| = |A|$, the degree of polarization after scattering of an initially unpolarized beam equals the analyzing power, i.e., the maximum scattering asymmetry which can be obtained with a totally polarized beam. The same is true in Mott scattering. However, the direction of the polarization P' is no longer identical with that of A, which gives, with respect to a scattering plane spanned by $\hat{e}_1$ and $\hat{e}_2$, the spin direction in a totally polarized beam for which maximum scattering asymmetry is observed. The difference is caused by the different signs of the interference term of g and h in (3.91 and 92).

The change of an electron polarization vector caused by scattering is obtained by insertion of the scattering matrix (3.90) into (3.74). The tedious multiplication is carried out in Problem 3.9. The formula for P' given there also contains the interference terms between h and the amplitudes f and g and is a generalization of (3.77).

Problem 3.8. Calculate the scattering cross section of a polarized electron beam if parity is not conserved.

Solution. With the scattering matrix $S = f\mathbf{1} + \boldsymbol{\sigma}\cdot\boldsymbol{b}$ one obtains the cross section

$$\tfrac{1}{2}\,\mathrm{tr}\{S(1+\boldsymbol{P}\cdot\boldsymbol{\sigma})S^\dagger\} = \tfrac{1}{2}\,\mathrm{tr}\{1+\boldsymbol{P}\cdot\boldsymbol{\sigma})(f^*\mathbf{1}+\boldsymbol{\sigma}\cdot\boldsymbol{b}^*)(f\mathbf{1}+\boldsymbol{\sigma}\cdot\boldsymbol{b})\}$$
$$= \tfrac{1}{2}\,\mathrm{tr}\{(1+\boldsymbol{P}\cdot\boldsymbol{\sigma})\,[|f|^2\mathbf{1}+f^*\boldsymbol{\sigma}\cdot\boldsymbol{b}+f\boldsymbol{\sigma}\cdot\boldsymbol{b}^*+|b|^2\mathbf{1}+\mathrm{i}\boldsymbol{\sigma}\cdot(\boldsymbol{b}^*\times\boldsymbol{b})]\},$$

where use has been made of

$$\mathrm{tr}\{M\cdot N\} = \mathrm{tr}\{N\cdot M\}, \quad \boldsymbol{\sigma}^\dagger = \boldsymbol{\sigma}, \quad \text{and} \quad (\boldsymbol{\sigma}\cdot\boldsymbol{A})(\boldsymbol{\sigma}\cdot\boldsymbol{B}) = \boldsymbol{A}\cdot\boldsymbol{B}\mathbf{1} + \mathrm{i}\boldsymbol{\sigma}\cdot(\boldsymbol{A}\times\boldsymbol{B}).$$

Making further use of the latter relation yields

$$\tfrac{1}{2}\,\mathrm{tr}\{|f|^2\mathbf{1} + \boldsymbol{\sigma}\cdot(f^*\boldsymbol{b}+f\boldsymbol{b}^*) + |b|^2\mathbf{1} + \mathrm{i}\boldsymbol{\sigma}(\boldsymbol{b}^*\times\boldsymbol{b}) + |f|^2\boldsymbol{P}\cdot\boldsymbol{\sigma}$$
$$+f^*[\boldsymbol{P}\cdot\boldsymbol{b}\mathbf{1}+\mathrm{i}\boldsymbol{\sigma}\cdot(\boldsymbol{P}\times\boldsymbol{b})]+f[\boldsymbol{P}\cdot\boldsymbol{b}^*\mathbf{1}+\mathrm{i}\boldsymbol{\sigma}\cdot(\boldsymbol{P}\times\boldsymbol{b}^*)]+|b|^2\boldsymbol{P}\cdot\boldsymbol{\sigma}$$
$$+\mathrm{i}\boldsymbol{P}\cdot(\boldsymbol{b}^*\times\boldsymbol{b})\mathbf{1}-\boldsymbol{\sigma}\cdot[\boldsymbol{P}\times(\boldsymbol{b}^*\times\boldsymbol{b})]\}.$$

Since $\mathrm{tr}\{\boldsymbol{\sigma}\} = 0$, see (2.2), all the terms of the form "(nonmatrix)$\cdot\boldsymbol{\sigma}$" do not give a contribution so that the cross section is

$$|f|^2 + |b|^2 + \boldsymbol{P}\cdot[f^*\boldsymbol{b}+f\boldsymbol{b}^* +\mathrm{i}(\boldsymbol{b}^*\times\boldsymbol{b})].$$

For $b = -ig\hat{n} - ih\hat{e}_1$ as in (3.90) one obtains the cross section

$$|f|^2 + |g|^2 + |h|^2 + [i(fg^* - f^*g)\hat{n} + i(fh^* - f^*h)\hat{e}_1 + i(g^*h - gh^*)\hat{e}_2] \cdot P$$
$$= (|f|^2 + |g|^2 + |h|^2)\left(1 + 2\frac{\mathrm{Im}\{f^*g\hat{n} + f^*h\hat{e}_1 - g^*h\hat{e}_2\}}{|f|^2 + |g|^2 + |h|^2} \cdot P\right).$$

Problem 3.9. Given an incident electron beam of arbitrary polarization P, how is P changed by scattering if parity is not conserved?

Solution. The polarization formula (3.74) has to be evaluated with the scattering matrix $S = f\mathbf{1} + \boldsymbol{\sigma} \cdot \boldsymbol{b}$. Since the denominator D is identical with the expression evaluated in Problem 3.8, only the numerator N has to be calculated:

$$N = \tfrac{1}{2}\mathrm{tr}\{S(1 + P \cdot \sigma)S^\dagger\sigma\} = \tfrac{1}{2}\mathrm{tr}\{(1 + P \cdot \sigma)(f^*1 + \sigma \cdot b^*)\sigma(f1 + \sigma \cdot b)\}$$
$$= \tfrac{1}{2}\mathrm{tr}\{|f|^2\sigma + f^*\sigma\sigma \cdot b + \sigma \cdot b^*\sigma f + \sigma \cdot b^*\sigma\sigma \cdot b$$
$$+ P \cdot \sigma|f|^2\sigma + P \cdot \sigma f^*\sigma\sigma \cdot b + P \cdot \sigma\sigma \cdot b^*\sigma f + P \cdot \sigma\sigma \cdot b^*\sigma\sigma \cdot b\}.$$

Utilizing the general rules applied in Problem 3.8 and, in addition, the rule $\tfrac{1}{2}\mathrm{tr}\{\sigma v \cdot \sigma\} = v$, where v is a three-dimensional vector, one has

$$N = f^*b + fb^* + \tfrac{1}{2}\mathrm{tr}\{\sigma[|b|^2 + i\sigma \cdot (b \times b^*)]\} + |f|^2P$$
$$+ \tfrac{1}{2}\mathrm{tr}\{f^*\sigma[b \cdot P + i\sigma \cdot (b \times P)] + f\sigma[P \cdot b^* + i\sigma \cdot (P \times b^*)]$$
$$+ \sigma[b \cdot P + i\sigma \cdot (b \times P)]\sigma \cdot b^*\}.$$

Transforming the last term into $\tfrac{1}{2}\mathrm{tr}\{i\sigma\{(b \times P) \cdot b^* + i\sigma \cdot [(b \times P) \times b^*]\}\}$ one obtains

$$N = f^*b + fb^* + i(b \times b^*) + |f|^2P + if^*(b \times P) - if(b^* \times P) + b \cdot Pb^* - (b \times P) \times b^*. \quad (3.93)$$

The evaluation of the first three terms with $b = -ig\hat{n} - ih\hat{e}_1$ can be seen in Problem 3.8 whereas the last five terms yield

$$|f|^2(P \cdot \hat{n}\hat{n} + P \cdot \hat{e}_1\hat{e}_1 + P \cdot \hat{e}_2\hat{e}_2) + (f^*g + fg^*)\hat{n} \times P + (f^*h + fh^*)\hat{e}_1 \times P$$
$$+ |g|^2\hat{n} \cdot P\hat{n} + |h|^2\hat{e}_1 \cdot P\hat{e}_1 + g^*h\hat{e}_1 \cdot P\hat{n} + gh^*\hat{n} \cdot P\hat{e}_1$$
$$- |g|^2(\hat{n} \times P) \times \hat{n} - |h|^2(\hat{e}_1 \times P) \times \hat{e}_1 - g^*h(\hat{e}_1 \times P) \times \hat{n} - gh^*(\hat{n} \times P) \times \hat{e}_1.$$

Using the vector relations given just before (3.78), one has

$$(\hat{n} \times P) \times \hat{n} = P_p = P \cdot \hat{e}_1\hat{e}_1 + P \cdot \hat{e}_2\hat{e}_2, \quad (3.94a)$$
$$(\hat{e}_1 \times P) \times \hat{e}_1 = P - P \cdot \hat{e}_1\hat{e}_1 = P \cdot \hat{n}\hat{n} + P \cdot \hat{e}_2\hat{e}_2,$$
$$(\hat{e}_1 \times P) \times \hat{n} = -P \cdot \hat{n}\hat{e}_1,$$
$$(\hat{n} \times P) \times \hat{e}_1 = -P \cdot \hat{e}_1\hat{n},$$

so that with

$$\hat{n} \times P = \hat{n} \times (P \cdot \hat{n}\hat{n} + P \cdot \hat{e}_1\hat{e}_1 + P \cdot \hat{e}_2\hat{e}_2) = P \cdot \hat{e}_1\hat{e}_2 - P \cdot \hat{e}_2\hat{e}_1, \quad (3.94b)$$
$$\hat{e}_1 \times P = -P \cdot \hat{n}\hat{e}_2 + P \cdot \hat{e}_2\hat{n},$$

one obtains

$$N = 2\,\mathrm{Im}\{f^*g\hat{n} + f^*h\hat{e}_1 + g^*h\hat{e}_2\} + (|f|^2 + |g|^2 - |h|^2)P \cdot \hat{n}\hat{n}$$
$$+ (|f|^2 - |g|^2 + |h|^2)P \cdot \hat{e}_1\hat{e}_1 + (|f|^2 - |g|^2 - |h|^2)P \cdot \hat{e}_2\hat{e}_2$$

$$+2 \operatorname{Re} \{f^*g\} (\mathbf{P} \cdot \hat{e}_1 \hat{e}_2 - \mathbf{P} \cdot \hat{e}_2 \hat{e}_1) + 2 \operatorname{Re} \{f^*h\} (\mathbf{P} \cdot \hat{e}_2 \hat{n} - \mathbf{P} \cdot \hat{n} \hat{e}_2)$$
$$+2 \operatorname{Re} \{g^*h\} (\mathbf{P} \cdot \hat{e}_1 \hat{n} + \mathbf{P} \cdot \hat{n} \hat{e}_1).$$

Consequently, the polarization $\mathbf{P}'$ of the scattered beam is given by $\mathbf{P}' = N/D$ with

$$N = [2 \operatorname{Im} \{f^*g\} + (|f|^2 + |g|^2 - |h|^2) \mathbf{P} \cdot \hat{n} + 2 \operatorname{Re} \{g^*h\} \mathbf{P} \cdot \hat{e}_1 + 2 \operatorname{Re} \{f^*h\} \mathbf{P} \cdot \hat{e}_2] \hat{n}$$
$$+ [2 \operatorname{Im} \{f^*h\} + 2 \operatorname{Re} \{g^*h\} \mathbf{P} \cdot \hat{n} + (|f|^2 - |g|^2 + |h|^2) \mathbf{P} \cdot \hat{e}_1 - 2 \operatorname{Re} \{f^*g\} \mathbf{P} \cdot \hat{e}_2] \hat{e}_1$$
$$+ [2 \operatorname{Im} \{g^*h\} - 2 \operatorname{Re} \{f^*h\} \mathbf{P} \cdot \hat{n} + 2 \operatorname{Re} \{f^*g\} \mathbf{P} \cdot \hat{e}_1 + (|f|^2 - |g|^2 - |h|^2) \mathbf{P} \cdot \hat{e}_2] \hat{e}_2$$

and

$$D = (|f|^2 + |g|^2 + |h|^2) \left(1 + 2 \frac{\operatorname{Im} \{f^*g \hat{n} + f^*h \hat{e}_1 - g^*h \hat{e}_2\}}{|f|^2 + |g|^2 + |h|^2} \cdot \mathbf{P} \right).$$

If parity is conserved ($h = 0$) one obtains with (3.94) the former result (3.78).

3.6 Quantitative Results

Numerical values are given for the quantities characterizing the scattering process. They show that if the atomic number of the target and the scattering angle are not too small, significant polarization effects occur. At energies of less than 100 eV the quantitative results are still incomplete and not very accurate.

Until now we have shown only which basic phenomena arise in elastic scattering and have said nothing about their magnitude. All that one needs to assess them are the complex amplitudes f and g since the scattering process is completely determined by these amplitudes if effects of parity violation (which are conceivable but very small) are neglected. Apart from depending on the scattering angle θ, the amplitudes f and g depend also on the energy of the incident electrons and on the scatterer (i.e., essentially on the scattering potential).

3.6.1 Coulomb Field

Even for the simplest case, the scattering in the pure Coulomb field of the atomic nucleus, the calculation of f and g is tedious. The Dirac equation for the Coulomb field can be exactly solved so that the scattering phases η_l and η_{-l-1} in the infinite series for f and g, (3.51), can be given exactly. However, closed expressions for the series cannot be obtained. Therefore one finds, apart from approximation formulae (e.g., from McKinley and Feshbach for the scattering cross section) only numerical tables for the quantities of interest [3.9].

As measurements in the past decades have shown, these theoretical results for the pure Coulomb field are very reliable. They are only valid, however, when the electron energies and scattering angles are not too small. This can be easily visualized: Slow electrons are strongly deflected even by the relatively weak forces at large distances from the nucleus. At these large distances the screening of the nuclear Coulomb field by the electron cloud is very effective. Also for the scattering of fast electrons at small angles, the relatively weak forces at large distances from the nucleus play the major role. In order to be scattered through large angles, however, fast electrons must come very close to the atomic nucleus, so that in the region of a few hundred keV with scattering angles of more than 45°, the description of the scattering process by the Coulomb field is a good approximation.

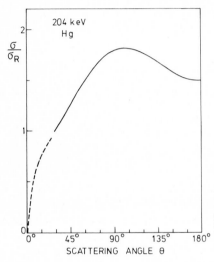

Fig. 3.18. Ratio of the cross section for elastic scattering to the Rutherford cross section for 204-keV electrons scattered by Hg [3.10]

As an example, Fig. 3.18 shows the cross section for the scattering of unpolarized 204-keV electrons by mercury, divided by the well-known Rutherford cross section. If the Rutherford formula were exact for electron scattering, one would obtain a horizontal straight line with the ordinate 1 for this scattering process. The actual scattering cross section deviates from this straight line at angles above about 30° (solid curve) because the Rutherford formula neglects the influence of the spin-orbit interaction. The very strong decrease of the cross section at small angles (dashed curve) is due to the screening of the nuclear Coulomb field by the elctron cloud of the atom.

Figure 3.19 shows the Sherman function in the region where scattering by the Coulomb potential is a very good approximation. It can be seen that with suitable electron energies and large scattering angles, polarizations of 40–50 % occur when an unpolarized beam is scattered by gold. The fairly large values of $S(\theta)$ also imply that the left-right asymmetry in scattering of polarized electrons is easily detectable. Such appreciable values occur only with elements having

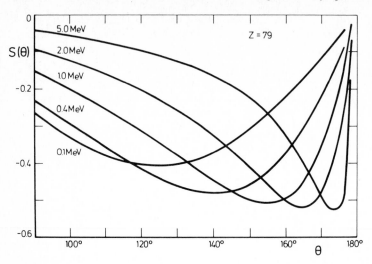

Fig. 3.19. Sherman function $S(\theta)$ for scattering by gold at various energies [3.9]

high atomic numbers since only then is the spin-orbit coupling large enough to evoke significant spin effects. With light elements the polarization effects are vanishingly small. A complete survey of theoretical and experimental work in the higher range of energies discussed here was given by *Überall* [3.11].

3.6.2 Screened Coulomb Field

The theoretical calculation of the scattering amplitudes f and g for low energies is much more tedious as it can no longer be based on the Coulomb potential. Very reliable results were obtained by using Hartree-type potentials and evaluating f and g by machine computation. We will now discuss these results.

The typical behavior of the cross sections in this energy region was already shown in Fig. 3.9 for 300-eV electrons. The curves are not smooth as in the 100-keV region but have an interference structure as λ is of the same order as the atomic radius. As explained in Sect. 3.4.2, the polarization curves, i.e., $S(\theta)$, are closely connected to the cross-section curves. They therefore also have an oscillating character and are not as smooth as in the region of a few hundred keV (Fig. 3.19). As an example, Fig. 3.20 shows the Sherman function for the scattering of 900-eV electrons by Hg in the angular range where high values of $S(\theta)$ occur. Small angles can always by disregarded when considering the polarization effects since the Sherman function is then practically zero. Since the cross sections are very large at small angles, this is again in accordance with our rule of Sect. 3.4.2 that high values of the Sherman function occur where the cross section is small and vice versa. It is also to be intuitively expected that there are no high polarization effects at small angles: Electrons that are scattered at small

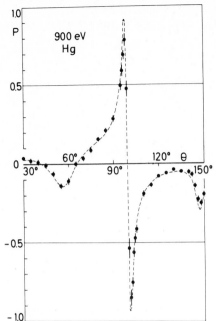

angles have passed through the relatively weak atomic field at fairly large distances from the nucleus. There the spin-orbit interaction which causes the polarization effects is very small, since, according to Sect. 3.1, it decreases quickly with increasing distance from the nucleus.

A good survey of polarization effects is given by the contours $S(\theta, E) = $ const. Figure 3.21 shows that for certain energy-angle combinations high values ($|S| > 0.8$) of the Sherman function occur. The positions of the extrema of $S(\theta, E)$ have been determined by *Walker* [3.13] using a method given by *Bühring* [3.14]. Such high values are of interest when scattering is used as a polarizer or when polarization is analyzed by measuring the left-right asymmetry. One must, however, remember that these "favorable" parameter combinations lie near the cross-section minima. One therefore obtains small scattering intensities and thus little efficiency (cf. Sects. 8.1.2 and 8.2). Furthermore, the extrema with values of $|S|$ near 1 are extremely narrow (see Fig. 3.20). Accordingly, $|S| \approx 1$ can be realized experimentally only if one works with very good angular resolution, which again means a reduction of intensity.

In Fig. 3.22 the corresponding contours for $U(\theta, E)$ are shown. The function U describes the rotation of the polarization vector out of its initial plane (cf. Sect. 3.3.3). The nearer $|U|$ is to 1, the stronger is this rotation of $\boldsymbol{P}$ during the scattering . The figure shows that for certain energy-angle combinations very strong rotation occurs.

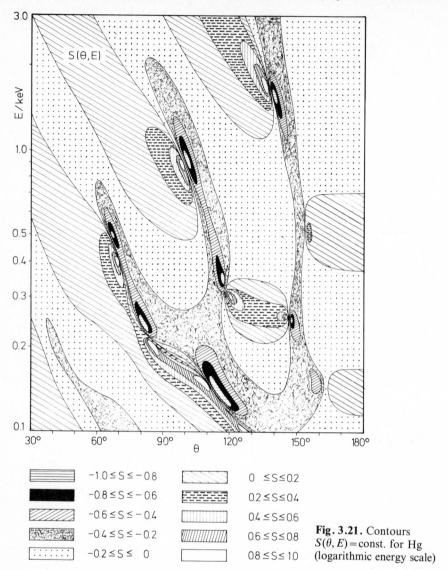

Fig. 3.21. Contours $S(\theta, E) = \text{const.}$ for Hg (logarithmic energy scale)

Figure 3.23 shows the corresponding contours for the function $T(\theta, E)$, which according to (3.77) describes the reduction of the polarization component $\boldsymbol{P}_p$ parallel to the scattering plane. The more $T(\theta, E)$ deviates from $+1$, the stronger is this reduction; $T(\theta, E) < 1$ means a reversal of $\boldsymbol{P}_p$. Again, strong changes of the polarization for certain parameter combinations can be seen.

A comparison of theoretical and experimental results shows that the theoretical prediction of the polarization effects is reliable down to the region of about 100 eV. The Sherman function for scattering angles between $30°$ and $150°$

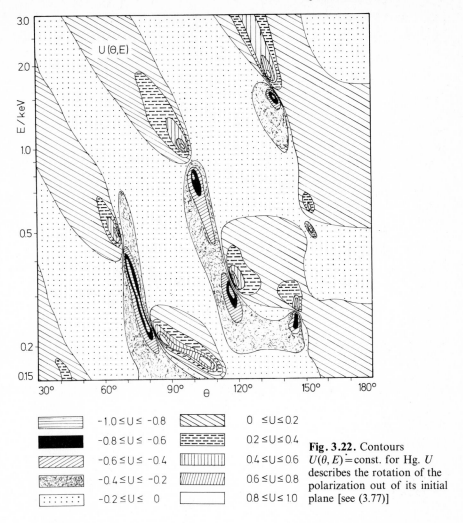

Fig. 3.22. Contours $U(\theta, E) = $ const. for Hg. U describes the rotation of the polarization out of its initial plane [see (3.77)]

Legend:
- $-1.0 \leq U \leq -0.8$
- $-0.8 \leq U \leq -0.6$
- $-0.6 \leq U \leq -0.4$
- $-0.4 \leq U \leq -0.2$
- $-0.2 \leq U \leq 0$
- $0 \leq U \leq 0.2$
- $0.2 \leq U \leq 0.4$
- $0.4 \leq U \leq 0.6$
- $0.6 \leq U \leq 0.8$
- $0.8 \leq U \leq 1.0$

was determined experimentally at many energies in the region shown in Fig. 3.21 so that the errors of these curves are well known. The error limits, even for the extrema, are usually only a few percent, and for the smaller energies shown occasional small angular shifts of up to 2° near the extrema cannot be excluded. With energies smaller than those shown in Fig. 3.21, deviations between theory and experiment continually increase, a statement which holds true not only for mercury but also for the other elements studied [3.13, 15–22].

The uncertainty of the theoretical values at low energies is due to the fact that with decreasing energy, processes become important which are difficult to describe theoretically. First, a charge distortion (and thus a potential distortion) arises in the atomic electron cloud during the scattering process, because slow

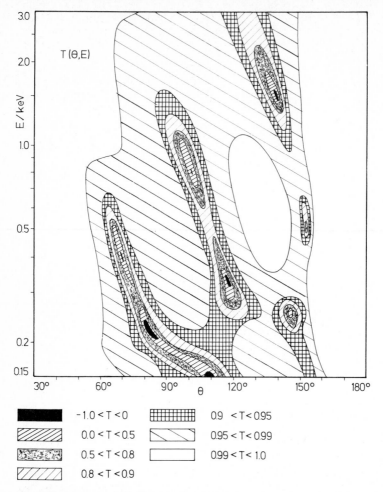

Fig. 3.23. Contours $T(\theta, E) = $ const. for Hg. T describes the reduction of the polarization component $\boldsymbol{P}_\mathrm{p}$ due to scattering [see (3.77)]

electrons stay near the atom long enough to give rise to such a charge-cloud polarization (not to be confused with spin polarization). Second, at low energies exchange processes of the incident electrons with the atomic electrons are important.

The results for xenon shown in Fig. 3.24 shed light on the general situation found at low energies. None of the existing theoretical approximations gives a completely satisfactory description of the experimental results. At energies below about 100 eV, the effort to treat the complications just mentioned [3.13, 17–21] has not yet led to a consistent theory which is reliable for all elements and at all energies studied experimentally. Further work remains to be done.

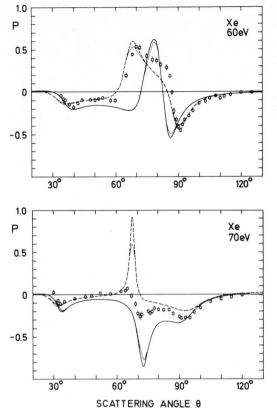

Fig. 3.24. Measured polarization (□) of electrons scattered elastically from xenon at energies of 60 and 70 eV [3.22] compared with theoretical curves: *Walker* with exchange plus charge-cloud polarization (——) [3.13]; *Feder's* $\bar{\alpha}$ approximation (– – –) [3.20]. Convolution of the theoretical curves with the angular resolution of $\pm 2°$ yields the dotted lines

SCATTERING ANGLE θ

Experimental studies of the functions T and U have been made only recently. Apart from a pioneer experiment in the late sixties [3.23], the available data have been taken in the present decade [3.24, 25]. Figure 3.25 compares experimental and theoretical results for xenon, showing good agreement between the experimental data and the two theoretical curves at the higher energies. As the energy decreases, discrepancies between the three sets of data arise which are considerable at the lowest energies shown. These uncertainties which are about the same in the other elements studied are the reason why Figs. 3.21–23 have not been continued to low energies even though experimental and theoretical data exist.

It has been discussed at the end of Sect. 3.3.3 how the measurements of S, T, U, and of the scattering cross section can be utilized to determine experimentally the complex scattering amplitudes f and g by which the scattering process is completely described. Such an evaluation is shown in Fig. 3.26 for xenon at $\theta = 60°$. Complete experiments of this kind which yield the maximum possible information on the scattering process are an outcome of the progress which has been made in the past few years with the production and analysis of polarized electrons.

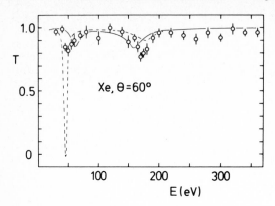

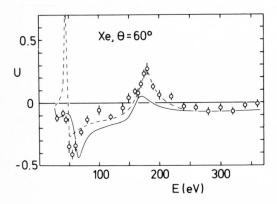

Fig. 3.25. Energy dependence of the parameters T and U at scattering angle $\theta = 60°$. The experimental points are compared with relativistic calculations by *Walker* including exchange (———) and charge-cloud polarization plus exchange (– – –) [3.24]

Among the other cases where polarization has been studied both in experiment and theory are the heavy elements krypton and cesium; but light elements like argon and helium, where the smaller spin-orbit interaction causes less pronounced polarization effects, have also been studied [3.26–30].

Due to the introductory character of this book we restrict ourselves to a few typical examples and refer to earlier review articles from the theoretical standpoint by *Walker* [3.13] and from the experimental standpoint with a view to the historical development by the author [3.31]. A considerable amount of material is also given in a review by *Eckstein* [3.32]. Tables of the relevant functions have been published by *Holzwarth* and *Meister* [3.33] and by *Fink* et al. [3.34].

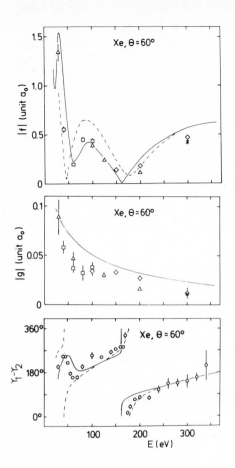

Fig. 3.26. Evaluation of the complex scattering amplitudes $f=|f|\exp(i\gamma_1)$, $g=|g|\exp(i\gamma_2)$ for $\theta=60°$. Different experimental results are due to different experimental cross sections used for the evaluation. Theoretical curves as in the preceding figure. For details see [3.24]

3.7 Experimental Setups

Typical examples of experimental setups for double and triple scattering experiments are given, and the main difficulties in carrying out such measurements (elimination of unwanted electrons, plural scattering, small intensity) are discussed.

3.7.1 Double Scattering Experiments

In polarization experiments one frequently does not have just a single scattering process as in the measurements of cross sections. According to Sect. 3.3.4, one can for example measure the Sherman function $S(\theta)$ by the following method. An unpolarized beam is scattered through the angle θ' (Fig. 3.27). The scattered beam which has the polarization $P = S(\theta')$ then hits a second target of the same kind. The respective numbers of electrons N_l and N_r scattered through the same angle θ' to the left and to the right are measured. From (3.84) it follows that

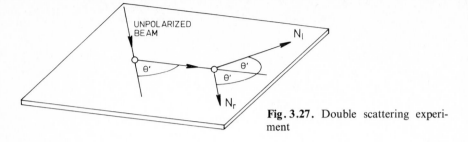

UNPOLARIZED
BEAM

N_l

θ'

θ'

θ'

N_r

Fig. 3.27. Double scattering experiment

$$\frac{N_l - N_r}{N_l + N_r} = S^2(\theta')$$

so that one obtains the value of $|S|$ for the angle θ'.

The angular dependence $S(\theta)$ can be found if one varies the first scattering angle and leaves the scattering angle at the second target unchanged. According to (3.84) one then measures

$$\frac{N_l - N_r}{N_l + N_r} = S(\theta)S(\theta').$$

As $|S(\theta')|$ is known from the first experiment, this measurement yields $S(\theta)$. Strictly speaking, the sign of $S(\theta)$ is not determined because only the magnitude of $S(\theta')$ is known. By using a source that produces polarized electrons of known spin direction, one can, however, determine the sign of $S(\theta')$ by measuring the sign of the left-right asymmetry which arises in the scattering of these electrons through θ' [see (3.70)].

Figure 3.28 shows an example of a practical setup for a double scattering experiment which has been used for a large number of measurements. The electron gun yields a well-collimated beam of unpolarized electrons of fixed energy E_0. The energy E_0 lies between 25 eV and a few keV. The electrons are fired at a mercury-vapor target and the polarization $P(\theta, E_0)$ of the scattered electrons is measured. As we are considering the scattering of an initially unpolarized beam, we have $P(\theta, E_0) = S(\theta, E_0)$. Thus the measurement produces the Sherman function. The scattering angle θ is varied by rotating the gun about the axis of the Hg beam.

The scattered beam passes through a filter lens, which removes inelastically scattered electrons, because in the present example elastic scattering is to be studied. The filter lens not only removes those electrons that excited the Hg atoms into higher energy states, but also removes unwanted electrons that hit the walls of the vacuum chamber (not shown) and were reflected into the direction of observation, having lost some of their energy.

The electrons that leave the filter lens are accelerated to 120 keV and then hit the second target, a gold foil. The electrons scattered through 120° to the right and left are counted. From the left-right asymmetry

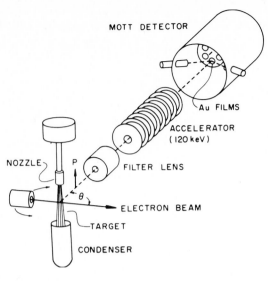

MOTT DETECTOR

Fig. 3.28. Experimental setup of a double scattering experiment [3.12]

Au FILMS

ACCELERATOR
(120 keV)

NOZZLE   P

FILTER LENS

θ

ELECTRON BEAM

TARGET

CONDENSER

$$\frac{N_l - N_r}{N_l + N_r} = S(\theta, E_0)\, S(120°,\ 120\ \text{keV}) \tag{3.95}$$

one obtains the quantity $S(\theta, E_0) = P(\theta, E_0)$ (i.e., the polarization after the first scattering process) if $S\,(120°,\ 120\ \text{keV})$ is known. Accelerating the electrons to 120 keV after the first scattering process has the advantage that at this energy the Sherman function is particularly well known. The theoretical values are very reliable as the approximation of scattering in the pure Coulomb field of the nucleus is very good in this energy region (see Sect. 3.6). Furthermore, the experimental determination of the Sherman function has been carried out particularly accurately at this energy [3.35]. This is partly because extensive Mott-scattering investigations of electron polarization in β decay took place in this energy region.

We emphasize the main problems of such experiments because these problems have not always been fully appreciated, the experiments thus having led to erroneous results.

Since the intensity of the electrons that are to be detected is small, the background of unwanted electrons must be carefully suppressed. This background appears because neither the electrons of the primary beam nor scattered electrons that hit the walls of the scattering chamber or of the polarization analyzer are completely absorbed there. They are instead reflected at the walls, and an appreciable portion of them, if not sufficiently suppressed, arrives at the counters and affects the measurements.

Electrons that are reflected into the direction of observation by plural scattering[11] in the target must also be suppressed. If, for example, one makes a

---

[11]   Scattering in which more than one deflection takes place while not enough are involved to give the characteristic Gaussian distribution of multiple scattering.

measurement at 120°, one does not find there only electrons that have been once scattered through 120°. One also finds electrons that have been scattered once through the angle α and then through $120° - \alpha$; or electrons that have reached the resulting angle 120° by more than two consecutive processes. The probability of such plural processes increases as the number of atoms in the target increases. This means that there are limits for the density of the Hg target and the thickness of the gold foil in the analyzer. For this reason, the density of the Hg beam in the experiment described had to correspond to a pressure which was considerably less than 10^{-3} mbar [3.36]. Likewise, the gold foil had an area density of ~ 200 µg/cm², i.e., a thickness of approximately 100 nm (Sect. 8.1.2).

Another important error source in polarization measurements is spurious asymmetries caused by the apparatus. Their elimination will be discussed in detail in Sect. 8.1.2.

3.7.2 Triple Scattering Experiments

By measuring the cross section in single scattering experiments and the Sherman function in double scattering experiments, one still has not found all the quantities that are required for a complete description of electron scattering. In addition, one needs the quantities $T(\theta)$ and $U(\theta)$ which describe the change of the polarization vector in a scattering process.

To measure these quantities by scattering only, one needs three consecutive scattering processes: In the first scattering the unpolarized electron beam becomes polarized. The second scattering process causes the change of the polarization vector, which is the object of the investigation. To be able to analyze this change of the polarization vector a third scattering is required – an asymmetry measurement with a Mott detector.

The problems that arise in double scattering naturally become much more pronounced in triple scattering. This is why only one measurement of $T(\theta)$ and $U(\theta)$ has been made with this method [3.23]. This pioneer experiment has been made with a gold target and electrons of 261 keV, an energy where no theoretical problems arise since at the angles studied it is the pure nuclear Coulomb field which causes the scattering process. An extension of T and U measurements to smaller energies, which pose more challenging theoretical problems, and to other elements, became possible only after the development of efficient sources of polarized electrons by which the first scattering process could be replaced. Although, with this technique, one no longer needs three scattering targets, we will retain the historical term "triple scattering experiment".

In the experiment depicted in Fig. 3.29 the polarized electrons are produced by photoemission from a GaAs crystal which is irradiated with circularly polarized light. The method is described in Sect. 8.2. The initial polarization was chosen to lie in the scattering plane. Accordingly one has $P = P_p$, $P_n = 0$, so that one obtains from (3.77) for the polarization P' after scattering the simple expression

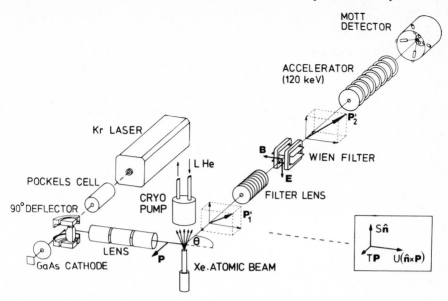

Fig. 3.29. Experimental setup for measurement of the quantities T and U describing the change of the polarization vector caused by scattering [3.24]

$$P' = S\hat{n} + TP + U(\hat{n} \times P).$$

Magnetic spin rotators not shown in the figure have oriented the initial polarization P along the scattering direction so that TP is the longitudinal polarization component of the scattered beam whereas $U(\hat{n} \times P)$ is its transverse component in the scattering plane. Measurement of these components yields the quantities T and U, if the initial polarization is known. Since the Mott detector can analyze transverse components only, a Wien filter (Sect. 8.1.1) is needed. If the Wien filter is switched off, the transverse components $S\hat{n}$ and $U(\hat{n} \times P)$ are analyzed simultaneously by the two pairs of counters in the Mott detector. If the Wien filter is switched on, the two polarization components perpendicular to its magnetic field B are rotated through $90°$ ($P'_1 \rightarrow P'_2$) so that the longitudinal component TP becomes transverse and can be measured. For measurement of the initial polarization P, the primary beam is deflected into the Mott detector. This can be done either by means of an electrostatic field or by scattering the electron beam from the graphite-coated surface of the atomic beam capillary. Both methods yield the same results since scattering from a low-Z material with its small spin-orbit interaction does not alter the polarization. In order to eliminate instrumental asymmetries when measuring TP and $U(\hat{n} \times P)$, these components are reversed by producing electrons of polarization $-P$ which was achieved by reversing the circular light polarization as described in Sect. 8.2.

Measurements of T and U have been made for Hg and Xe at several energies and angles. An example has been shown in Fig. 3.25.

3.8 Resonance Scattering

Cross section and polarization curves may have significant resonance features. These arise when the scattered electron attaches itself temporarily to a target atom. The origin of polarization resonances is explained.

Significant polarization effects may occur in resonance scattering [3.37–39], a process in which the electron does not simply pass through the potential field of the target atom as we have assumed so far, but instead attaches itself to the atom forming a temporary negative ion or compound state. After a short lifetime (10^{-13} s or less) the compound state decays by emission of the electron.

A typical mechanism for electron trapping is illustrated by the following example. We consider the excited state $1s^2 2s^2 2p^5 3s$ of the neon atom lying approximately 16.7 eV above the ground state $1s^2 2s^2 2p^6$. The excited state (which is called "parent" in Fig. 3.30) can bind an electron, the binding energy being about 0.6 eV. The negative ion thus formed has the structure $2p^5 3s^2$ of the outer shells and an energy of approximately $16.7 - 0.6 = 16.1$ eV. If an electron of this energy is fired on neon atoms it has just the right energy to form the negative ion state. After the short lifetime the temporary ion decays to the ground state emitting the electron with its full primary energy, so that an observer finds an electron which has been elastically scattered by the neon atom. If one measures the energy dependence of the scattering cross section one will find at 16.1 eV a "resonance" structure. It is caused by a shift in the phase of the scattered electron that is much larger than at the other energies where formation of a negative ion is energetically forbidden.

Since there is a great number of parent states which are able to bind electrons, one finds many resonances if the energy is varied. Compound states of high enough energy need not decay into the ground state of the atom; instead the atom may be left in an excited state so that the emitted electron does not have the full primary energy. In such cases the resonance is found when inelastically scattered electrons are observed. In general the compound state has various channels of decay so that the same resonance can be observed in several scattering channels.

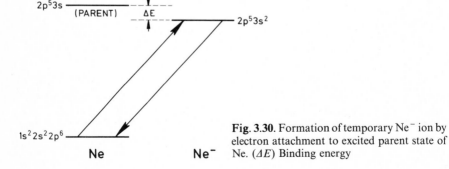

Fig. 3.30. Formation of temporary Ne⁻ ion by electron attachment to excited parent state of Ne. (ΔE) Binding energy

The mechanism described here produces the so-called Feshbach resonances which lie energetically below their parents. For completeness we mention that this is not the only origin of resonances. There are also "shape resonances" which lie above their parents. In this case the attractive potential of the parent, in conjunction with the centrifugal potential of the trapped electron, forms a well with a penetrable barrier that traps the electron for a limited time (usually much shorter than for Feshbach resonances). The properties of this resonance depend on the shape of the potential.

After these introductory remarks, let us discuss how resonances can give rise to polarization effects. Returning to our example of the neon resonance, we will now take spin-orbit interaction into account. From the configurations given, the excited states of Fig. 3.30 are clearly seen to be P states because they have one hole in the p shell. Spin-orbit coupling causes the P state of the Ne$^-$ to split into two states with configurations $(2p^5 3s^2)P_{1/2}$ and $(2p^5 3s^2)P_{3/2}$. They differ in energy by about 0.1 eV, an amount by which the energies mentioned above need correction. Accordingly, one expects the cross section for elastic electron scattering to have two resonance features close to 16 eV which differ by 0.1 eV. They can be seen in the upper part of Fig. 3.31, while the lower part shows a positive and a negative polarization peak which are clearly related to the cross-section peaks.

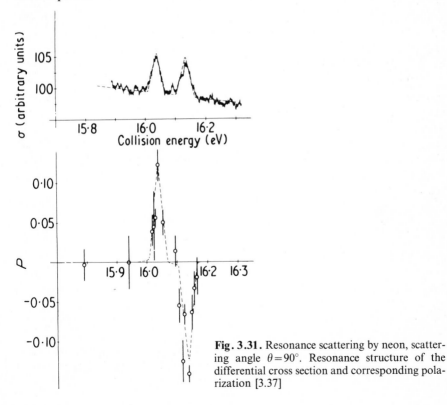

Fig. 3.31. Resonance scattering by neon, scattering angle $\theta = 90°$. Resonance structure of the differential cross section and corresponding polarization [3.37]

The polarization peaks can be explained as follows. The process

$$e + Ne \rightarrow Ne^- \rightarrow e + Ne$$

must obey the general conservation laws for energy, total angular momentum J, and parity π. Since the compound states of Ne^- which are considered here have $J = 1/2$ and $3/2$ and $\pi = -1$, the input channel must have the same quantum numbers. Since the Ne is in its ground state, $^1S_0$, the incident electron must have $j = 1/2$ or $3/2$, respectively, in order to excite the $^2P_{1/2}$ or $^2P_{3/2}$ compound state. Its quantum number l must be odd in order to make $\pi = (-1)^l$ odd. Accordingly, only electrons described by the partial wave with $l = 1$ and $j = 1/2$ or $3/2$ can excite the respective ion states $^2P_{1/2}$ and $^2P_{3/2}$. In the resonance maxima of the cross section these partial waves are particularly accentuated. Let us pick out the $^2P_{3/2}$ resonance at 16.04 eV and visualize its polarization by the model used in Sect. 3.4.2. In the $l = 1$ wave of an unpolarized incident beam, the electrons with spins parallel to the orbital angular momentum have the proper $j = 3/2$ to excite the $^2P_{3/2}$ ion state, whereas electrons with opposite spins $(j = 1/2)$ cannot go through this resonant state. The preferential scattering of the $j = 3/2$ wave (i.e., of the e↑) which is indicated by the cross section maximum at 16.04 eV results in a positive polarization as illustrated by Fig. 3.32. At 16.14 eV, the energy of the $^2P_{1/2}$ compound state, it is the $j = 1/2$ wave containing the e↓ that is preferentially scattered which results in a negative polarization of the scattered beam. Quantitatively, the polarization is given by the general expression (3.73) with f and g from (3.51). Near the resonance energies, the $l = 1$ partial wave plays a dominant role in the scattering amplitudes: the two phases corresponding to $j = 1 + 1/2$ and $j = 1 - 1/2$, respectively, change rapidly by π radians as the energy traverses the two resonances, thus affecting f and g (and therewith cross section and polarization) strongly. The polarization is significant since at such small primary energies only a few partial waves of low l contribute to scattering. Thus the dominant scattering of the $l = 1$ wave which produces the resonant polarization clearly stands out against the background scattering of the other partial waves.

The compound states of Ne^- discussed so far have a simple electronic structure, so that their classification followed soon after discovery. In many other cases, the identification of compound states is less obvious. In these circumstances, measurement of the resonant polarization can help considerably in classifying the negative ion states: The partial wave dominating near

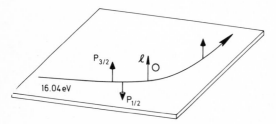

Fig. 3.32. Resonance scattering of 16.04 eV electrons via the $(2p^5 3s^2)P_{3/2}$ compound state of Ne^-. The conservation laws enforce preferential scattering of e↑ which results in positive polarization

resonance determines the angular dependence of the scattering amplitudes through the Legendre polynomials $P_l(\cos\theta)$ and $P_l^1(\cos\theta)$ (3.51). As a consequence, one has an angular distribution of the cross section and of the polarization which depends near resonance characteristically on the dominant partial wave. Measurement of the angular dependence of the resonances therefore allows the dominant partial wave to be identified and yields its total angular momentum and parity. This, in conjunction with the quantum numbers of the atomic state and the conservation laws, gives J and π for the compound state. In this way, some controversial energy levels of Hg^- ions have been classified which are formed when an electron attaches itself to the $6\,^3P$ excited states of mercury [3.40] (see also Sect. 4.7).

3.9 Inelastic Scattering

Spin-orbit coupling produces polarization effects in inelastic electron scattering. They are similar to those found in elastic scattering as long as inelastic scattering may be described as a two-step process. The relation $P = A$ does not generally hold in inelastic scattering so that asymmetry and polarization measurements yield different information. Separation of electrons that have been scattered from different fine-structure levels yields polarization effects which cannot be explained by spin-orbit interaction alone.

So far our attention has been focused on elastic scattering. Polarization effects do, however, also occur in inelastic electron scattering. They have been studied, in particular, for mercury targets. Figure 3.33 shows the polarization of 180-eV electrons which have excited the $6\,^1P_1$ state (energy loss 6.7 eV) of Hg atoms and have thereby been scattered through the angle θ. If one compares the polarization curve with that for elastic scattering of 180-eV electrons from

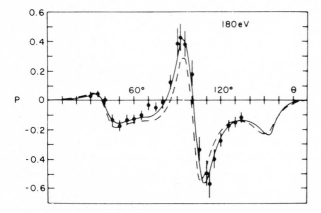

Fig. 3.33. Polarization of electrons scattered inelastically by Hg atoms (excitation of the $6\,^1P_1$ state; energy loss 6.7 eV). Experimental [3.41] and theoretical [3.42] values calculated in the distorted wave approximation using different potentials

mercury [3.41] one finds that the curves are very similar. The similarity decreases as the electron energy is reduced. Below 25 eV the measured polarization for the inelastic process lies mostly below 10% whereas for elastic scattering the polarization is still significant as at the higher energies [3.15, 16].

The similarity between elastic and inelastic polarization at higher energies indicates that, at these energies, inelastic scattering may be considered as a two-step process [3.43], a model which is corroborated by the fact that the cross sections also look very similar. An electron can be thought of as first losing energy in exciting the atom, a process usually resulting in little deviation, and then undergoing scattering in the field of the excited atom. Similarly, it might first be scattered without energy loss and then suffer the inelastic transition. If the field of the excited atom does not differ very much from that of the normal atom, and the energy loss suffered in the excitation process is small compared with the incident energy, one would expect the outcome of elastic and inelastic scattering to be almost the same. For more detailed discussion of this coupling between elastic and inelastic channels see [3.41–45]. It is worth noting that such two-step processes have also been observed in inelastic low-energy electron diffraction (cf. [3.46]).

Beyond the similarities there are fundamental differences of the polarization effects in elastic and inelastic processes. A major difference is that the equality of polarizing and analyzing power (which has been shown in Sect. 3.4.5 to exist in Mott scattering) does not necessarily hold in inelastic scattering. The reason why the arguments of Sect. 3.4.5 can, in general, not be transferred to the inelastic case may be visualized as follows. Referring to Fig. 3.17, let the inelastic scattering intensities still be proportional to $1 + S$ and $1 - S$ for totally polarized incident beams of opposite directions of polarization. The scattering asymmetry A then still equals S. The essential difference is now that inelastic scattering may be associated with transfer of angular momentum to the atom, which would change the angular momentum of the incident electrons. Due to their spin-orbit interaction in the atomic field, this may affect both their orbital and spin angular momentum. Although each of the beams in Fig. 3.17 is totally polarized normal to the scattering plane, when inelastic scattering plays a role their polarization may change. Scattering of the e↑ (or of the e↓) produces therefore e↑ as well as e↓ so that the relations $N_\uparrow \propto 1 + S$, $N_\downarrow \propto 1 - S$ for the respective numbers of e↑ and e↓ found after the scattering in one direction are no longer valid in general. Thus the polarization obtained by inelastic scattering cannot be found by the simple method of (3.87). It is no longer given by S, but instead the exact value is determined by the dynamics of the inelastic process.[12]

---

[12] The case of spin transfer to the atom owing to exchange interaction between the incident and atomic electrons will be treated in the following chapter. It is obvious from the above argument that $P = A$ is then broken not only in inelastic but also in elastic scattering from atoms which allow such a spin transfer without violation of the Pauli principle (atoms with unsaturated spins, cf. Problem 4.1).

When $P = A$ does not apply, as in inelastic scattering, measurements of both quantities are meaningful because they yield different information. In special cases, such as the two-step process discussed above where the elastic step is responsible for the inelastic polarization, $P = A$ can be valid whereas in other cases differences of the two quantities may reveal a different type of process. There are not many measurements of polarization effects in inelastic scattering because of the experimental difficulties. The cross sections for inelastic processes are about two orders of magnitude smaller than for elastic ones. Asymmetry measurements of inelastically scattered polarized electrons have therefore only been made since the development of efficient polarized-electron sources.

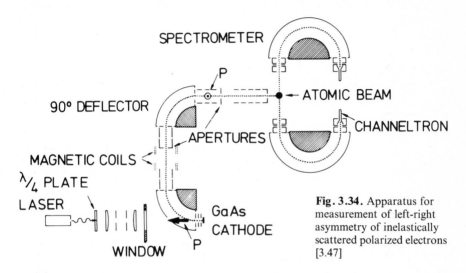

Fig. 3.34. Apparatus for measurement of left-right asymmetry of inelastically scattered polarized electrons [3.47]

Figure 3.34 shows the experimental arrangement of an asymmetry measurement with polarized electrons which have excited the $6\,^1P_1$ state and triplet fine-structure states of Hg. A GaAs source, to be described in Sect. 8.2, produces longitudinally polarized electrons. After several deflections of the beam, and transformation of the polarization by means of magnetic coils to the transverse direction, the electrons hit the mercury target (currents at the target 3–10 nA). A pair of energy analyzers positioned at the scattering angle of 90° selects the inelastic channel to be studied. The overall energy resolution of the experiment is about 200 meV. The left-right asymmetry $A = (N_l - N_r)/(N_l + N_r)$ of the inelastically scattered polarized electrons is detected by a pair of channel-trons (N_l, N_r = number of electrons scattered to left and right, respectively). Instrumental asymmetries can easily be corrected by reversal of the electron polarization.

Figure 3.35 shows the observed asymmetry A normalized to the incident polarization which has been determined by elastic scattering from mercury where the Sherman function is well known. Particularly within a few eV above

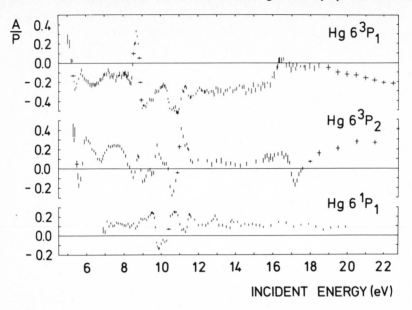

Fig. 3.35. Left-right asymmetry A (normalized to incident electron polarization P) of inelastically scattered polarized electrons. The energy losses are 4.89, 5.46, and 6.7 eV corresponding to excitations of the $6\,^3P_1$, $6\,^3P_2$, and $6\,^1P_1$ states of Hg [3.47]

threshold, pronounced structures with large values of A/P are observed. This is the energy range where resonance structures in the cross sections exist due to formation of negative ion states. From what has been said in the last section we are prepared for the appearance of these resonances also in the polarization effects. We have seen that only electrons of well-defined energy and angular momentum j may excite a certain compound state. Picking out the electrons of the dominant partial wave l, we must take into account the fact that transversely polarized electrons which are scattered to the left have total angular momenta j different from those of the electrons scattered to the right, since the orbital angular momenta are opposite to each other. Accordingly, either the electrons scattered to the right or to the left, but not both, may have j suitable to excite the compound state and to exhibit a resonance in the cross section. As a result one has a significant difference of the scattering intensities to the right and to the left, i.e., a large asymmetry.

A closer look at the asymmetry curves for $6\,^3P_1$ and $6\,^3P_2$ excitation shows that they tend to differ in sign and reach larger values than for $6\,^1P_1$ excitation. The same is true for the polarization curves of (initially unpolarized) inelastically scattered electrons which have excited the triplet fine-structure levels as shown in Fig. 3.36 for the scattering angle $\theta = 60°$ (the angular resolution in this experiment did not quite reach that of the asymmetry measurement of Fig. 3.35, but excitation of the $6\,^3P_2$ and $6\,^3P_{1,0}$ levels could be well separated). This behavior of the results for fine-structure excitation cannot be explained by spin-

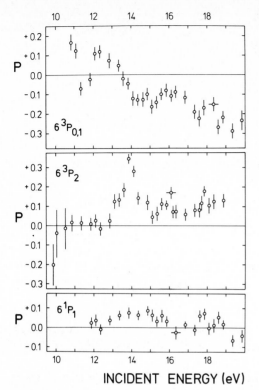

Fig. 3.36. Polarization of electrons which have excited $6\,^3P$ fine-structure levels of Hg and have thereby been scattered through $\theta = 60°$. Analogous results for $6\,^1P_1$ excitation for comparison [3.48]

orbit interaction alone. In Sect. 4.5.2 we will see how the exchange interaction plays a crucial role in excitation processes.

Let us conclude this section with the remark that the polarization effects described here for inelastic scattering also occur in the time-reversed process, in superelastic scattering. The reciprocity following from time-reversal invariance implies that the polarization P_{in} arising from inelastic scattering of an unpolarized beam equals the asymmetry A_{sup} found in superelastic scattering of a polarized beam. Vice versa one has $P_{sup} = A_{in}$ [3.49]. A first polarization measurement in superelastic scattering has been reported [3.50].

4. Polarization Effects Caused by Exchange Processes in Electron-Atom Scattering

4.1 The Polarization Mechanism in Exchange Scattering

Exchange scattering can be separated from direct scattering processes by using polarized electrons and/or atoms and observing the polarization of the scattered electrons or of the recoil atoms. The mechanism underlying the polarization effects is discussed.

In the last chapter we saw that the spin-orbit interaction can give rise to a spin polarization in electron scattering. Another mechanism that can cause a polarization of the scattered electrons is the exchange interaction. Consider, for example, scattering from a target of alkali atoms whose valence electrons all have the same spin direction. If exchange processes occur between the valence electrons and the free electrons, one obtains polarized electrons in the scattered beam.

In this section we will explain the polarization mechanism by dealing with the simple case of elastic exchange scattering of electrons from atoms with one valence electron (hydrogen, alkalis). The polarization effects to be discussed are a consequence of the Pauli principle rather than of explicit spin-dependent forces between the colliding electrons. The dipole-dipole interaction between the incident and atomic electrons is small compared with the Coulomb interaction for all energies considered here; its influence on the scattering is completely masked by the influence of the Coulomb interaction. Furthermore we make the assumption that the spin-orbit interaction of the scattered electron, the mechanism causing the polarization effects in Mott scattering, can be neglected. This is a good approximation for light atoms or small scattering angles.

Let the scattering take place on an atom of arbitrary spin direction which we denote by $A\uparrow$. For the scattering of electrons with spins parallel or antiparallel to the atomic spin the following possibilities are conceivable:

$$e\downarrow + A\uparrow \rightarrow e\downarrow + A\uparrow \tag{4.1}$$

$$e\downarrow + A\uparrow \rightarrow e\uparrow + A\downarrow \tag{4.2}$$

$$e\uparrow + A\uparrow \rightarrow e\uparrow + A\uparrow. \tag{4.3}$$

In the first two processes the two colliding electrons can be distinguished. Since explicit spin-dependent forces were excluded, each electron retains its spin direction during the scattering process and is thus marked by this spin direction.

This chapter will show that this possibility of labeling electrons by their spin direction is of central importance for experimental studies of exchange scattering.

We recall from elementary scattering theory that the scattering amplitude can be expressed by

$$-\frac{m}{2\pi\hbar^2} \langle\psi_f|T|\psi_i\rangle, \tag{4.4}$$

where, in the first Born approximation, the transition operator T equals the scattering potential and ψ_i and ψ_f are the wave functions of the initial and the final state. In the scattering process discussed here, two electrons are involved: the incident electron and the valence electron. In order to describe these identical particles properly, one has to use antisymmetric wave functions in the above formula, thus taking the Pauli principle into account. We therefore write

$$\psi_i = \frac{1}{\sqrt{2}} [e^{i k\cdot r_1}\eta(1)u(r_2)\chi(2) - e^{i k\cdot r_2}\eta(2)u(r_1)\chi(1)] \quad \text{and} \tag{4.5a}$$

$$\psi_f = \frac{1}{\sqrt{2}} [e^{i k'\cdot r_1}\eta'(1)u'(r_2)\chi'(2) - e^{i k'\cdot r_2}\eta'(2)u'(r_1)\chi'(1)], \tag{4.5b}$$

where u and u' are the atomic wave functions in the initial and the final state; k and k' are the wave vectors of the incident and the scattered electrons; η, χ and η', χ' are the spin functions of the free and the bound electron in the initial and the final state, respectively.

With these wave functions we obtain the scattering amplitude

$$-\frac{m}{2\pi\hbar^2} \langle\psi_f|T|\psi_i\rangle = \tfrac{1}{2} \{ f(k',k) [\langle\eta'(1)|\eta(1)\rangle \langle\chi'(2)|\chi(2)\rangle$$

$$+ \langle\eta'(2)|\eta(2)\rangle \langle\chi'(1)|\chi(1)\rangle] - g(k',k) [\langle\eta'(1)|\chi(1)\rangle \langle\chi'(2)|\eta(2)\rangle$$

$$+ \langle\eta'(2)|\chi(2)\rangle \langle\chi'(1)|\eta(1)\rangle]\}, \quad \text{where} \tag{4.6}$$

$$f(k',k) = -\frac{m}{2\pi\hbar^2} \langle e^{i k'\cdot r_\lambda}u'(r_\mu)|T|e^{i k\cdot r_\lambda}u(r_\mu)\rangle, \tag{4.7a}$$

$$g(k',k) = -\frac{m}{2\pi\hbar^2} \langle e^{i k'\cdot r_\mu}u'(r_\lambda)|T|e^{i k\cdot r_\lambda}u(r_\mu)\rangle, \tag{4.7b}$$

with $\lambda, \mu = 1, 2$ or $2, 1$. The scattering amplitude g obviously describes a process in which the incident electron is captured by the atom and the atomic electron is ejected. One therefore calls g the exchange amplitude, whereas f is called the direct amplitude. Needless to say, the exchange amplitude has nothing to do with the amplitude g which we introduced in the treatment of Mott scattering. As a consequence of this fundamental difference there is no left-right asymmetry of the scattered intensity, since in the approximation made in this section (negligible spin-orbit interaction) there is no potential which depends on whether the electron passes the atom on the right or on the left. The cross sections can therefore be considered to have no azimuthal dependence.

Let us describe the spin directions ↑ and ↓ in (4.1–3) by the respective spin functions $\alpha = \begin{pmatrix} 1 \\ 0 \end{pmatrix}$ and $\beta = \begin{pmatrix} 0 \\ 1 \end{pmatrix}$. Then we have for the process (4.1)

$$\eta = \beta, \quad \chi = \alpha, \quad \eta' = \beta, \quad \chi' = \alpha,$$

so that (4.6) yields the scattering amplitude

$$\tfrac{1}{2} \{ f[1+1] - g \cdot 0 \} = f.$$

In the process (4.2) we have $\eta = \beta$, $\chi = \alpha$, $\eta' = \alpha$, $\chi' = \beta$, resulting in the scattering amplitude $-g$. In (4.3) the spin functions are $\eta = \chi = \eta' = \chi' = \alpha$, so that (4.6) yields the scattering amplitude $f - g$.

Since the cross sections are the squares of the scattering amplitudes we can summarize our results as follows:

Process	Amplitude	Cross section	
$e\downarrow + A\uparrow \rightarrow e\downarrow + A\uparrow$	f	$\lvert f \rvert^2$	(4.8)
$e\downarrow + A\uparrow \rightarrow e\uparrow + A\downarrow$	$-g$	$\lvert g \rvert^2$	(4.9)
$e\uparrow + A\uparrow \rightarrow e\uparrow + A\uparrow$	$f - g$	$\lvert f - g \rvert^2$.	(4.10)

These relations, which can be easily generalized to inelastic scattering (Sect. 4.5.2), demonstrate the obvious physical meaning of the "direct" and the "exchange" cross section. The corresponding results for $A\downarrow$ are

$e\uparrow + A\downarrow \rightarrow e\uparrow + A\downarrow$	f	$\lvert f \rvert^2$	(4.11)
$e\uparrow + A\downarrow \rightarrow e\downarrow + A\uparrow$	$-g$	$\lvert g \rvert^2$	(4.12)
$e\downarrow + A\downarrow \rightarrow e\downarrow + A\downarrow$	$f - g$	$\lvert f - g \rvert^2$.	(4.13)

By scattering electrons and atoms with well-defined spin states on each other, and by analyzing the spin directions of the observed electrons and/or atoms, one could determine each of the cross sections listed above.

Needless to say, such experiments are very difficult. Usually unpolarized particles are used for scattering experiments and a spin analysis after scattering is not made. In this case one measures the sum of the cross sections given above and the information on the individual contributions is lost. If, for example, one scatters an unpolarized electron beam by an $A\uparrow$ target and observes all the scattered electrons independent of their spin direction, it follows from (4.8–10) that the cross section is

$$\frac{d\sigma}{d\Omega} \equiv \sigma_0(\theta) = \tfrac{1}{2}(|f|^2 + |g|^2) + \tfrac{1}{2}|f-g|^2$$

$$= \tfrac{1}{4}|f+g|^2 + \tfrac{3}{4}|f-g|^2. \tag{4.14}$$

The factor $1/2$ is due to the fact that the cross sections given in (4.8–13) are valid for totally polarized beams; for an unpolarized primary beam, which may be considered to be made up of equal numbers of e↑ and e↓, the scattering intensities in the individual channels are only half as large as for a totally polarized beam. It is obvious that we would have obtained the same result if we had considered scattering of an unpolarized electron beam by an A↓ target (opposite spin direction). Thus the cross section (4.14) also describes the scattering of an unpolarized beam by an unpolarized target made up of equal numbers of A↑ and A↓.[1]

The validity of the last equality in (4.14) follows immediately from elementary rules for complex number calculation. The physical meaning of this form of the cross section can be seen as follows: When an unpolarized electron beam is scattered by an atom with one outer electron, the two colliding electrons may form either a triplet state $S=1$ with the symmetric spin functions

$$\chi_S = \begin{cases} \alpha(1)\alpha(2) & \text{describing the substate } S_z = 1 & (4.15a) \\ \dfrac{1}{\sqrt{2}}[\alpha(1)\beta(2) + \beta(1)\alpha(2)] & \text{describing the substate } S_z = 0 & (4.15b) \\ \beta(1)\beta(2) & \text{describing the substate } S_z = -1 & (4.15c) \end{cases}$$

or a singlet state $S=0$ with the antisymmetric spin function

$$\chi_A = \frac{1}{\sqrt{2}}[\alpha(1)\beta(2) - \beta(1)\alpha(2)]. \tag{4.16}$$

The straightforward evaluation of the scattering amplitude (4.4) with the antisymmetric wave functions

$$\psi_i = \frac{1}{\sqrt{2}}[e^{ik \cdot r_1}u(r_2) \pm e^{ik \cdot r_2}u(r_1)]\chi_{A,S}, \tag{4.17a}$$

$$\psi_f = \frac{1}{\sqrt{2}}[e^{ik' \cdot r_1}u'(r_2) \pm e^{ik' \cdot r_2}u'(r_1)]\chi_{A,S} \tag{4.17b}$$

yields $f-g$ for the three symmetric spin states and $f+g$ for the antisymmetric spin state. The differential cross section must be computed with the former term in three-quarters of the collisions, and with the latter term in one-quarter of the cases. We thus obtain the last expression of (4.14).

---

[1] For later purposes we state here that in this case neither the scattered electron beam nor the recoil atoms are, of course, polarized, since for every possible scattering process there exists an analogous process with opposite spins. Scattered electrons and atoms with each of the two spin directions occur with the same cross section $\sigma(\theta)/2$.

For the observation of the individual cross sections listed in (4.8–13) it is not necessary to use both polarized electrons and polarized atoms, as these equations may suggest.[2] It suffices to make a simpler experiment in which either the electrons or the atoms are initially unpolarized. If, for example, one scatters totally polarized electrons by unpolarized atoms then from (4.10–12) one has

Process   Cross section

$$e\uparrow + \begin{pmatrix} A\uparrow \\ A\downarrow \end{pmatrix}$$

$$\nearrow \quad e\uparrow + A\uparrow \qquad \tfrac{1}{2}|f-g|^2 \qquad (4.18)$$

$$\rightarrow \quad e\uparrow + A\downarrow \qquad \tfrac{1}{2}|f|^2 \qquad (4.19)$$

$$\searrow \quad e\downarrow + A\uparrow \qquad \tfrac{1}{2}|g|^2. \qquad (4.20)$$

Here, as in (4.14), it has been taken into account that for scattering by an unpolarized target (equal numbers of $A\uparrow$ and $A\downarrow$) the cross sections given in (4.10–12), which are valid for a totally polarized target, have to be multiplied by the factor $1/2$. As it is not possible to select scattered electrons of one spin direction for the analysis of these processes, one measures the polarization of the scattered electron beam,

$$P_e'(\theta) = \frac{\sigma_e^\uparrow(\theta) - \sigma_e^\downarrow(\theta)}{\sigma_e^\uparrow(\theta) + \sigma_e^\downarrow(\theta)} = \frac{\tfrac{1}{2}|f-g|^2 + \tfrac{1}{2}|f|^2 - \tfrac{1}{2}|g|^2}{\tfrac{1}{2}|f-g|^2 + \tfrac{1}{2}|f|^2 + \tfrac{1}{2}|g|^2}. \qquad (4.21)$$

Here $\sigma_e^\uparrow(\theta)$ and $\sigma_e^\downarrow(\theta)$ are the cross sections for the occurrence of $e\uparrow$ and $e\downarrow$, respectively, in the scattered beam. Their values have been taken from (4.18–20). According to (4.14), the denominator of (4.21) is none other than $\sigma_0(\theta)$, the cross section for the scattering of unpolarized electrons by unpolarized atoms. If $\sigma_0(\theta)$ is known, the measurement of P_e' yields $\tfrac{1}{2}|f-g|^2 + \tfrac{1}{2}|f|^2 - \tfrac{1}{2}|g|^2$ and by subtraction of this quantity from $\sigma_0(\theta) = \tfrac{1}{2}|f-g|^2 + \tfrac{1}{2}|f|^2 + \tfrac{1}{2}|g|^2$ one obtains $|g|^2$.

If one observes in this experiment the recoil atoms $A\downarrow$, then according to (4.19) one obtains $|f|^2$. (Contrary to what we have discussed in Sect. 1.2 for the case of electrons, atoms of a certain spin direction can be selected, e.g., by a Stern-Gerlach type magnet; see also Sect. 4.2.) With the help of three measurements (cross section for scattering of unpolarized particles, polarization of scattered electrons, fraction of the recoil atoms with certain spin direction) one thus can determine the quantities $|f|^2$, $|g|^2$, and $|f-g|^2$. But still this is not quite sufficient to obtain all the information which is hidden in the scattering amplitudes, as will be shown in Sect. 4.3.

In practice one cannot use totally polarized electron beams as we have assumed up to now. In principle this does not, however, change anything we have considered. A partially polarized electron beam with polarization P_e can be considered to be split up into two fractions in the ratio $P_e/(1 - P_e)$ with total or zero polarizations, respectively (see Sect. 2.3). Accordingly, if we scatter a

---

[2] A "perfect" experiment as discussed in Sect. 4.3 needs, however, both polarized electrons and polarized atoms.

partially polarized electron beam by unpolarized atoms the cross section for occurrence of $e\uparrow$ in the scattered beams is

$$\sigma_e^\uparrow(\theta) = P_e(\tfrac{1}{2}|f-g|^2 + \tfrac{1}{2}|f|^2) + (1-P_e)\frac{\sigma_0(\theta)}{2},$$

where $\sigma_0(\theta)$ is again the cross section for the scattering of unpolarized electrons by unpolarized atoms. Here use has been made of the fact that the unpolarized fraction of the incident electron beam produces equal numbers of $e\uparrow$ and $e\downarrow$ in the scattered beam (see footnote on p. 87). Correspondingly one has

$$\sigma_e^\downarrow(\theta) = P_e\tfrac{1}{2}|g(\theta)|^2 + (1-P_e)\frac{\sigma_0(\theta)}{2}.$$

Thus the polarization of the scattered electrons is

$$P_e'(\theta) = \frac{\sigma_e^\uparrow(\theta) - \sigma_e^\downarrow(\theta)}{\sigma_e^\uparrow(\theta) + \sigma_e^\downarrow(\theta)} = \frac{P_e}{\sigma_0(\theta)}[\sigma_0(\theta) - |g(\theta)|^2],$$

where use has been made of (4.14). One therefore obtains

$$|g(\theta)|^2 = \sigma_0(\theta)(1 - P_e'(\theta)/P_e). \tag{4.22}$$

Hence, by a measurement of the electron polarization after the scattering, $|g(\theta)|$ can be determined if the electron polarization before the scattering and the cross section for the scattering of unpolarized electrons are known.

One can also observe the recoil atoms in this experiment. $A\uparrow$ emerge with the cross section

$$\sigma_A^\uparrow(\theta) = P_e(\tfrac{1}{2}|f-g|^2 + \tfrac{1}{2}|g|^2) + (1-P_e)\frac{\sigma_0(\theta)}{2}.$$

Similarly

$$\sigma_A^\downarrow(\theta) = P_e\tfrac{1}{2}|f|^2 + (1-P_e)\frac{\sigma_0(\theta)}{2}.$$

If we introduce the polarization of the atoms after the scattering

$$P_A'(\theta) = \frac{\sigma_A^\uparrow(\theta) - \sigma_A^\downarrow(\theta)}{\sigma_A^\uparrow(\theta) + \sigma_A^\downarrow(\theta)} = \frac{P_e}{\sigma_0(\theta)}[\sigma_0(\theta) - |f(\theta)|^2]$$

we obtain

$$|f(\theta)|^2 = \sigma_0(\theta)(1 - P_A'(\theta)/P_e), \tag{4.23}$$

which shows that a measurement of the polarization of the recoil atoms yields $|f(\theta)|$.

Thus we see that by scattering a partially polarized electron beam from unpolarized atoms all the individual cross sections listed in (4.8–10) can be obtained. Equivalent experiments can be made starting with unpolarized electrons and polarized atoms. In complete analogy to the above treatment, one then obtains

$$|f(\theta)|^2 = \sigma_0(\theta)\,(1 - P_e'(\theta)/P_A) \quad \text{and} \tag{4.24}$$
$$|g(\theta)|^2 = \sigma_0(\theta)\,(1 - P_A'(\theta)/P_A). \tag{4.25}$$

Hence one again obtains all the individual cross sections by measuring the electron polarization P_e' and the atomic polarization P_A' after scattering, if P_A, the polarization of the atoms before scattering, and $\sigma_0(\theta)$ are known.

We point out that the quantities $|f(\theta)|^2/\sigma_0(\theta)$ and $|g(\theta)|^2/\sigma_0(\theta)$ can actually be larger than 1. [Example: if $g = f/2$ then from (4.14) one has $|f|^2/\sigma_0 = 4/3$]. This means that the polarization of the scattered particles in (4.22–25) can also be antiparallel to that of the particles which cause the polarization.

Since the purpose of this section was to introduce the polarization mechanism that is effective in exchange scattering, the discussion has been limited to the situation of greatest conceptual simplicity where the spins of electrons and atoms were either parallel or antiparallel to one another. For treatment of the general case we will take advantage of density matrix techniques. For a summary of earlier literature on electron polarization in exchange scattering we refer to [4.1].

Problem 4.1. In Sect. 3.4.5 it was proved that for Mott scattering one has $P = A$, where P is the polarizing power (for the scattering of an unpolarized beam) and A is the analyzing power (= asymmetry for polarization measurement on a totally polarized beam). Are these two quantities also equal in exchange scattering? Compare the intensity asymmetries in exchange scattering and Mott scattering.

Solution. We first consider the case in which an unpolarized electron beam is polarized by scattering from totally polarized atoms A↑. From (4.9, 10) one obtains scattered e↑ with the cross section $\frac{1}{2}|f-g|^2 + \frac{1}{2}|g|^2$, and from (4.8) scattered e↓ with cross section $\frac{1}{2}|f|^2$. Thus

$$P_e' = \frac{\frac{1}{2}|f-g|^2 + \frac{1}{2}|g|^2 - \frac{1}{2}|f|^2}{\frac{1}{2}|f-g|^2 + \frac{1}{2}|g|^2 + \frac{1}{2}|f|^2} = \frac{\sigma_0 - |f|^2}{\sigma_0} = 1 - \frac{|f|^2}{\sigma_0}.$$

We now consider the scattering of a totally polarized electron beam by totally polarized atoms A↑. If the direction of polarization of the primary electrons is the same as that of the atoms, then according to (4.10) the scattering intensity is determined by the cross section $|f-g|^2$. According to (4.8, 9), reversal of the polarization of the primary beam yields the cross section $|f|^2 + |g|^2$. The polarization analysis can be made by measuring the scattering intensity in a certain direction for the polarization directions ↑ and ↓ of the primary beam. The relative intensity difference gives us the analyzing power

$$A = \frac{|f-g|^2 - |f|^2 - |g|^2}{|f-g|^2 + |f|^2 + |g|^2} = 1 - \frac{|f|^2 + |g|^2}{\sigma_0},$$

which is different from the polarization calculated above.

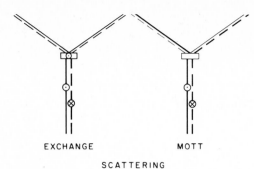

Fig. 4.1. Spin-up-down asymmetry and left-right asymmetry (intensities indicated by line thickness)

EXCHANGE   MOTT

SCATTERING

The intensity asymmetries discussed here and those in Mott scattering are compared in Fig. 4.1. In exchange scattering of polarized electrons by polarized atoms there is, in the approximation made in this chapter, no left-right asymmetry since the forces do not depend on whether an electron passes the atom on the right or on the left. However, the number of electrons scattered into a certain direction depends on their spin direction. So we do not have a "left-right" asymmetry, but we do have a "spin-up-down" asymmetry as illustrated in the left-hand part of Fig. 4.1. In Mott scattering there is a spin-up-down asymmetry as well as a left-right asymmetry.

4.2 The Composite Spin Space of Electron and Target

The density matrix formalism for the composite spin space of an electron and a spin one-half atom is introduced.

In order to generalize the special cases discussed in the preceding section we will now describe elastic exchange scattering by means of the density matrix formalism. Particularly in cases more complicated than heretofore discussed, where beam polarization and target polarization have arbitrary directions, density matrices turn out to be very useful.

Limiting ourselves to the important case of scattering by spin one-half atoms we compose a four-dimensional spin space from the two-dimensional spin spaces of the free and the target electron. For this purpose we recall the direct product (or tensor product) of matrices which, for 2×2 matrices, is defined by

$$\begin{pmatrix} m_{11} & m_{12} \\ m_{21} & m_{22} \end{pmatrix} \otimes \begin{pmatrix} n_{11} & n_{12} \\ n_{21} & n_{22} \end{pmatrix} = \begin{pmatrix} m_{11}n_{11} & m_{11}n_{12} & m_{12}n_{11} & m_{12}n_{12} \\ m_{11}n_{21} & m_{11}n_{22} & m_{12}n_{21} & m_{12}n_{22} \\ m_{21}n_{11} & m_{21}n_{12} & m_{22}n_{11} & m_{22}n_{12} \\ m_{21}n_{21} & m_{21}n_{22} & m_{22}n_{21} & m_{22}n_{22} \end{pmatrix}$$

producing a 4×4 matrix [4.2]. From the spin functions $\chi^e = \begin{pmatrix} a_1^e \\ a_2^e \end{pmatrix}$ of the free electron and $\chi^A = \begin{pmatrix} a_1^A \\ a_2^A \end{pmatrix}$ of the atom we define a spin function of the system electron-plus-target by

$$\chi^e \otimes \chi^A = \begin{pmatrix} a_1^e a_1^A \\ a_1^e a_2^A \\ a_2^e a_1^A \\ a_2^e a_2^A \end{pmatrix}, \tag{4.26}$$

and as spin operators in the composite spin space we use

$$\sigma_\mu^e = \sigma_\mu \otimes \mathbf{1} \quad (\mu = x, y, z)$$

for the electron and

$$\sigma_\mu^A = \mathbf{1} \otimes \sigma_\mu \tag{4.27}$$

for the atom. The σ_μ are the Pauli spin matrices (2.2) and $\mathbf{1}$ is the 2×2 unit matrix. Explicitly we have

$$\sigma_x^e = \begin{pmatrix} 0 & 0 & 1 & 0 \\ 0 & 0 & 0 & 1 \\ 1 & 0 & 0 & 0 \\ 0 & 1 & 0 & 0 \end{pmatrix}, \quad \sigma_y^e = \begin{pmatrix} 0 & 0 & -i & 0 \\ 0 & 0 & 0 & -i \\ i & 0 & 0 & 0 \\ 0 & i & 0 & 0 \end{pmatrix}, \quad \sigma_z^e = \begin{pmatrix} 1 & 0 & 0 & 0 \\ 0 & 1 & 0 & 0 \\ 0 & 0 & -1 & 0 \\ 0 & 0 & 0 & -1 \end{pmatrix},$$

$$\sigma_x^A = \begin{pmatrix} 0 & 1 & 0 & 0 \\ 1 & 0 & 0 & 0 \\ 0 & 0 & 0 & 1 \\ 0 & 0 & 1 & 0 \end{pmatrix}, \quad \sigma_y^A = \begin{pmatrix} 0 & -i & 0 & 0 \\ i & 0 & 0 & 0 \\ 0 & 0 & 0 & -i \\ 0 & 0 & i & 0 \end{pmatrix}, \quad \sigma_z^A = \begin{pmatrix} 1 & 0 & 0 & 0 \\ 0 & -1 & 0 & 0 \\ 0 & 0 & 1 & 0 \\ 0 & 0 & 0 & -1 \end{pmatrix}. \tag{4.28}$$

Let us verify by a few examples that by means of these definitions we can properly describe the spin of the system electron-plus-target. For the state $e{\uparrow}A{\downarrow}$ (electron spin $+\hbar/2$ and atom spin $-\hbar/2$ in the z direction), which is represented by

$$a_1^e = 1, \quad a_2^e = 0, \quad a_1^A = 0, \quad a_2^A = 1,$$

one obtains from (4.26) the spin function

$$\chi_{\uparrow}^e \otimes \chi_{\downarrow}^A = \begin{pmatrix} 0 \\ 1 \\ 0 \\ 0 \end{pmatrix}.$$

and one immediately finds from (4.28) that this is an eigenfunction of σ_z^e with eigenvalue $+1$ and of σ_z^A with eigenvalue -1. This is what one expects for the state $e{\uparrow}A{\downarrow}$. Similarly one sees that the states $e{\uparrow}A{\uparrow}$, $e{\downarrow}A{\uparrow}$, $e{\downarrow}A{\downarrow}$ have the spin functions

$$\begin{pmatrix} 1 \\ 0 \\ 0 \\ 0 \end{pmatrix}, \quad \begin{pmatrix} 0 \\ 0 \\ 1 \\ 0 \end{pmatrix}, \quad \begin{pmatrix} 0 \\ 0 \\ 0 \\ 1 \end{pmatrix},$$

respectively, which are eigenfunctions both of σ_z^e and σ_z^A with the eigenvalues symbolized by the arrows.

More generally, one finds by straightforward matrix multiplication that, analogous to what has been said in Sect. 2.2, the spin function (4.26) with

$$a_1^e = \cos\frac{\vartheta^e}{2}, \quad a_2^e = \sin\frac{\vartheta^e}{2}\exp(i\varphi^e), \quad a_1^A = \cos\frac{\vartheta^A}{2}, \quad a_2^A = \sin\frac{\vartheta^A}{2}\exp(i\varphi^A)$$

$$(4.29)$$

is a simultaneous eigenfunction of the operators $\boldsymbol{\sigma}^e \cdot \hat{e}^e$ and $\boldsymbol{\sigma}^A \cdot \hat{e}^A$ with eigenvalue $+1$, if $\hat{e}^e$ and $\hat{e}^A$ are unit vectors in the directions ϑ^e, φ^e and ϑ^A, φ^A, respectively. In other words: this function represents an electron spin with direction ϑ^e, φ^e and an atom spin with direction ϑ^A, φ^A. If we have, for example, the electron spin in the x direction and the target spin in the y direction ($\vartheta^e = \vartheta^A = \pi/2$, $\varphi^e = 0$, $\varphi^A = \pi/2$), the spin function is

$$\begin{pmatrix} 1/2 \\ i/2 \\ 1/2 \\ i/2 \end{pmatrix}.$$

Before discussing the polarization of mixed states let us first consider the density matrix for the trivial case of total polarization of electrons and atoms (pure state). Recalling (4.26) and (2.17) we have

$$\varrho = \begin{pmatrix} a_1^e a_1^A \\ a_1^e a_2^A \\ a_2^e a_1^A \\ a_2^e a_2^A \end{pmatrix} (a_1^{e*} a_1^{A*} \quad a_1^{e*} a_2^{A*} \quad a_2^{e*} a_1^{A*} \quad a_2^{e*} a_2^{A*})$$

$$= \begin{pmatrix} |a_1^e|^2|a_1^A|^2 & |a_1^e|^2 a_1^A a_2^{A*} & a_1^e a_2^{e*}|a_1^A|^2 & a_1^e a_2^{e*} a_1^A a_2^{A*} \\ |a_1^e|^2 a_2^A a_1^{A*} & |a_1^e|^2|a_2^A|^2 & a_1^e a_2^{e*} a_2^A a_1^{A*} & a_1^e a_2^{e*}|a_2^A|^2 \\ a_2^e a_1^{e*}|a_1^A|^2 & a_2^e a_1^{e*} a_1^A a_2^{A*} & |a_2^e|^2|a_1^A|^2 & |a_2^e|^2 a_1^A a_2^{A*} \\ a_2^e a_1^{e*} a_2^A a_1^{A*} & a_2^e a_1^{e*}|a_2^A|^2 & |a_2^e|^2 a_2^A a_1^{A*} & |a_2^e|^2|a_2^A|^2 \end{pmatrix} = \varrho^e \otimes \varrho^A. \quad (4.30)$$

Keeping to the general definition $\boldsymbol{P} = \langle\boldsymbol{\sigma}\rangle$ given in Chap. 2 we have

$$\boldsymbol{P}_e = \text{tr}\{\varrho\boldsymbol{\sigma}^e\}, \quad \boldsymbol{P}_A = \text{tr}\{\varrho\boldsymbol{\sigma}^A\}, \quad (4.31)$$

if the spin functions from which ϱ is formed are normalized.

Checking these relations for two examples we find from (4.28, 30),

$$\text{tr}\{\varrho\sigma_z^e\} = |a_1^e|^2|a_1^A|^2 + |a_1^e|^2|a_2^A|^2 - |a_2^e|^2|a_1^A|^2 - |a_2^e|^2|a_2^A|^2 = |a_1^e|^2 - |a_2^e|^2,$$

since $|a_1^A|^2 + |a_2^A|^2 = 1$. From (2.14) we know that $|a_1^e|^2 - |a_2^e|^2 = P_{ez}$ so that $P_{ez} = \text{tr}\{\varrho\sigma_z^e\}$ in agreement with (4.31).

As a second example we calculate

$$\text{tr}\{\varrho\sigma_x^A\} = |a_1^e|^2(a_1^A a_2^{A*} + a_2^A a_1^{A*}) + |a_2^e|^2(a_1^A a_2^{A*} + a_2^A a_1^{A*}).$$

Since $|a_1^e|^2 + |a_2^e|^2 = 1$, we have $\text{tr}\{\varrho\sigma_x^A\} = a_1^A a_2^{A*} + a_2^A a_1^{A*} = P_{Ax}$ where (2.14) has been used.

Expressing the elements of the density matrix (4.30) in terms of the polarization one finds that, instead of relation (2.19) for the spinless target, one obtains for the density matrix of a state in which both target and projectile electrons are totally polarized

$$\varrho = \frac{1}{4}\left(1 + \boldsymbol{P}_e\cdot\boldsymbol{\sigma}^e + \boldsymbol{P}_A\cdot\boldsymbol{\sigma}^A + \sum_{\mu,\nu} P_{e\mu}P_{A\nu}\sigma_\mu^e\sigma_\nu^A\right), \quad \left.\begin{array}{c}\mu\\\nu\end{array}\right\} = x, y, z. \qquad (4.32\text{a})$$

This is the direct product of the density matrices (2.19) for two spin one-half particles. For a verification of (4.32) see Problem 4.2.

Our assumption of total polarization of both target and projectile electrons means that we regard the initial polarizations of atoms and electrons as independent of each other, which is consistent with the spin functions we used. If one describes the system by spin functions like those given by (4.15b or 16), which include correlation of the two electrons, the factor

$$P_{e\mu}P_{A\nu} = \langle\sigma_\mu^e\rangle\langle\sigma_\nu^A\rangle \text{ has to be replaced by } Q_{\mu\nu} = \langle\sigma_\mu^e\sigma_\nu^A\rangle \text{ [4.3, 4]}. \qquad (4.32\text{b})$$

If there are no correlations between the electrons, (4.32a) is seen to be valid also for a state which is a mixture of states $\chi^{(n)e}\otimes\chi^{(m)A}$ in each of which projectile and target electrons are totally polarized. One has this situation in two noninteracting beams of partially polarized electrons and atoms. When $w^{(n)}$ and $w^{(m)}$ are the normalized statistical weights of $\chi^{(n)}$ and $\chi^{(m)}$, respectively, the density matrix of the mixed state is the sum of the pure-state density matrices $\varrho^{(nm)}$:

$$\varrho = \sum_{n,m} w^{(n)}w^{(m)}\varrho^{(nm)}$$

$$= \frac{1}{4}\sum_{n,m} w^{(n)}w^{(m)}\left(1 + \boldsymbol{P}_e^{(n)}\cdot\boldsymbol{\sigma}^e + \boldsymbol{P}_A^{(m)}\cdot\boldsymbol{\sigma}^A + \sum_{\mu,\nu} P_{e\mu}^{(n)}P_{A\nu}^{(m)}\sigma_\mu^e\sigma_\nu^A\right)$$

$$= \frac{1}{4}\left(1 + \sum_n w^{(n)}\boldsymbol{P}_e^{(n)}\cdot\boldsymbol{\sigma}^e + \sum_m w^{(m)}\boldsymbol{P}_e^{(m)}\cdot\boldsymbol{\sigma}^A + \sum_{\mu,\nu}\sum_n w^{(n)}P_{e\mu}^{(n)}\sum_m w^{(m)}P_{A\nu}^{(m)}\sigma_\mu^e\sigma_\nu^A\right),$$

where

$$\sum_n w^{(n)} = \sum_m w^{(m)} = 1$$

has been used. Recalling (2.16), we see that this is identical with (4.32a).

If there are correlations between the electrons, one obtains with (4.32b)

$$\varrho = \frac{1}{4}\left(1 + \boldsymbol{P}_e \cdot \boldsymbol{\sigma}^e + \boldsymbol{P}_A \cdot \boldsymbol{\sigma}^A + \sum_{\mu,\nu} Q_{\mu\nu}\sigma_\mu^e\sigma_\nu^A\right) \quad \left.\begin{matrix}\mu\\ \nu\end{matrix}\right\} = x, y, z$$

where $Q_{\mu\nu} = \sum_{m,n} w^{(n)}w^{(m)}Q_{\mu\nu}^{(nm)}$.

In this section we have so far assumed that the spin functions are normalized. If one starts from unnormalized spin functions, the density matrix must be normalized to have trace unity before applying (4.31) and (4.32a): as in (2.21, 23) ϱ should be replaced by $\varrho/\mathrm{tr}\{\varrho\}$.

In order to be able to calculate the scattering cross section, or the polarization of a scattered beam, one needs to know the spin scattering matrix. We have explained before that the connection between the basic spin states and their spin functions is

$$e{\uparrow}A{\uparrow} \triangleq \begin{pmatrix} 1 \\ 0 \\ 0 \\ 0 \end{pmatrix}, \quad e{\uparrow}A{\downarrow} \triangleq \begin{pmatrix} 0 \\ 1 \\ 0 \\ 0 \end{pmatrix}, \quad e{\downarrow}A{\uparrow} \triangleq \begin{pmatrix} 0 \\ 0 \\ 1 \\ 0 \end{pmatrix}, \quad e{\downarrow}A{\downarrow} \triangleq \begin{pmatrix} 0 \\ 0 \\ 0 \\ 1 \end{pmatrix}.$$

The transitions which can occur between the various spin states have been listed in (4.8–13). Process (4.9), for instance, leads from the initial state

$$\chi_i = \begin{pmatrix} 0 \\ 0 \\ 1 \\ 0 \end{pmatrix} \quad \text{to the final state} \quad \chi_f = \begin{pmatrix} 0 \\ 1 \\ 0 \\ 0 \end{pmatrix}$$

and has the transition amplitude $-g$. The scattering matrix S must therefore

contain the element $\begin{pmatrix} \cdot & \cdot & \cdot & \cdot \\ \cdot & \cdot & -g & \cdot \\ \cdot & \cdot & \cdot & \cdot \\ \cdot & \cdot & \cdot & \cdot \end{pmatrix}$ so that one has the transition

matrix element

$$\langle \chi_f|S|\chi_i\rangle = (0\,1\,0\,0) \begin{pmatrix} \cdot & \cdot & \cdot & \cdot \\ \cdot & \cdot & -g & \cdot \\ \cdot & \cdot & \cdot & \cdot \\ \cdot & \cdot & \cdot & \cdot \end{pmatrix} \begin{pmatrix} 0 \\ 0 \\ 1 \\ 0 \end{pmatrix} = (0\,1\,0\,0) \begin{pmatrix} \cdot \\ -g \\ \cdot \\ \cdot \end{pmatrix} = -g.$$

The same reasoning applied to the other five processes shows that the complete scattering matrix must be

$$
S = \begin{pmatrix}
f-g & 0 & 0 & 0 \\
0 & f & -g & 0 \\
0 & -g & f & 0 \\
0 & 0 & 0 & f-g
\end{pmatrix},
\tag{4.33}
$$

having six nonzero elements which describe the six conceivable processes (4.8–13).

Knowing the scattering matrix we are now in a position to calculate the polarizations P'_e and P'_A of the electrons and atoms in the final state according to (4.31):

$$
P'_e = \frac{\operatorname{tr}\{\varrho'\boldsymbol{\sigma}^e\}}{\operatorname{tr}\{\varrho'\}}, \quad
P'_A = \frac{\operatorname{tr}\{\varrho'\boldsymbol{\sigma}^A\}}{\operatorname{tr}\{\varrho'\}}
\tag{4.34}
$$

where ϱ' is the (unnormalized) density matrix of the final state which has been shown in Sect. 3.3 to be $\varrho' = S\varrho S^\dagger$. The density matrix ϱ of the initial state is given by (4.32a). The scattering cross section follows from

$$
\sigma(\theta) = \operatorname{tr}\{\varrho'\}/\operatorname{tr}\{\varrho\}
$$

which has also been derived in Sect. 3.3.

Problem 4.2. Verify the relation (4.32a) between the density matrix (4.30) and the polarizations of projectile and target electrons.

Solution. Multiplication of the spin matrices with the polarizations in (4.32a) yields the explicit form of the density matrix:

$$
\frac{1}{4}
\begin{pmatrix}
1+P_{ez}+P_{Az}+P_{ez}P_{Az} & (1+P_{ez})(P_{Ax}-iP_{Ay}) & (P_{ex}-iP_{ey})(1+P_{Az}) & (P_{ex}-iP_{ey})(P_{Ax}-iP_{Ay}) \\
(1+P_{ez})(P_{Ax}+iP_{Ay}) & 1+P_{ez}-P_{Az}-P_{ez}P_{Az} & (P_{ex}-iP_{ey})(P_{Ax}+iP_{Ay}) & (P_{ex}-iP_{ey})(1-P_{Az}) \\
(P_{ex}+iP_{ey})(1+P_{Az}) & (P_{ex}+iP_{ey})(P_{Ax}-iP_{Ay}) & 1-P_{ez}+P_{Az}-P_{ez}P_{Az} & (1-P_{ez})(P_{Ax}-iP_{Ay}) \\
(P_{ex}+iP_{ey})(P_{Ax}+iP_{Ay}) & (P_{ex}+iP_{ey})(1-P_{Az}) & (1-P_{ez})(P_{Ax}+iP_{Ay}) & 1-P_{ez}-P_{Az}+P_{ez}P_{Az}
\end{pmatrix}.
\tag{4.35}
$$

From (4.31) one has the connection between the polarization components and the elements of the spin functions, so that (4.32a) and (4.30) can easily be seen to be identical. We verify this for an arbitrary element, say ϱ_{34}. Starting from (4.31), we have explicitly shown in the main body of this section that $P_{ez} = |a_1^e|^2 - |a_2^e|^2$, $P_{Ax} = a_1^A a_2^{A*} + a_2^A a_1^{A*}$. In the same way one obtains $P_{Ay} = i(a_1^A a_2^{A*} - a_2^A a_1^{A*})$ so that with $|a_1^e|^2 + |a_2^e|^2 = 1$ one has

$$
\varrho_{34} = \tfrac{1}{4}(1-P_{ez})(P_{Ax}-iP_{Ay}) = \tfrac{1}{4}\cdot 2|a_2^e|^2 \cdot 2\, a_1^A a_2^{A*}
$$

which is identical to the expression for ϱ_{34} in (4.30).

4.3 Cross Section and Polarization in Elastic Exchange Scattering

The general results of the preceding section are applied to elastic exchange scattering in the case where the electron polarization has an arbitrary direction with respect to the target polarization. Measurement of the cross section and the change of the electron polarization and/or target polarization (including the rotation of the polarization out of its initial plane) constitutes a "perfect" scattering experiment which yields all the attainable information on the scattering amplitudes.

We can now apply the density matrix formalism to treat the general case of elastic exchange scattering of an electron beam whose polarization has an arbitrary direction with respect to the target polarization P_A. By proper choice of the z axis one obtains $P_A = P_{Az}\hat{e}_z$, $P_{Ax} = P_{Ay} = 0$. The density matrix ϱ of the initial state is (cf. Problem 4.2)

$$
\varrho = \frac{1}{4}
\begin{pmatrix}
(1+P_{ez})(1+P_A) & 0 & (P_{ex}-iP_{ey})(1+P_A) & 0 \\
0 & (1+P_{ez})(1-P_A) & 0 & (P_{ex}-iP_{ey})(1-P_A) \\
(P_{ex}+iP_{ey})(1+P_A) & 0 & (1-P_{ez})(1+P_A) & 0 \\
0 & (P_{ex}+iP_{ey})(1-P_A) & 0 & (1-P_{ez})(1-P_A)
\end{pmatrix}.
$$

By multiplication with the scattering matrix (4.33) we find the final-state density matrix $\varrho' = S\varrho S^\dagger$ whose elements are

$\varrho'_{11} = \frac{1}{4}(1+P_{ez})(1+P_A)|f-g|^2,$

$\varrho'_{12} = -\frac{1}{4}(P_{ex}-iP_{ey})(1+P_A)(f-g)g^*,$

$\varrho'_{13} = \frac{1}{4}(P_{ex}-iP_{ey})(1+P_A)(f-g)f^*,$

$\varrho'_{14} = 0,$

$\varrho'_{21} = -\frac{1}{4}(P_{ex}+iP_{ey})(1+P_A)g(f^*-g^*),$

$\varrho'_{22} = \frac{1}{4}[(1+P_{ez})(1-P_A)|f|^2 + (1-P_{ez})(1+P_A)|g|^2],$

$\varrho'_{23} = -\frac{1}{4}[(1+P_{ez})(1-P_A)fg^* + (1-P_{ez})(1+P_A)f^*g],$

$\varrho'_{24} = \frac{1}{4}(P_{ex}-iP_{ey})(1-P_A)(f^*-g^*)f,$

$\varrho'_{31} = \frac{1}{4}(P_{ex}+iP_{ey})(1+P_A)f(f^*-g^*),$

$\varrho'_{32} = -\frac{1}{4}[(1+P_{ez})(1-P_A)f^*g + (1-P_{ez})(1+P_A)fg^*],$

$\varrho'_{33} = \frac{1}{4}[(1+P_{ez})(1-P_A)|g|^2 + (1-P_{ez})(1+P_A)|f|^2],$

$\varrho'_{34} = -\frac{1}{4}(P_{ex}-iP_{ey})(1-P_A)g(f^*-g^*),$

$\varrho'_{41} = 0,$

$\varrho'_{42} = \frac{1}{4}(P_{ex}+iP_{ey})(1-P_A)(f-g)f^*,$

$\varrho'_{43} = -\frac{1}{4}(P_{ex}+iP_{ey})(1-P_A)(f-g)g^*,$

$\varrho'_{44} = \frac{1}{4}(1-P_{ez})(1-P_A)|f-g|^2.$

This density matrix yields the cross section

$$\sigma(\theta) = \frac{\mathrm{tr}\{\varrho'\}}{\mathrm{tr}\{\varrho\}} = \frac{1}{2}[(1+P_{ez}P_A)|f-g|^2 + (1-P_{ez}P_A)(|f|^2+|g|^2)]. \tag{4.36}$$

Introducing the cross section $\sigma_0(\theta)$ from (4.14) for zero electron and/or target polarization, (4.36) may be written

$$\sigma(\theta) = \sigma_0 \left[1 + P_{ez}P_A \left(1 - \frac{|f|^2+|g|^2}{\sigma_0}\right)\right]. \tag{4.37}$$

This shows that with arbitrary spin orientation it is only the electron-polarization component along the direction of the atomic polarization which contributes to the spin dependence of the cross section. The relative cross-section difference for scattering with antiparallel ($P_{ez}P_A < 0$) and parallel ($P_{ez}P_A > 0$) spins is

$$\frac{\sigma_{\uparrow\downarrow} - \sigma_{\uparrow\uparrow}}{\sigma_{\uparrow\downarrow} + \sigma_{\uparrow\uparrow}} = P_{ez}P_A A \tag{4.38}$$

with the asymmetry

$$A = -\frac{|f-g|^2 - |f|^2 - |g|^2}{|f-g|^2 + |f|^2 + |g|^2} = \frac{fg^* + f^*g}{2\sigma_0}. \tag{4.39}$$

The polarization of the scattered electrons becomes

$$P'_{ex} = \frac{\mathrm{tr}\{\varrho'\sigma^e_x\}}{\mathrm{tr}\{\varrho'\}} = \frac{1}{2\sigma(\theta)} [P_{ex}(2|f|^2 - fg^* - f^*g) - iP_{ey}P_A(fg^* - f^*g)]$$

$$P'_{ey} = \frac{\mathrm{tr}\{\varrho'\sigma^e_y\}}{\mathrm{tr}\{\varrho'\}} = \frac{1}{2\sigma(\theta)} [iP_{ex}P_A(fg^* - f^*g) + P_{ey}(2|f|^2 - fg^* - f^*g)] \tag{4.40}$$

$$P'_{ez} = \frac{\mathrm{tr}\{\varrho'\sigma^e_z\}}{\mathrm{tr}\{\varrho'\}} = \frac{1}{2\sigma(\theta)} [(P_{ez} + P_A)|f-g|^2 + (P_{ez} - P_A)(|f|^2 - |g|^2)].$$

For a discussion of this result it is expedient to resolve the electron polarization P_e into components parallel and perpendicular to the atomic polarization (cf. Fig. 4.2),

$$P_e = P_{ez} + P_{ep}.$$

From

$$P_{ep} = P_{ex}\hat{e}_x + P_{ey}\hat{e}_y \quad \text{and} \quad \hat{e}_z \times \hat{e}_x = \hat{e}_y \quad \text{(etc., cyclic)},$$

one has

$$\hat{e}_z \times P_{ep} = P_{ex}\hat{e}_y - P_{ey}\hat{e}_x,$$

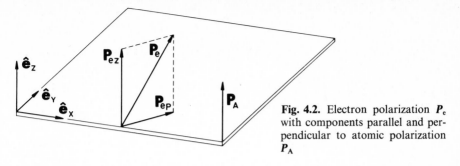

Fig. 4.2. Electron polarization P_e with components parallel and perpendicular to atomic polarization P_A

so that one obtains from (4.40)

$$P'_e = \left[(P_{ez} + P_A) \frac{|f-g|^2}{2\sigma(\theta)} + (P_{ez} - P_A) T'(\theta) \right] \hat{e}_z$$

$$+ \left[\frac{|f|^2}{\sigma(\theta)} - U'(\theta) \right] P_{ep} + S'(\theta) P_A (\hat{e}_z \times P_{ep}). \tag{4.41}$$

Here, for reasons of comparison with (3.77), we have set

$$S'(\theta) = i \frac{fg^* - f^*g}{2\sigma(\theta)}, \quad T'(\theta) = \frac{|f|^2 - |g|^2}{2\sigma(\theta)}, \quad U'(\theta) = \frac{fg^* + f^*g}{2\sigma(\theta)}.$$

Just like (3.77) in the case of Mott scattering, (4.41) enables one to see very easily how the scattering process affects the polarization. Let us, however, emphasize the essential differences compared to Mott scattering: The physical origin and the meaning of the scattering amplitudes g in exchange scattering are quite different. Furthermore, the vector $\hat{n}$ of (3.77) which is determined by the directions of incident and scattered electrons no longer plays a role here; it is replaced by $\hat{e}_z$, determined by the atomic polarization. Consequently, in contrast to Mott scattering, reversal of the incident direction affects neither the change of the polarization components in the initial plane ($\hat{e}_z$, P_{ep}), nor the rotation of the polarization vector out of this plane. The latter is described by the additional component $S'(\theta) P_A (\hat{e}_z \times P_{ep})$. The angle α of this rotation is seen from Fig. 4.3 and (4.41) to be given by

$$\tan \alpha = \frac{i(fg^* - f^*g)}{2|f|^2 - (fg^* + f^*g)}, \tag{4.42}$$

which shows that such a rotation occurs only if the exchange amplitude is different from zero. The angle of rotation α vanishes also in the special case where $f = |f| \exp(i\gamma_1)$ and $g = |g| \exp(i\gamma_2)$ have the same phase:

$$i(fg^* - f^*g) = -2|f||g| \sin(\gamma_1 - \gamma_2) = 0 \quad \text{for} \quad \gamma_1 = \gamma_2.$$

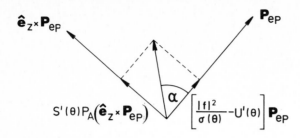

Fig. 4.3. Angle of rotation of the polarization component $\boldsymbol{P}_{ep}$

In order to become familiar with the physics of exchange scattering we will now discuss a few simple cases.

If $g=0$, there is no change of the polarization at all: in this case one has $\sigma(\theta)=|f|^2$ so that (4.41) yields $\boldsymbol{P}'_e = P_{ez}\hat{\boldsymbol{e}}_z + \boldsymbol{P}_p = \boldsymbol{P}_e$. It is quite obvious that there can be no change of the polarization when there is no exchange scattering causing target electrons to go into the scattered beam. Since we have excluded explicit spin-dependent forces there is, with vanishing exchange scattering, no way of affecting the electron polarization so that the polarization of the scattered electrons must be identical to that of the incident beam.

If, on the other hand, $f=0$ so that no direct processes contribute to scattering, one has $\sigma=|g|^2$ and (4.41) yields

$$\boldsymbol{P}'_e = [\tfrac{1}{2}(P_{ez}+P_A)-\tfrac{1}{2}(P_{ez}-P_A)]\hat{\boldsymbol{e}}_z = P_A\hat{\boldsymbol{e}}_z.$$

This is what one expects, since now all the scattered electrons have been in the target before the scattering. No explicit spin-dependent forces being present, these electrons retain their spin direction during the exchange process so that the atomic polarization is transferred to the scattered beam.

In the case $P_A=0$ which is of practical importance, (4.41) yields

$$\boldsymbol{P}'_e = \frac{1}{2\sigma_0}\,[(|f-g|^2+|f|^2-|g|^2)\boldsymbol{P}_{ez}+(2|f|^2-fg^*-f^*g)\boldsymbol{P}_{ep}]$$

$$= \frac{|f-g|^2+|f|^2-|g|^2}{|f-g|^2+|f|^2+|g|^2}\,\boldsymbol{P}_e \tag{4.43}$$

which describes the fact that there is a partial depolarization of the electron beam due to exchange processes with the unpolarized target. The electron polarization retains its direction (the depolarization is the same for each component) but is reduced due to the admixture of target electrons.

If, on the other hand, $\boldsymbol{P}_e=0$, (4.41) yields

$$\boldsymbol{P}'_e = \frac{|f-g|^2-|f|^2+|g|^2}{|f-g|^2+|f|^2+|g|^2}\,\boldsymbol{P}_A \tag{4.44}$$

which describes a polarization transfer from the target to the electrons due to exchange scattering.

From (4.41) one also sees that, with $P_A \neq 0$, there is one special case where only the magnitude but not the axis of P_e is changed by the scattering; this occurs when $P_{ep} = 0$, i.e., when P_e is parallel or antiparallel to the target polarization. P'_e has then a z component only. This is the situation which has been used in Sect. 4.1 for introducing the basic facts of exchange scattering. The formulas given there follow immediately as special cases of (4.41).

In order to obtain complete information on the scattering amplitudes it does not suffice to measure $|f|^2$, $|g|^2$, and $|f-g|^2 = |f|^2 + |g|^2 - fg^* - f^*g = |f|^2 + |g|^2 - 2|f||g| \cos(\gamma_1 - \gamma_2)$ by means of the experiments discussed in Sect. 4.1, since the difference of γ_1 and γ_2 [which are defined by $f = |f| \exp(i\gamma_1)$, $g = |g| \exp(i\gamma_2)$] cannot be obtained from $\cos(\gamma_1 - \gamma_2)$ alone. For a "perfect" scattering experiment which fully determines the scattering amplitudes, an investigation of the rotation of the polarization is also necessary. Such an additional measurement yields, according to (4.42), $i(fg^* - f^*g) = -2|f||g| \sin(\gamma_1 - \gamma_2)$ so that, together with a measurement yielding $\cos(\gamma_1 - \gamma_2)$, the phase difference $\gamma_1 - \gamma_2$ can be unambiguously determined. Consequently, an experiment where the rotation of the polarization has not been measured cannot be called "perfect" as is sometimes done in literature [4.5]. The situation is quite similar to Mott scattering where we have seen that rotation experiments with the polarization are required in order to fully determine the scattering amplitudes. This is not surprising because of the formal similarity of the expressions (4.41) and (3.77) which describe the behavior of the polarization in scattering.

Before discussing the change of the atomic polarization by exchange scattering, let us point out that the scattering amplitudes can be fully determined by studies of the scattered electrons alone. Apart from the set of measurements we have explicitly discussed, there are several other combinations of measurements which represent such a perfect scattering experiment.

In order to find the polarization of the scattered atoms we have to evaluate the expression

$$P'_{Ai} = \frac{\text{tr}\{\varrho' \sigma_i^A\}}{\text{tr}\{\varrho'\}} \qquad (i = x, y, z).$$

Using the σ_i^A given in (4.28) one obtains, by proceeding in the same way as in the evaluation of P'_e,

$$P'_A = \left[(P_{ez} + P_A) \frac{|f-g|^2}{2\sigma(\theta)} + (P_A - P_{ez}) T'(\theta) \right] \hat{e}_z$$
$$+ \left[\frac{|g|^2}{\sigma(\theta)} - U'(\theta) \right] P_{ep} - S'(\theta) P_A (\hat{e}_z \times P_{ep}), \qquad (4.45)$$

which can easily be seen to follow from (4.41) by interchanging f and g. The formula shows that the polarization of the recoil atoms has three components, in the directions $\hat{e}_z$, P_{ep}, and $\hat{e}_z \times P_{ep}$, although the initial target polarization was assumed to have a component in the z direction only. Comparison with (4.41) shows that the components of P'_A and P'_e in the direction $\hat{e}_z \times P_{ep}$ are opposite to each other. In the special case $P_{ep} = 0$ where the formulae of Sect. 4.1 containing P'_A are seen to follow from (4.45), the polarization of the recoil atoms has a z component only.

We will refrain from illustrating (4.45) in detail as we did with (4.41) for the scattered electrons, since such a discussion is quite analogous for the two cases. Let us just mention a few facts: For $g = 0$ one finds $P'_A = P_A$, i.e., no change of the target polarization because there is no exchange scattering. For $f = 0$ one finds $P'_A = P_e$, i.e., complete transfer of the electron polarization to the scattered atoms because there is exchange scattering only. If unpolarized electrons are scattered by a polarized target, the depolarization of the recoil atoms is described by the same factor as in (4.43). Similarly, the polarization transfer from polarized electrons to unpolarized atoms is given by the multiplicative factor in (4.44).

A perfect scattering experiment yielding the full information on the scattering amplitudes can be made by studies of the recoil atoms alone. By suitable choice of the initial polarizations and the measured polarization components of the recoil atoms the quantities $|f|$, $|g|$, and $\gamma_1 - \gamma_2$ can be obtained as in the case of the scattered electrons. Consequently, in order to make a perfect scattering experiment, one has the choice of observing either the recoil atoms or the scattered electrons or both. One can therefore achieve a highly desirable redundancy in the determination of f and g so that it is possible to check the reliability of these difficult experiments.

4.4 Polarization Experiments in Elastic Exchange Scattering

Measurements of quantities characteristic of exchange scattering are discussed. In early experiments, unpolarized electrons were scattered from polarized atoms and the polarization of either the recoil atoms or the scattered electrons was measured. More recently, scattering of polarized electrons from polarized atoms has been started.

In the preceding sections we saw that for atoms with one outer electron there are several experimental possibilities for determining the quantities that are essential in elastic exchange scattering. In the first investigations made, polarized atoms were crossed by unpolarized electrons [4.6–9], since polarized electrons of sufficient intensity could not be produced at that time. We shall first discuss an experiment [4.7] in which the change of the atomic spin by the scattering process was measured according to (4.25), which is a special case of (4.45).

Figure 4.4 is a schematic diagram of the experiment in which a spin analysis of the recoil atoms was made. A beam of potassium atoms is sent through a Stern-Gerlach magnet which selects atoms of a certain spin direction and energy range. Slow electrons are fired across the polarized atom beam. Their energy is in the 1-eV region and is thus insufficient to excite the K atoms, so that only elastic collisions between the K atoms and the electrons occur. Both collision partners change their momentum direction during the scattering, the scattering angle of

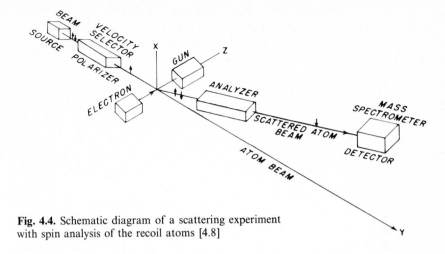

Fig. 4.4. Schematic diagram of a scattering experiment with spin analysis of the recoil atoms [4.8]

the heavy atoms being much smaller ($\lesssim 1°$) than that of the electrons owing to the large ratio of atomic masses to electron mass.

Due to the exchange interaction, some of the recoil atoms change their spin direction during the scattering process. They are selected by the spin analyzer if it is set so that only atoms with reversed spin direction are transmitted. This spin filter consists of a strong inhomogeneous magnetic field on which a strong inhomogeneous electric field is superimposed. The electric field is such that the force exerted on the induced electric dipole moments of the atoms has the same magnitude but the opposite direction as the magnetic force on the atoms with the desired spin direction. These atoms therefore pass through the analyzer without being deflected, while the atoms with the opposite spin direction are removed from the beam.

By rotating the analyzer-detector assembly about the scattering center, the angular distribution of the scattered atoms can be measured. From this, the angular distribution of the scattered electrons can be calculated using energy-momentum conservation.

If the spin filter is switched off, all atoms scattered within the angular range subtended by the detector are observed and not just those which have experienced a spin flip. In this case one obtains the full differential cross section $\sigma_0(\theta)$. In the present experiment the ratio $|g(\theta)|^2/\sigma_0(\theta)$ was first measured by using the two analyzer settings mentioned. Then, in order to find $|g(\theta)|^2$, the quantity $Q = 2\pi \int \sigma_0(\theta) \sin \theta \, d\theta$ was determined from the measured angular distribution and fitted to the total cross section Q measured in a separate experiment. This yields the absolute values of $\sigma_0(\theta)$ and thus of $|g(\theta)|^2$.

Needless to say, there were deviations from the ideal conditions described here. For example, the imperfect transmission and polarization of analyzer and polarizer had to be determined by test measurements. The influence of the nuclear spins on the atomic polarization was avoided by using an additional strong magnetic field in the scattering region (see Sect. 5.1). The two spin systems were thereby decoupled so that the change of the atomic polarization may be assumed to be due solely to the scattering. The detection of the weak atomic-beam intensities was accomplished by using the lock-in technique: The electron beam was modulated with a certain frequency. This frequency was superimposed by the scattering process on the recoil atoms which reached the detector. The phasesensitive narrow-band amplifier for the detector signal was locked to the oscillator controlling the electron-beam modulation and therefore responded only to the modulation frequency. In this way all disturbing background frequencies were cut out.

The main difference between this measurement and the next experiment to be mentioned [4.9] is that instead of the spins of the recoil atoms, those of their collision partners, the scattered electrons, are analyzed. This yields $|f(\theta)|^2/\sigma_0(\theta)$ according to (4.24). The electrons scattered by the polarized potassium atoms pass through a filter lens which removes unwanted inelastically scattered electrons. They are then accelerated to 100 keV so that their polarization can be measured with a Mott detector, just as in the experiments described in Sect. 3.7.

In this experiment it was not possible to decouple electron and nuclear spins in the scattering region by using a strong magnetic field. This would have had too strong an effect on the path of the electrons whose scattering angle is one of the quantities to be measured. Accordingly, the polarization of the atoms is reduced by the hyperfine coupling to rather small values ($<20\%$). In turn, the electron polarization is small, so that very low scattering asymmetries had to be measured. Furthermore, the experiment suffered from the lack of intensity, which is typical of many double scattering experiments.

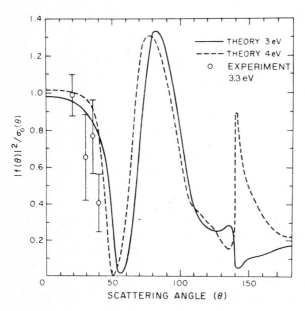

Fig. 4.5. Experimental [4.9] and theoretical [4.10] values of $|f(\theta)|^2/\sigma_0(\theta)$ for potassium

These difficulties explain the rather large error bars on the results shown in Fig. 4.5. The measurements could be carried out only at relatively small scattering angles because only there were the cross sections large enough to produce usable intensities. Despite all the problems, the measurements show that calculations made with the close-coupling approximation [4.10] appear to be quite reliable. This is particularly brought out in the first experiment discussed where $|g(\theta)|^2$ was determined in a large angular range at several energies with an error of $< 30\%$ (see Fig. 4.6).

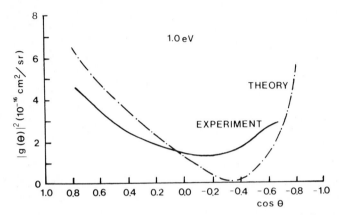

Fig. 4.6. Experimental [4.8] and theoretical [4.10] values of $|g|^2$ for potassium

The rather large cross sections $|g(\theta)|^2$ of more than 10^{-16} cm^2/sr suggest the use of exchange scattering of slow electrons for building a source of polarized electrons. We shall treat this and give further references in Sect. 8.2.

The experiments discussed so far were made at times when efficient sources of polarized electrons were not yet available. Now the situation has changed and first measurements have been reported where polarized electrons are scattered from polarized atoms. It is seen from (4.38) that such an experiment yields the scattering asymmetry A provided that the polarizations of the incident electrons and of the target are known. With the help of (4.14) the asymmetry A given by (4.39) may be rewritten in the form

$$A = \frac{\text{Re}\{f^*g\}}{\sigma_0} = \frac{|f+g|^2 - |f-g|^2}{|f+g|^2 + 3|f-g|^2} \tag{4.46}$$

as is done by the authors of an experiment in which polarized electrons are scattered from polarized hydrogen atoms [4.11]. From (4.46) the lower bound of A is seen to be $-1/3$ which is reached for $|f+g| = 0$, i.e., for the case of pure triplet scattering explained in Sect. 4.1.

The description of the apparatus used for the polarized-electron–polarized-hydrogen experiment will be postponed to the next section, since it has also been

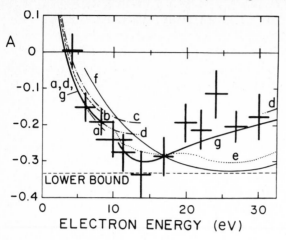

Fig. 4.7. Measured values of asymmetry at $\theta = 90°$ for elastic scattering of polarized electrons from polarized hydrogen atoms with various theoretical curves. For details see [4.11]

used for extensive asymmetry measurements of inelastic exchange scattering which will be discussed there. The essential result of the elastic scattering experiment, the asymmetry observed at the scattering angle $\theta = 90°$ ($\Delta\theta = 24°$), is shown in Fig. 4.7 together with a number of theoretical curves. It is interesting to note that the measured value for 13.8 eV is consistent with $A = -1/3$ which corresponds to pure triplet scattering.

Although electron-hydrogen elastic scattering is the simplest of all electron-atom collision problems, the agreement between the experimental and theoretical data is in most cases not satisfactory at the higher energies shown where calculations become increasingly difficult. In fact, agreement is good with only one of the theoretical predictions that is based on the close-coupling approximation. This is quite a different situation than is found for theoretical results on the spin-averaged cross section in this region, which mostly agree with the data obtained in conventional experiments with unpolarized electrons. Such agreement does not guarantee that calculated values of the scattering amplitudes f and g are correct as is demonstrated by the present example and by many other similar cases in electron-atom scattering. That is why measurements of polarization effects put theory to a more stringent test than do cross-section measurements alone, in particular if the angular variation of the asymmetry is also studied [4.12].

The evaluation of all of the above measurements has been made under the assumption that the role of spin-orbit coupling can be neglected. Since light atoms were used as targets and the experimental error bars are large, this is certainly a reasonable approach. When one is able to increase the accuracy of the experiments, it will be necessary to consider, quantitatively, how far the polarization effects can be influenced by spin-orbit interaction, particularly if heavier atoms and fairly large scattering angles are studied. The theoretical treatment of this problem has been initiated [4.13–15].

4.5 Inelastic Exchange Processes with One-Electron Atoms

Spin-dependent impact ionization and excitation are discussed. Investigations analogous to those of elastic scattering can be made. Additional information on the cross sections for excitation of the various sublevels can be obtained by polarization analysis of the emitted light. Electrons exciting fine-structure levels become polarized by exchange scattering even from an unpolarized target.

4.5.1 Spin-Dependent Electron-Impact Ionization

When now dealing with inelastic scattering, let us first assume that enough energy is transferred from the incident electron to the atom for impact ionization to take place. The results that have been derived in this chapter are then still applicable if f and g are taken to be the amplitudes for inelastic direct and exchange scattering leading to impact ionization of the atom. There are several experimental studies of the spin dependence of electron-impact ionization. Let us pick out here the polarized-electron–polarized-hydrogen experiment we have already met in the preceding section.

The experimental layout is illustrated in Fig. 4.8. Longitudinally polarized electrons ($P_e \approx 70\%$) intersect perpendicularly a beam of thermally dissociated

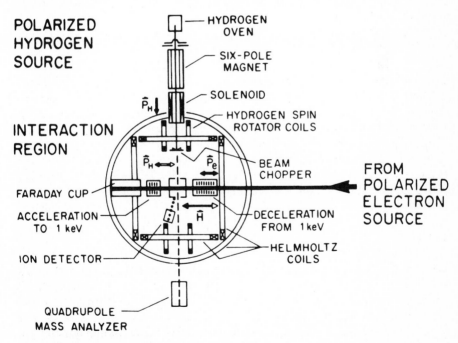

Fig. 4.8. Schematic diagram of the scattering chamber for polarized-electron–polarized-hydrogen experiment [4.16]

polarized hydrogen atoms ($P_A = 50\%$) whose polarization vector is oriented either antiparallel or parallel to that of the incident electrons in accordance with a 100-mG magnetic field in the interaction region. The methods used for producing the polarized electrons (Fano effect) and the polarized atoms (six-pole magnet) will be discussed in Chap. 5. By counting the protons produced in e-H collisions the relative ionization cross sections for antiparallel and parallel spins of the colliding beams are determined. A mass analyzer monitors the relative amounts of atomic and molecular hydrogen so that the ion counts originating from molecular hydrogen can be corrected for. The measurements have been made over the energy range 13.8 to 197 eV. The experimental details of such ambitious experiments are too sophisticated to be described here. Some of them (like atomic-beam modulation) have been outlined in the preceding section; for others we have to refer to the original papers [4.16].

The measurements yield the asymmetry A given in (4.46), where f and g are now the amplitudes for inelastic scattering. Since the ionization processes are detected regardless of the scattering angles and final energies of the scattered electrons, the terms in the expression for A must be integrated over these variables.

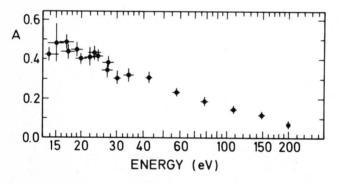

Fig. 4.9. Measured values of asymmetry vs. incident electron energy for impact ionization of polarized hydrogen atoms by polarized electrons [4.16]

The experimental results are shown in Fig. 4.9. They are in substantial disagreement with the theoretical predictions between threshold and 50 eV confirming that the electron-hydrogen scattering problem cannot be said at the present time to have been mastered, even though it is the most fundamental and least complicated of all electron collision problems.

Such investigations of impact ionization of polarized atoms by polarized electrons have become quite popular now that the techniques for producing polarized electrons and atoms are well established. Among the elements studied are potassium, lithium, and sodium [4.17–20]. The polarized-electron sources that have been employed are based on various techniques (cf. Sect. 8.2) like scattering, field emission from EuS, or photoemission from GaAs, whereas the polarized atoms are produced either by means of six-pole magnets or by optical pumping with circularly polarized light, a method whose principle will be discussed in the next chapter. One of the reasons for the interest in such experiments is the problem of the threshold law, i.e., the relation $\sigma \propto E^m$ between the ionization cross section and the final energy E of the escaping electrons. The measurements were of considerable help in clarifying such controversial questions as whether there is a drastic difference of the exponents m for triplet and singlet excitation or whether the exponential relation even needs modification [4.20–22].

4.5.2 Spin-Dependent Effects in Electron-Impact Excitation

In the rest of this section and in the following two sections we shall see that polarization studies are a powerful means for exploring the details of excitation processes. Inelastic electron scattering may raise the atoms to discrete states which differ in their energies and their angular momenta. A separation of transitions into different angular momentum states cannot be achieved by energy analysis if the states are energetically degenerate; it can, however, be performed by the methods discussed below.

If one wants to describe inelastic processes leading to various degenerate angular momentum states, one needs more scattering amplitudes than in the cases treated before but there are also more observable quantities. A direct spin analysis of the excited atoms is not easy because their lifetimes are generally short. One can, however, observe the polarization of the atoms after their return to the ground state [4.23, 24]. A direct spin analysis of the excited atoms has been made for metastable hydrogen which has a sufficiently long lifetime [4.25]. Further information on inelastic exchange scattering is obtained by analyzing the polarization of the light produced in the decay of the various excited states [4.26] and by observing the polarization of the scattered electrons as in the elastic case. Another observable is the scattering asymmetry obtained with polarized electrons. The technique of scattering polarized electrons from polarized atoms has recently been applied to asymmetry measurements in electron-impact excitation, analogous to those discussed above for impact ionization [4.27].

In line with our major topic, we shall focus attention on processes where polarized electrons are involved. In this subsection, we treat the simplest case of atoms with a single valence electron in an s state and consider excitations from the ground state into the resonance states, that is, transitions $S \to P$. As in the case of elastic exchange scattering, we neglect the spin-orbit interaction of the unbound electrons that causes the polarization effects in Mott scattering.

For the P state there are, with respect to some reference direction, three possible orientations, specified by $m_l = 0, \pm 1$. Accordingly, three scattering amplitudes f_0, f_1, and f_{-1} are needed to describe the inelastic direct scattering leading to the excitation of one of these states. Correspondingly, three scattering amplitudes g_0, g_1, and g_{-1} are needed to describe the exchange scattering. From symmetry considerations, one has $|f_1|^2 = |f_{-1}|^2$ and $|g_1|^2 = |g_{-1}|^2$. If this were not the case, the two states $m_l = \pm 1$ would obtain different populations even when excited by ordinary collisions with unpolarized collision partners. The resulting angular momentum in the direction of quantization, together with the components in this direction of the momenta occurring in the scattering process, would define a screw sense which would be reversed in the mirror image of the

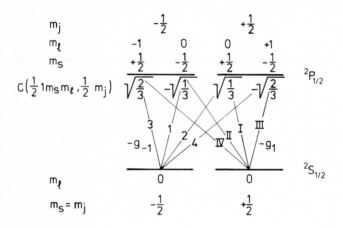

Transition		Cross section	Spin direction of scattered electron		Circular polarization of light emitted in transitions $\Delta m_j = +1$ or -1	
a)	b)		a)	b)	a)	b)
I	1	$\frac{1}{3}|f_0 - g_0|^2$	↑	↓	σ^+	σ^-
II	2	0				
III	3	0				
IV	4	$\frac{2}{3}|f_1 - g_1|^2$	↑	↓	σ^-	σ^+
c)	d)		c)	d)	c)	d)
I	1	$\frac{1}{3}|f_0|^3$	↓	↑	σ^+	σ^-
II	2	$\frac{1}{3}|g_0|^2$	↑	↓	σ^-	σ^+
III	3	$\frac{2}{3}|g_1|^2$	↑	↓	σ^+	σ^-
IV	4	$\frac{2}{3}|f_1|^2$	↓	↑	σ^-	σ^+

Fig. 4.10. Transitions to $^2P_{1/2}$ with totally polarized collision partners. a) to d) denote the following processes:

a) $e\uparrow + A\uparrow \rightarrow A(^2P_{1/2})$ b) $e\downarrow + A\downarrow \rightarrow A(^2P_{1/2})$
c) $e\downarrow + A\uparrow \rightarrow A(^2P_{1/2})$ d) $e\uparrow + A\downarrow \rightarrow A(^2P_{1/2})$

The amplitudes for transitions 3 and III are indicated for illustration experiment. Thus – as explained in more detail in Sect. 3.4.4 – parity conservation would be violated.

Applying to the inelastic case the results summarized in (4.8–13), we obtain

Process	Amplitudes	Cross section	
$e{\downarrow}+A{\uparrow} \rightarrow e{\downarrow}+A(^2P){\uparrow}$ $e{\uparrow}+A{\downarrow} \rightarrow e{\uparrow}+A(^2P){\downarrow}$	f_0, f_1, f_{-1}	$\lvert f_0\rvert^2 + \lvert f_1\rvert^2 + \lvert f_{-1}\rvert^2 = \lvert f_0\rvert^2 + 2\lvert f_1\rvert^2,$	(4.47)
$e{\downarrow}+A{\uparrow} \rightarrow e{\uparrow}+A(^2P){\downarrow}$ $e{\uparrow}+A{\downarrow} \rightarrow e{\downarrow}+A(^2P){\uparrow}$	$-g_0, -g_1, -g_{-1}$	$\lvert g_0\rvert^2 + 2\lvert g_1\rvert^2,$	(4.48)
$e{\uparrow}+A{\uparrow} \rightarrow e{\uparrow}+A(^2P){\uparrow}$ $e{\downarrow}+A{\downarrow} \rightarrow e{\downarrow}+A(^2P){\downarrow}$	$f_0-g_0, f_1-g_1, f_{-1}-g_{-1}$	$\lvert f_0-g_0\rvert^2 + 2\lvert f_1-g_1\rvert^2,$	(4.49)

where $A(^2P)$ represents an atom in the excited ^2P state and the tabulated quantities apply to the individual processes to their left (the kinematical factor k'/k of the inelastic cross sections has been omitted). Analogous to the elastic case of (4.14), the differential cross section for the situation where at least one of the colliding beams is unpolarized is thus

$$\sigma_0(^2P) = \frac{k'}{k}\,(\tfrac{1}{2}\lvert f_0\rvert^2 + \lvert f_1\rvert^2 + \tfrac{1}{2}\lvert g_0\rvert^2 + \lvert g_1\rvert^2 + \tfrac{1}{2}\lvert f_0-g_0\rvert^2 + \lvert f_1-g_1\rvert^2). \qquad (4.50)$$

For later discussion of light emission and electron scattering with high energy resolution we have to take the fine structure of the P levels into account. The transitions to $^2P_{1/2}$ which occur with various initial polarizations are illustrated by Fig. 4.10. A characterization of the excited states by the quantum numbers m_s and m_l is only possible when the excitation time is short compared with the spin-orbit relaxation time so that the spin-orbit-coupled states need not be considered during excitation. This is justified for light alkali atoms, where the relaxation time is of order 10^{-12} s, whereas the excitation time may be considered to be of order 10^{-15} s. For all processes that need $\sim 10^{-12}$ s or more, we have to take into account the fact that due to spin-orbit coupling m_s and m_l are not good quantum numbers; from the quantum numbers of the excited states in Fig. 4.10, only j and m_j represent constants of the motion.

Using the Clebsch-Gordan coefficients $C(slm_sm_l, jm_j)$ tabulated in text-books on quantum mechanics, a state with certain values j, m_j can be written as a superposition of states[3] $\lvert m_s, m_l\rangle$ which have fixed values of m_s and m_l, where $m_s+m_l=m_j$. For example, the coupled wave function for the $m_j=\tfrac{1}{2}$ substate of $^2P_{1/2}$ is given by[4]

$$\sqrt{\tfrac{1}{3}}\lvert\tfrac{1}{2}, 0\rangle - \sqrt{\tfrac{2}{3}}\lvert-\tfrac{1}{2}, 1\rangle, \qquad (4.51)$$

---

[3] Since we consider states of fixed s and l here, we dispense with these quantum numbers in the state vectors.

[4] As to the sign of the Clebsch-Gordan coefficients, see [Ref. 4.28, p. 123].

which is a superposition of the wave functions corresponding to $m_s = +\frac{1}{2}$, $m_l = 0$, and $m_s = -\frac{1}{2}$, $m_l = 1$. The Clebsch-Gordan coefficients $\sqrt{\frac{1}{3}}$ and $-\sqrt{\frac{2}{3}}$ are chosen so that (4.51) is an eigenfunction not only of j_z but also of j^2, since the total angular momentum also is a constant of the motion (see also Sect. 5.2.1 and Problem 5.1).

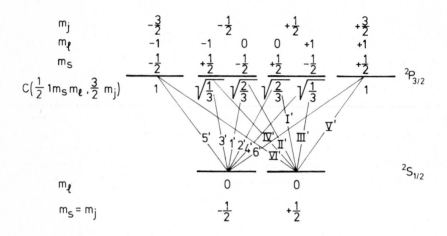

Transition		Cross section	Spin direction of scattered electron		Circular polarization of light emitted in transitions $\Delta m_j = +1$ or -1	
a)	b)		a)	b)	a)	b)
I'	1'	$\frac{2}{3}\lvert f_0 - g_0 \rvert^2$	↑	↓	σ^+	σ^-
II'	2'	0				
III'	3'	0				
IV'	4'	$\frac{1}{3}\lvert f_1 - g_1 \rvert^2$	↑	↓	σ^-	σ^+
V'	5'	$\lvert f_1 - g_1 \rvert^2$	↑	↓	σ^+	σ^-
VI'	6'	0				
c)	d)		c)	d)	c)	d)
I'	1'	$\frac{2}{3}\lvert f_0 \rvert^2$	↓	↑	σ^+	σ^-
II'	2'	$\frac{2}{3}\lvert g_0 \rvert^2$	↑	↓	σ^-	σ^+
III'	3'	$\frac{1}{3}\lvert g_1 \rvert^2$	↑	↓	σ^+	σ^-
IV'	4'	$\frac{1}{3}\lvert f_1 \rvert^2$	↓	↑	σ^-	σ^+
V'	5'	$\lvert f_1 \rvert^2$	↓	↑	σ^+	σ^-
VI'	6'	$\lvert g_1 \rvert^2$	↑	↓	σ^-	σ^+

Fig. 4.11. Transitions to $^2P_{3/2}$ with totally polarized collision partners. a) to d) denote the following processes:

a) $e\uparrow + A\uparrow \rightarrow A(^2P_{3/2})$ b) $e\downarrow + A\downarrow \rightarrow A(^2P_{3/2})$

c) $e\downarrow + A\uparrow \rightarrow A(^2P_{3/2})$ d) $e\uparrow + A\downarrow \rightarrow A(^2P_{3/2})$

Using the coupled atomic wave functions, one finds from (4.5–7) that one must multiply the amplitudes for transitions to the uncoupled states by the Clebsch-Gordan coefficients, also given in Fig. 4.10, in order to find the contribution of these transitions to the excitation of the coupled state $|jm_j\rangle$. This explains the numerical factors of the cross sections in Fig. 4.10, which apply to the basic processes listed in the figure caption, where e↑ and e↓ refer to the states $\begin{pmatrix}1\\0\end{pmatrix}$ and $\begin{pmatrix}0\\1\end{pmatrix}$, respectively. The last column concerns the polarization of emitted radiation which will be discussed later.

The transitions into the $^2P_{3/2}$ levels are illustrated by Fig. 4.11 in the same way.

We now consider the experimental possibilities for determining the quantities introduced here and choose as an example the excitation of unpolarized atoms by polarized electrons e↑ with polarization P_e. Let us assume that the quantity to be measured is the polarization P'_e of the scattered electrons after the excitation process. We will first consider the case where the fine-structure splitting is not resolved. Since half the collisions take place with atoms A↑ and the other half with A↓ (factor $\frac{1}{2}$), we obtain, from the relevant processes in (4.47, 49) the cross section for the appearance of e↑ :

$$\sigma_e^{\uparrow}(^2P)=\frac{k'}{k}\,P_e(\tfrac{1}{2}|f_0|^2+|f_1|^2+\tfrac{1}{2}|f_0-g_0|^2+|f_1-g_1|^2)$$

$$+\tfrac{1}{2}(1-P_e)\sigma_0(^2P) \tag{4.52}$$

with $\sigma_0(^2P)$ from (4.50). Here the partially polarized primary beam has again been split into a totally polarized and an unpolarized part as in the derivation of (4.22). Similarly, it follows from the lower process in (4.48) that the cross section for appearance of e↓ is

$$\sigma_e^{\downarrow}(^2P)=\frac{k'}{k}\,P_e(\tfrac{1}{2}|g_0|^2+|g_1|^2)+\tfrac{1}{2}(1-P_e)\sigma_0(^2P). \tag{4.53}$$

Thus

$$P'_e=\frac{\sigma_e^{\uparrow}-\sigma_e^{\downarrow}}{\sigma_e^{\uparrow}+\sigma_e^{\downarrow}}=P_e\,\frac{\sigma_0(^2P)-\dfrac{k'}{k}(|g_0|^2+2|g_1|^2)}{\sigma_0(^2P)}. \tag{4.54}$$

Hence, by measuring $P'_e(\theta)$, one obtains the quantity

$$\frac{k'}{k}(|g_0(\theta)|^2+2|g_1(\theta)|^2)=\sigma_0(^2P)\left(1-\frac{P'_e}{P_e}\right), \tag{4.55}$$

if the initial polarization P_e and the differential excitation cross section $\sigma_0(^2P)$ are known.

By observing the electron polarization P_e' after the inelastic scattering of initially unpolarized electrons by polarized atoms with polarization P_A, one obtains

$$\frac{k'}{k}\left(|f_0(\theta)|^2 + 2|f_1(\theta)|^2\right) = \sigma_0(^2P)\left(1 - \frac{P_e'}{P_A}\right), \tag{4.56}$$

which can be derived in the same way as (4.55).

From the results obtained so far it seems as though exchange scattering of unpolarized electrons from unpolarized targets cannot produce spin polarization. This is, however, not generally true! If the electrons which have excited different fine-structure levels are separated from each other, they may have significant polarization even making the assumption of this chapter that spin-orbit interaction in the continuum state (the origin of polarization in Mott scattering) disappears [4.29, 30]. For a straightforward calculation of the polarization one needs only to apply the formalism we have presented. Let us settle on the case that unpolarized electrons are inelastically scattered from unpolarized atoms, thereby exciting the $^2P_{1/2}$ state. Since direct polarization analysis of the short-lived excited atoms is difficult, we will restrict our considerations to the polarization of the electrons and therefore treat the problem in the 2×2 electron spin space rather than in the composite spin space of electron and target, which would be more rigorous but more lengthy. Let us take the direction of incidence z as the quantization axis and the $x-z$ plane as the scattering plane.

In order to apply our formalism we need to know the scattering matrices for the processes shown in Fig. 4.10. The scattering amplitudes can be read off from this figure by taking account of the Clebsch-Gordan coefficients given there. For transition $A\uparrow \rightarrow A(^2P_{1/2}, m_j = 1/2)$ the scattering matrix is seen to be

$$S = \frac{1}{\sqrt{3}}\begin{pmatrix} f_0 - g_0 & \sqrt{2}g_1 \\ 0 & f_0 \end{pmatrix}, \tag{4.57}$$

since the process may go via paths I and III and may have $e\uparrow$ and $e\downarrow$ contributions as given by reactions a) and c). In finding (4.57) we have recalled (cf. Sects. 3.2, 3) that the arrangement of the matrix elements is

$$\begin{pmatrix} S_{\uparrow\uparrow} & S_{\uparrow\downarrow} \\ S_{\downarrow\uparrow} & S_{\downarrow\downarrow} \end{pmatrix}$$

where $S_{\uparrow\downarrow}$ denotes a spin flip of the scattered electron from $\downarrow$ to $\uparrow$, etc.

A pure inital state $\begin{pmatrix} a_1 \\ a_2 \end{pmatrix}$ is transformed by the scattering matrix to the pure final state

$$\begin{pmatrix} a_1' \\ a_2' \end{pmatrix} = S\begin{pmatrix} a_1 \\ a_2 \end{pmatrix} = \frac{1}{\sqrt{3}}\begin{pmatrix} (f_0 - g_0)a_1 + \sqrt{2}g_1 a_2 \\ f_0 a_2 \end{pmatrix}.$$

This differs from the situation in scattering without fine-structure analysis, where the scattered beam is a mixture of electrons with different spin directions, because it contains electrons whose spins have flipped by exchange and others which have not flipped. In the present case, however, it cannot be said whether the transition goes via path I or III, since the excited state has no well-defined magnetic spin quantum number m_s. The atoms are all excited to the same pure state m_j. Consequently, since electrons and atoms were initially in a pure state the scattered electrons are also in a pure state [4.2].

The scattering matrices for the remaining transitions

$$A\uparrow \rightarrow A(^2P_{1/2}, m_j = -\tfrac{1}{2}), \quad A\downarrow \rightarrow A(^2P_{1/2}, m_j = \tfrac{1}{2}),$$
$$A\downarrow \rightarrow A(^2P_{1/2}, m_j = -\tfrac{1}{2})$$

are given in Problem 4.3 where the polarization of an inelastically scattered beam which has excited the $^2P_{1/2}$ state is found to be

$$P^{(1/2)} = \frac{i\sqrt{2}}{3} \frac{(f_1 g_0^* - f_1^* g_0) - (f_0 g_1^* - f_0^* g_1)}{|f_1 - g_1|^2 + \tfrac{1}{3}|f_1 + g_1|^2 + \tfrac{1}{2}|f_0 - g_0|^2 + \tfrac{1}{6}|f_0 + g_0|^2} \hat{n} \qquad (4.58)$$

if the incident beam is unpolarized ($\hat{n}$ = unit vector normal to scattering plane).

The analogous calculation for $^2P_{3/2}$ excitation yields

$$P^{(3/2)} = -\tfrac{1}{2} P^{(1/2)}.$$

A numerical evaluation of these quantities as a function of the scattering angle θ has been made for excitation of the $6\,^2P_{1/2}$ and $6\,^2P_{3/2}$ states of cesium [4.31]. The ratio of the cross sections for scattering with $^2P_{3/2}$ and $^2P_{1/2}$ excitation is $\sigma_{3/2}/\sigma_{1/2} = 2$, which is seen by adding up the cross sections given in Figs. 4.10 and 11. Hence the polarization of initially unpolarized electrons which have excited the unresolved ^2P state disappears in the approximation made here:

$$P = \frac{\sigma_{1/2} P^{(1/2)} + \sigma_{3/2} P^{(3/2)}}{\sigma_{1/2} + \sigma_{3/2}} = 0$$

in accordance with (4.54). This shows that the fine-structure splitting caused by spin-orbit coupling of the atomic electrons is crucial for the polarization to appear. The polarization is a cooperative result of exchange scattering and spin-orbit interaction in the bound state; both are required to produce the spin orientation of the scattered electrons.

From the results given, one may easily derive the other polarization effects in inelastic scattering with excitation of a fine-structure level. Using in Problem 4.3 the density matrix ϱ (2.19) for a polarized beam one may obtain (by the same straightforward calculation we made earlier for other scattering processes) the scattering asymmetry [4.30] and the change of the polarization vector caused by the excitation process. The observables discussed may be utilized to disentangle the various scattering amplitudes.

Polarization effects caused by fine-structure interaction are not restricted to one-electron atoms. In fact, experimental studies have mostly been made with mercury where the fine-structure splitting is sufficiently large that the scattered electrons may be separated by spectrometers. In Figs. 3.35 and 36 of Sect. 3.9 one can clearly see the polarization effects related to fine-structure splitting and compare them with those caused by spin-orbit interaction in the unbound electron state alone, which are seen in 6^1P_1 excitation. Although the latter mechanism is also effective when the triplet levels are excited, the fine-structure effects are evident because the results obtained for different fine-structue levels differ in sign. Polarization phenomena which are caused by the fine-structure interaction alone cancel if the fine structure is not resolved. At the energies chosen in the experiment, the fine-structure effect is obviously an essential polarization mechanism [4.32].

In superelastic scattering (the time-reversed process of inelastic scattering) from laser-excited sodium atoms the "fine-structure polarization" of the scattered electrons has also been observed to yield opposite polarizations for deexcitation of $3^2P_{1/2}$ and $3^2P_{3/2}$ levels [4.33].

Problem 4.3. Find the polarization of an initially unpolarized electron beam after excitation of the $^2P_{1/2}$ level of an unpolarized target of one-electron atoms. Spin-orbit interaction of the unbound electron, the mechanism effective in Mott scattering, is to be neglected.

Solution. The scattering matrix for the transition from the A↑ ground state to the $m_j = 1/2$ excited state is given by (4.57). Like this matrix, those for the three transitions A↑ → A($^2P_{1/2}, m_j = -1/2$), A↓ → A($^2P_{1/2}, m_j = 1/2$), A↓ → A($^2P_{1/2}, m_j = -1/2$) can be read off from Fig. 4.10 to be

$$\frac{1}{\sqrt{3}}\begin{pmatrix} \sqrt{2}(f_{-1} - g_{-1}) & g_0 \\ 0 & \sqrt{2}f_{-1} \end{pmatrix}, \quad \frac{1}{\sqrt{3}}\begin{pmatrix} -\sqrt{2}f_1 & 0 \\ -g_0 & -\sqrt{2}(f_1 - g_1) \end{pmatrix},$$

$$\frac{1}{\sqrt{3}}\begin{pmatrix} -f_0 & 0 \\ -\sqrt{2}g_{-1} & -(f_0 - g_0) \end{pmatrix}.$$

With the relations $f_{-1} = -f_1$, $g_{-1} = -g_1$, which follow from parity conservation [4.34], and $\varrho' = 1/2\, SS^\dagger \, \mathrm{tr}\{\varrho\}$ from (3.71) for vanishing polarization of the incident electrons, one obtains the density matrices of the scattered electrons,

$$\varrho_1' = \frac{1}{12}\begin{pmatrix} |f_0 - g_0|^2 + 2|g_1|^2 & \sqrt{2}f_0^*g_1 \\ \sqrt{2}f_0 g_1^* & |f_0|^2 \end{pmatrix}, \quad \varrho_2' = \frac{1}{12}\begin{pmatrix} 2|f_1 - g_1|^2 + |g_0|^2 & -\sqrt{2}f_1^*g_0 \\ -\sqrt{2}f_1 g_0^* & 2|f_1|^2 \end{pmatrix},$$

$$\varrho_3' = \frac{1}{12}\begin{pmatrix} 2|f_1|^2 & \sqrt{2}f_1 g_0^* \\ \sqrt{2}f_1^*g_0 & 2|f_1 - g_1|^2 + |g_0|^2 \end{pmatrix}, \quad \varrho_4' = \frac{1}{12}\begin{pmatrix} |f_0|^2 & -\sqrt{2}f_0 g_1^* \\ -\sqrt{2}f_0^*g_1 & |f_0 - g_0|^2 + 2|g_1|^2 \end{pmatrix},$$

for the four respective transitions. An additional factor 1/2 has been included as before, cf. (4.52), since half of the collisions take place with A↑, and the other half with A↓. The scattered electron beam is an incoherent mixture of the four polarized beams resulting from the individual transitions because the atomic states involved are incoherent mixtures of well-defined angular momentum states. Which of the four paths was followed can in principle be determined for each scattering event. (A thorough treatment of coherent and incoherent transitions can be found in [4.2]). The density matrix of the scattered beam is therefore

$$\varrho' = \sum_{n=1}^{4} \varrho'_n.$$

The calculation of the polarization $\boldsymbol{P}' = \mathrm{tr}\{\varrho'\boldsymbol{\sigma}\}/\mathrm{tr}\{\varrho'\}$ is straightforward and yields $P_x = P_z = 0$. For P_y one obtains (4.58) where the expression for $\mathrm{tr}\{\varrho'\}$, which is the sum of the cross sections given in Fig. 4.10, has been rewritten in a convenient form. The fact that there is only a single polarization component normal to the scattering plane is a consequence of parity conservation as shown in Sect. 3.4.4.

4.5.3 Emission of Circularly Polarized Light Induced by Excitation with Polarized Electrons

Further information on the cross sections for excitation of the various substates is obtained by observing the circularly polarized light which is emitted by the atoms. The fine-structure splitting has to be taken into account because it is easily resolved by optical spectrometers. We assume the hyperfine interaction to be decoupled experimentally. Since the 10^{-8} s excited-state lifetime is large compared with the relaxation time of spin-orbit coupling, the analysis of the emitted radiation has to be based on the coupled states $|j, m_j\rangle$.

If one observes the light regardless of the angle at which the electrons are scattered in the excitation process, one does not obtain information on the differential cross sections but rather on the integral cross sections

$$|F_0|^2 = 2\pi \frac{k'}{k} \int_0^\pi |f_0(\theta)|^2 \sin\theta \, d\theta, \text{ etc.} \quad \text{and} \tag{4.59a}$$

$$Q = 2\pi \int_0^\pi \sigma(\theta) \sin\theta \, d\theta. \tag{4.59b}$$

The emitted light has linearly and circularly polarized components. Their intensity ratio I^π/I^σ for the transitions from the states $m_j = \pm\frac{1}{2}$ is $\frac{1}{2}$ for $^2P_{1/2} \to {}^2S_{1/2}$ and 2 for $^2P_{3/2} \to {}^2S_{1/2}$. This follows from the calculation of the corresponding transition matrix elements (see, for example, [Ref. 4.35, Sect. 48]).

Let us first focus attention on the transitions from $^2P_{1/2}$. If all excited atoms return to the ground state by light emission, the cross sections for production of polarized radiation are

$$I^\pi(^2P_{1/2}) = \tfrac{1}{3}Q(^2P_{1/2}), \tag{4.60}$$

$$I^\sigma(^2P_{1/2}) = I^{\sigma+}(^2P_{1/2}) + I^{\sigma-}(^2P_{1/2}) = \tfrac{2}{3}Q(^2P_{1/2}), \tag{4.61}$$

where $Q(^2P_{1/2})$ is the total cross section for excitation of $^2P_{1/2}$. The last column in Fig. 4.10 indicates the sense of circular polarization of the emitted light, considering that in transitions $\Delta m_j = -1$ and $+1$ from the levels $m_j = +\frac{1}{2}$ and $m_j = -\frac{1}{2}$ circularly polarized σ^+ and σ^- light is emitted along the quantization axis which again is defined as the direction z of the incident electron beam. We will deal here with circularly polarized light, since it is typical of excitation by polarized electrons [see (4.64)]: owing to parity conservation we know that it cannot be produced by unpolarized electrons in the experiments discussed here, since reflection would change its helicity without changing the initial electron state (cf. Sect. 4.6).

For impact excitation of unpolarized atoms by electrons totally polarized along the z axis, that is, for

$$e\uparrow + \frac{A\uparrow}{A\downarrow} \rightarrow A(^2P_{1/2})$$

which is described by processes a) and d) in Fig. 4.10, we see that

$$I^{\sigma +} = \tfrac{1}{9}|F_0 - G_0|^2 + \tfrac{1}{9}|G_0|^2 + \tfrac{2}{9}|F_1|^2 \tag{4.62}$$
$$I^{\sigma -} = \tfrac{2}{9}|F_1 - G_1|^2 + \tfrac{1}{9}|F_0|^2 + \tfrac{2}{9}|G_1|^2$$

[where the factors $\tfrac{2}{3}$ from (4.61) and $\tfrac{1}{2}$ due to the scattering on an unpolarized target have been included]. Thus for excitation with a partially polarized electron beam of polarization P_e, one has

$$I^{\sigma +} = \tfrac{1}{9} P_e(|F_0 - G_0|^2 + |G_0|^2 + 2|F_1|^2) + \tfrac{1}{2}(1 - P_e)\tfrac{2}{3} Q(^2P_{1/2}) \tag{4.63}$$
$$I^{\sigma -} = \tfrac{1}{9} P_e(2|F_1 - G_1|^2 + |F_0|^2 + 2|G_1|^2) + \tfrac{1}{2}(1 - P_e)\tfrac{2}{3} Q(^2P_{1/2}).$$

If the fraction of the emitted intensity observed along the z direction is Δ^σ, the circular polarization of the light emitted in the transition $^2P_{1/2} \rightarrow {}^2S_{1/2}$ is

$$P^\sigma = \frac{\Delta^{\sigma +} - \Delta^{\sigma -}}{\Delta^{\sigma +} + \Delta^{\sigma -}} = \frac{I^{\sigma +} - I^{\sigma -}}{I^{\sigma +} + I^{\sigma -}} \tag{4.64}$$
$$= P_e \frac{\tfrac{1}{9}(|F_0 - G_0|^2 + |G_0|^2 + 2|F_1|^2 - 2|F_1 - G_1|^2 - |F_0|^2 - 2|G_1|^2)}{\tfrac{2}{3} Q(^2P_{1/2})}.$$

Since according to (4.61 and 62)

$$Q(^2P_{1/2}) = \tfrac{1}{6}|F_0 - G_0|^2 + \tfrac{1}{3}|F_1 - G_1|^2 + \tfrac{1}{6}|F_0|^2 + \tfrac{1}{3}|F_1|^2 + \tfrac{1}{6}|G_0|^2 + \tfrac{1}{3}|G_1|^2 \tag{4.65}$$

it follows that

$$P^\sigma Q(^2P_{1/2}) = P_e[Q(^2P_{1/2}) - \tfrac{1}{3}|F_0|^2 - \tfrac{2}{3}|G_1|^2 - \tfrac{2}{3}|F_1 - G_1|^2] \tag{4.66}$$

or

$$\tfrac{1}{3}|F_0|^2 + \tfrac{2}{3}|G_1|^2 + \tfrac{2}{3}|F_1 - G_1|^2 = Q(^2P_{1/2})(1 - P^\sigma/P_e). \tag{4.67}$$

If we know the polarization P_e of the incident electrons, and the cross section $Q(^2P_{1/2})$ which is found by measuring the total light emitted in the transitions $^2P_{1/2} \rightarrow {}^2S_{1/2}$, then, by measuring the circular polarization of the emitted line, we can obtain the sum of three of the unknown cross sections. Thus the sum of the remaining terms in Q is also known: From (4.65 and 66) one has

$$P^\sigma Q(^2P_{1/2}) = P_e[\tfrac{2}{3}|F_1|^2 + \tfrac{1}{3}|G_0|^2 + \tfrac{1}{3}|F_0 - G_0|^2 - Q(^2P_{1/2})]$$

or

$$\tfrac{2}{3}|F_1|^2 + \tfrac{1}{3}|G_0|^2 + \tfrac{1}{3}|F_0 - G_0|^2 = Q(^2P_{1/2})(1 + P^\sigma/P_e). \tag{4.68}$$

Further relations between the cross sections can be obtained in a similar way from Fig. 4.11 by considering the circular polarization of the light emitted in the transitions from $^2P_{3/2}$ (see Problem 4.4).

In order to determine the various unknown cross sections experimentally, other independent combinations of these quantities are needed. Such relations can be obtained by considering the excitation at the threshold energy. The scattered electrons then leave with vanishing energy and hence vanishing orbital angular momentum. As the quantization axis has been chosen parallel to the incident electron beam, we can state that the incident electrons also have vanishing orbital angular momentum along the quantization axis. Hence the electrons neither transfer nor carry away angular momentum parallel to the quantization axis, so that $\Delta m_l = 0$. Consequently, at the threshold energy E_{th}, the transition from the ground state of the alkalis ($l = 0$, $m_l = 0$) can lead only to states with $m_l = 0$. This is different at incident energies well above the threshold energy. In this case the scattered electrons leave with considerable energy and thus with considerable orbital angular momentum. Unless the electrons are scattered at 0° or 180°, they do have orbital angular momentum components along the direction of quantization. This means that usually there is a transfer of orbital angular momentum parallel to the quantization axis at higher energies.

Accordingly, at threshold energy all the cross sections for $m_l = \pm 1$ vanish, whereas with increasing energy the cross sections for $m_l = 0$ tend to zero. For example, we obtain from (4.67 and 68)

$$\tfrac{1}{3}|F_0|^2 = \lim_{E \to E_{th}} Q(^2P_{1/2})(1 - P^\sigma/P_e)$$

$$\tfrac{2}{3}|F_1|^2 = \lim_{E \to \infty} Q(^2P_{1/2})(1 + P^\sigma/P_e) \tag{4.69}$$

for the $^2P_{1/2} \to {}^2S_{1/2}$ transition.

Further possibilities for determining the cross sections arise from measurements where both collision partners are polarized. Such experiments are difficult, but have the advantage that the number of possible transitions is further reduced so that the various terms can be more easily separated. Furthermore, one can observe the scattered electrons in coincidence with the emitted light as briefly considered in Sects. 4.6 and 7. Measurement of the electron polarization in such coincidence experiments is, however, not within the reach of present experimental techniques. But in principle the individual cross sections can all be determined [4.36] by combining the various possibilities whose results can be read from Figs. 4.10 and 11.

Problem 4.4. Equation (4.67) gives the combination of cross sections that is obtained by measuring the circular light polarization in the transition $^2P_{1/2} \to {}^2S_{1/2}$ if unpolarized atoms are excited by polarized electrons. Find the corresponding expression obtained by measuring the circular polarization in the transition $^2P_{3/2} \to {}^2S_{1/2}$.

Solution. In this process, transitions from $m_j = \pm\frac{3}{2}$ result in circularly polarized light only, whereas in transitions from $m_j = \pm\frac{1}{2}$ due to the relation $I^\pi/I^\sigma = 2$ only a third of the emitted light is circularly polarized. Thus one obtains, for totally polarized electrons e↑, from processes a) and d) of Fig. 4.11,

$$I^{\sigma+} = \tfrac{1}{9}|F_0 - G_0|^2 + \tfrac{1}{2}|F_1 - G_1|^2 + \tfrac{1}{9}|G_0|^2 + \tfrac{1}{18}|F_1|^2 + \tfrac{1}{2}|G_1|^2$$
$$I^{\sigma-} = \tfrac{1}{18}|F_1 - G_1|^2 + \tfrac{1}{9}|F_0|^2 + \tfrac{1}{18}|G_1|^2 + \tfrac{1}{2}|F_1|^2.$$

Hence for partially polarized electrons it follows that

$$I^{\sigma+} = P_e(\tfrac{1}{9}|F_0 - G_0|^2 + \tfrac{1}{2}|F_1 - G_1|^2 + \tfrac{1}{9}|G_0|^2 + \tfrac{1}{18}|F_1|^2 + \tfrac{1}{2}|G_1|^2) + \tfrac{1}{2}(1 - P_e)I^\sigma$$
$$I^{\sigma-} = P_e(\tfrac{1}{18}|F_1 - G_1|^2 + \tfrac{1}{9}|F_0|^2 + \tfrac{1}{18}|G_1|^2 + \tfrac{1}{2}|F_1|^2) + \tfrac{1}{2}(1 - P_e)I^\sigma,$$

where

$$I^\sigma = I^{\sigma+} + I^{\sigma-} = \tfrac{1}{9}|F_0 - G_0|^2 + \tfrac{5}{9}|F_1 - G_1|^2 + \tfrac{1}{9}|F_0|^2 + \tfrac{5}{9}|F_1|^2 + \tfrac{1}{9}|G_0|^2 + \tfrac{5}{9}|G_1|^2$$

is independent of the polarization of the incident electrons. This yields

$$P^\sigma = \frac{I^{\sigma+} - I^{\sigma-}}{I^{\sigma+} + I^{\sigma-}}$$

$$= \frac{P_e}{9 I^\sigma} (|F_0 - G_0|^2 + 4|F_1 - G_1|^2 + |G_0|^2 + 4|G_1|^2 - 4|F_1|^2 - |F_0|^2). \tag{4.70}$$

Contrary to what might be expected from (4.64), $I^\sigma(^2P_{3/2})$ and $Q(^2P_{3/2})$ are not connected by a fixed numerical factor. Reason: the fraction of excitations which lead to $m_j = \pm 3/2$ (σ light only) and to $m_j = \pm 1/2$ (fraction of σ light is $1/3$) is not specified by $Q(^2P_{3/2})$. It depends on the size of the individual terms of $Q(^2P_{3/2})$ which thus also determine the fraction of circularly polarized light. From (4.70) it follows that

$$P^\sigma I^\sigma = P_e(I^\sigma - \tfrac{1}{9}|F_1 - G_1|^2 - \tfrac{2}{9}|F_0|^2 - |F_1|^2 - \tfrac{1}{9}|G_1|^2) \quad \text{or}$$

$$\tfrac{1}{9}|F_1 - G_1|^2 + \tfrac{2}{9}|F_0|^2 + |F_1|^2 + \tfrac{1}{9}|G_1|^2 = I^\sigma \left(1 - \frac{P^\sigma}{P_e}\right).$$

To determine the left-hand side of the last equation one needs to measure P^σ, P_e, and the total cross section for producing circularly polarized light by excitation of the $^2P_{3/2}$ state.

Let us note that the fine structure need not be resolved in order to obtain circularly polarized light: Adding the above expressions for $I^{\sigma+}$ and $I^{\sigma-}$ to the corresponding expressions (4.62) for $^2P_{1/2}$ excitation one finds that $P^\sigma = (I^{\sigma+} - I^{\sigma-})/(I^{\sigma+} + I^{\sigma-})$ differs in general from zero.

4.6 General Laws for Polarization of Light Emitted After Excitation by Polarized Electrons

From symmetry arguments, general relations between the polarization of light and that of the electrons causing its emission are derived. The results depend on whether the photons are observed alone or in coincidence with the scattered electrons.

We will now show that the results on light polarization caused by polarized-electron-impact excitation are not restricted to the special case discussed in the Sect. 4.5.3. We will see from symmetry arguments that there are close general relations between the polarization of light and the polarization of the electrons causing its emission.

Let us first recall from textbooks of optics the set of four independent measurements which must be performed in order to completely determine the properties of a light beam [4.37, 38]. The direction of the photon detector is given by $\hat{n}$ in Fig. 4.12 where $\hat{n}$, $\hat{e}_1$, and $\hat{e}_2$ are three orthogonal vectors. Let $I(\alpha)$ denote the intensity transmitted through a linear-polarization filter oriented at an angle α with respect to $\hat{e}_1$. For complete specification of the light beam one must measure the "Stokes parameters"

1. total intensity of the beam observed along $\hat{n}$,
2. degree of linear polarization with respect to $\hat{e}_1$,

$$\eta_3 = \frac{I(0°) - I(90°)}{I}, \tag{4.71}$$

3. degree of linear polarization with respect to an axis oriented at 45° to $\hat{e}_1$,

$$\eta_1 = \frac{I(45°) - I(135°)}{I}, \tag{4.72}$$

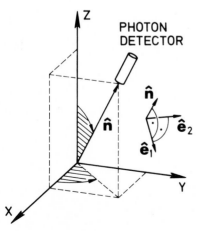

Fig. 4.12. Geometry for observation of light beam

4. degree of circular polarization,

$$\eta_2 = \frac{I^{\sigma+} - I^{\sigma-}}{I}, \tag{4.73}$$

where $I^{\sigma+}$ and $I^{\sigma-}$ are the light intensities transmitted through polarization filters for σ^+ and σ^- light. In the following discussion we will always assume that the primary electron beam propagates along the z direction and we define $\hat{e}_1$ to lie in the plane spanned by $\hat{n}$ and the z axis.

Let us first assume that the light is observed along the y direction. According to our definition, $\hat{e}_1$ then coincides with the z axis. It is easy to show that $\eta_1 = 0$ if the incident electron beam is unpolarized: For measurement of η_1 the polarization filter is oriented successively along the axes 1 and 2 indicated by the dashed lines of Fig. 4.13. Assume that $\eta_1 \neq 0$ because the emitted light is linearly polarized along axis 1. Reflection at the y-z plane leaves the initial state unchanged whereas the sign of η_1 would be reversed because the polarization along axis 1 would be transformed into polarization along axis 2. The different results obtained in the laboratory and in the mirror image would mean violation of parity as discussed in Sect. 3.4.4. Hence $\eta_1 = 0$. For the same reason one has, for an unpolarized incident beam, $\eta_2 = 0$. Otherwise the reflection at the y-z plane would change the sign of η_2 by changing the helicity, whereas the initial state remains unchanged. On the other hand, parity conservation permits $\eta_3 \neq 0$. In the present case, one has $\eta_3 = [I(z) - I(x)]/I$, where $I(z) = I(0°)$ and $I(x) = I(90°)$ are the light intensities measured with the orientation of the filter along the z and x axis, respectively. Unlike $I(45°)$ and $I(135°)$, $I(0°)$ and $I(90°)$ are not transformed into one another by reflection at the y-z plane so that η_3 remains unchanged. Consequently, parity conservation is compatible with $\eta_3 \neq 0$. Indeed, such a light polarization is quite common. It is caused by different excitation cross sections for different sublevels m_l, a situation discussed at the end of Sect. 4.5.3. When the various states m_l with their different populations decay, one obtains different intensities from the transitions with $\Delta m_l = 0$ and $\Delta m_l = \pm 1$ which yield, respectively, light polarization parallel and perpendicular to the z axis.

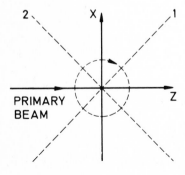

Fig. 4.13. Measurement of light polarization generated by electron beam incident along the z axis. Light detector normal to the x-z plane

If the primary electron beam has transverse polarization along the y direction, the arguments used are no longer valid because the initial state is then also changed by the reflection (cf. Sect. 3.4.4). The light observed along the y direction will therefore, in general, be polarized not only with respect to the directions (x, z) as expressed by $\eta_3 \neq 0$, but also with respect to $(1, 2)$ in Fig. 4.13 $(\eta_1 \neq 0)$ and it will also have circular polarization $(\eta_2 \neq 0)$.

For the same reason light polarization cannot be excluded by the parity arguments used if the electron beam is longitudinally polarized. However, axial symmetry around the z axis may then be exploited to show that $\eta_1 = \eta_2 = 0$ for dipole radiation.[5] Owing to the axial symmetry of the excitation process, the radiation field must also be axially symmetric so that the signal in the light detector cannot change if the detector is rotated around the z axis. Let us assume that $\eta_1 \neq 0$ because the dipole radiation emitted by the atom is linearly polarized along axis 1. The detector in the y direction with polarization filter along axis 1 then transmits a certain intensity. Rotation of the detector by $180°$ into the $-y$ direction interchanges the filter axes 1 and 2 so that the transmitted intensity is reduced. The measured intensity therefore fluctuates when the detector is rotated around the z axis, which means that emission of dipole radiation with $\eta_1 \neq 0$ leads to violation of axial symmetry. Consequently η_1 must be zero. For the same reason the circular light polarization η_2 disappears. Let us assume that $\eta_2 \neq 0$ because the angular momentum transferred to the photon by the atomic dipole transition has the sense of rotation indicated in Fig. 4.13. Light accepted by the detector in the y position will not be accepted in the $-y$ position, because the sense of rotation transmitted through the circular polarization filter is reversed as the light detector is rotated by $180°$. The fluctuation of the measured light intensity thus caused contradicts axial symmetry, so that η_2 must be zero. On the other hand, $\eta_3 \neq 0$ is compatible with axial symmetry, since the filter orientations $\alpha = 0°$ and $90°$ which coincide with the z and x axis, respectively, are not changed by rotation around the z axis.

The results for observation along the y axis obtained so far are summarized in Table 4.1 (upper third) together with those which we will derive now, more briefly, since the line of argumentation is basically the same. For an unpolarized or longitudinally polarized electron beam the light polarization observed in the x direction is the same as in the y direction since, due to axial symmetry, the two directions are equivalent. Hence $\eta_1 = \eta_2 = 0$, $\eta_3 \neq 0$. For the primary beam with transverse polarization P_y one finds by reflection at the x-z plane that $\eta_1 = \eta_2 = 0$ (cf. Fig. 4.14). The reflection does not change the initial excitation process, since the axial polarization vector represents a sense of rotation around the y axis. Accordingly, reflection cannot change the outcome of the experiment so that light polarization along axis 1 can be excluded since it would be transformed by the reflection into polarization along axis 2. Hence $\eta_1 = 0$. Similarly, the observed

---

[5] Since attention is focused on optical transitions, only dipole radiation needs to be considered here. Those results which cannot be transferred to higher multipole transitions (e.g., bremsstrahlung) due to their different polarization characteristics are indicated in Table 4.1.

Table 4.1. Connection between light polarization and electron polarization. The Stokes parameters η_1 and η_3 describe the two independent kinds of linear light polarization; η_2 describes circular light polarization. Electron beam is incident along z axis; direction of transverse electron polarization along y axis.

Electron polarization	Position of photon detector along direction	Light polarization		
		η_1	η_2	η_3
0		0	0	$\neq 0$
longitudinal	y	0^a	0^a	$\neq 0$
transverse		$\neq 0$	$\neq 0$	$\neq 0$
0		0	0	$\neq 0$
longitudinal	x	0^a	0^a	$\neq 0$
transverse		0	0	$\neq 0$
0		0	0	0
longitudinal	z	0	$\neq 0$	0
transverse		0	0	0

[a] dipole radiation assumed.

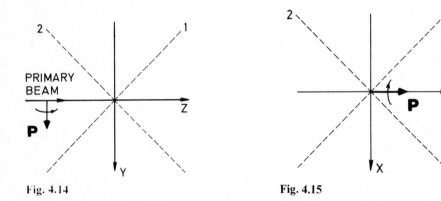

Fig. 4.14   Fig. 4.15

Fig. 4.14. Measurement of light polarization generated by electron beam incident along z axis and transversely polarized along y axis. Light detector in x direction

Fig. 4.15. Same as Fig. 4.14, but light detector in z direction

light cannot have a sense of rotation around the x axis since it would be transformed by the reflection into the opposite direction. Hence $\eta_2 = 0$. On the other hand, $\eta_3 \neq 0$ is allowed since the reflection does not interchange the y and z axes.[6]

---

[6] For observation in x direction, $I(0°)$ and $I(90°)$ denote, according to the definition of directions given at the beginning of this section, filter orientations along the z and y axis, respectively.

The same line of reasoning yields the results given in Table 4.1 for observation in z direction (Fig. 4.15). From axial symmetry one obtains, for unpolarized or longitudinally polarized electrons, $\eta_1 = \eta_3 = 0$. The quantity η_2 disappears for unpolarized electrons owing to the parity argument used above. Since this argument cannot be applied to longitudinally polarized electrons (cf. Sect. 3.4.4), η_2 may be different from zero in this case which has been treated in detail in Sect. 4.5.3. For the transversely polarized beam with polarization P_y (the special case indicated in Fig. 4.15) reflection at the x-z plane shows, as in the last paragraph, that $\eta_1 = \eta_2 = 0$. The fact that even η_3 disappears is seen by reflection at the y-z plane. The reflection does not change η_3, since it does not affect light polarization along x or y, but it changes the polarization of the incident electrons from P_y to $-P_y$. Because of parity conservation, the mirror image is also a process which can occur in nature. Consequently the same η_3 is observed, no matter whether the incident polarization is P_y or $-P_y$. This means that η_3 cannot differ from the value which is produced by an unpolarized beam made up of electrons with polarizations P_y and $-P_y$ and which we have seen before to be zero.

The results obtained here are more general than those of Sect. 4.5.3. They show that excitation by polarized electrons yields, in addition to the linear light polarization η_3 also obtained with unpolarized electrons, circular light polarization η_2 if the primary electrons have longitudinal polarization. If they have transverse polarization, the light may have circular polarization η_2 and also linear polarization η_1 which is independent of η_3.

From the symmetry arguments given, we see when the different kinds of light polarization may appear, but we do not see how large they are. This is determined by the dynamics of the particular process considered. Earlier in this book, we found a similar situation for the electron polarization after scattering of an unpolarized electron beam: symmetry permits polarization normal to the scattering plane, but the interaction decides whether the polarization appears and how large it is. Correspondingly, it has been shown that η_1 can be different from zero only if spin-orbit interaction plays a role in the excitation process [4.39]. Under the same condition, the light polarization η_3 produced by transversely polarized electrons is different from η_3 produced by unpolarized electrons (in order to see the difference, the light has to be observed in a direction different from x, y or z [4.40]). An important consequence is that measurement of the light polarization is one of the possibilities of getting insight into the interaction mechanism if the experimental results are compared to values computed from theoretical models.

Experiments exploiting this new source of information have been made only in recent years. Examples will be given in the next section. The information obtained can be enhanced if the light polarization is observed in coincidence with the scattered electron which has excited the photon emission (cf. Fig. 4.16). Since the primary electron beam and the direction of the scattered-electron detector now define a scattering plane (which we assume to be the x-z plane), a photon detector rotating around the z axis no longer observes axial symmetry.

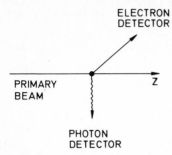

Fig. 4.16. Electron-photon coincidence

Table 4.2. Connection between light polarization and electron polarization if the photon is observed in coincidence with the scattered electron which has caused the photon emission (cf. Fig. 4.16). Electron scattering plane is x-z plane

Electron polarization	Position of photon detector along direction	Light polarization		
		η_1	η_2	η_3
0		$\neq 0$	$\neq 0$	$\neq 0$
$P_x\hat{e}_x + P_z\hat{e}_z$	y	$\neq 0$	$\neq 0$	$\neq 0$
$P_y\hat{e}_y$		$\neq 0^a$	$\neq 0^a$	$\neq 0^a$
0		0	0	$\neq 0$
$P_x\hat{e}_x + P_z\hat{e}_z$	x	$\neq 0^a$	$\neq 0^a$	$\neq 0$
$P_y\hat{e}_y$		0	0	$\neq 0^a$
0		0	0	0
$P_x\hat{e}_x + P_z\hat{e}_z$	z	0	$\neq 0^a$	0
$P_y\hat{e}_y$		0	0	0

[a] denotes that η differs from the value obtained with unpolarized electrons. (For arbitrary position of the photon detector all of the η obtained with polarized electrons differ from those obtained with unpolarized electrons.)

Furthermore, reflection symmetry with respect to the x-z plane is now broken not only by longitudinal polarization P_z of the primary electron beam, but also by transverse polarization P_x lying in the scattering plane (since the axes have now been defined by the scattering process we are no longer free to assume that the transverse polarization is directed along the y axis). This is an entirely different situation from that which we analyzed in Table 4.1, and many of the arguments which proved there that certain Stokes parameters vanished, do not apply here. Thus, in coincidence experiments the observed light is usually polarized as illustrated in Table 4.2, which may be easily verified by the reader using the symmetry arguments used above. If the light is not observed along one of the axes x, y, z, but along an arbitrary direction, all Stokes parameters η_i are generally different from zero and all of them depend on all components of the electron polarization. As a consequence, coincidence experiments allow more information about the dynamics of the excitation process to be extracted than non-coincidence experiments. An adequate and expedient description can be given by using the method of state multipoles introduced by *Fano* and *Macek* [4.41]. For a more detailed discussion of this point we must refer to the original literature [4.40].

4.7 Inelastic Exchange Processes with Two-Electron Atoms

Even though two-electron atoms with saturated spins cannot be polarized, exchange interaction can be studied by suitable polarized-electron experiments. A depolarization experiment which yields the energy dependence of exchange excitation and measurements of the light polarization characteristic of polarized-electron excitation are discussed. Such measurements allow us to disentangle the various atomic interactions (Coulomb-, exchange-, spin-orbit interaction) which are effective in the excitation process, since certain observables are sensitive to only one or the other of these forces.

Experiments with polarized atoms, such as those discussed earlier in this chapter, cannot be made with atoms that have two outer electrons and saturated spins (spin quantum number $S=0$). These atoms are in singlet states which cannot be polarized. Still, experiments with polarized electrons give valuable insights into exchange processes with such atoms. We have seen in Sect. 3.9 and 4.5.2 how both the polarization after scattering of an initially unpolarized electron beam and the scattering asymmetry of a polarized beam may be affected by exchange interaction in mercury atoms. We will now discuss measurements of other observables: the depolarization of an electron beam by excitation and the polarization of the light that is emitted after atomic collisions with polarized electrons.

Let us consider the excitation of a triplet state $(S=1)$ of a two-electron atom by polarized electrons e↑. As before, we will first exclude forces that are explicitly spin dependent which implies that the total spin quantum number and its projection along the quantization axis remain invariant in the collision. This is a good approximation for light atoms like helium. Then the excitation of a triplet state from the singlet ground state can occur only by exchange of the incident electron with one of the atomic electrons as shown in Fig. 4.17. We see from the figure that in this process the orientation quantum numbers have the following values:

	m_s(electron)	M_S(atom)
before the collision	$+\frac{1}{2}$	0
after the collision	either $+\frac{1}{2}$	0
	or $-\frac{1}{2}$	1

The polarization direction of the electrons has been chosen as the reference axis. We see that with e↑ the state $M_S= -1$ cannot be reached without violating the conservation of spin angular momentum.

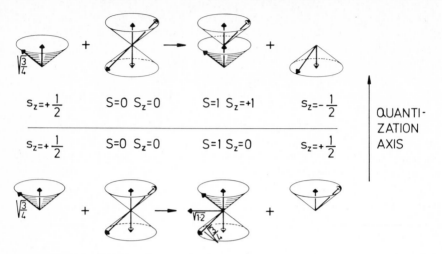

Fig. 4.17. Excitation of triplet states by exchange collisions

Let us now consider the polarization of the scattered electrons. The cross section for excitation of $M_S = 1$ is twice as large as that for excitation of $M_S = 0$. This can be seen from a simple calculation (see Problem 4.5) and is made plausible by Fig. 4.18, which represents the excitation by an unpolarized electron beam. The e↑ of this beam excite the levels $M_S = +1, 0$, whereas the e↓ excite $M_S = -1, 0$. Since $M_S = 0$ can be excited by the e↑ of the unpolarized beam as well as by the e↓, one would obtain a disparity in the populations of $M_S = 0$ and $M_S = \pm 1$ if the excitation cross sections for $M_S = +1$ and -1 were not twice as large as that for excitation of $M_S = 0$. There is no reason why such an alignment of the spin directions should occur by excitation with an unpolarized beam. We have seen in Fig. 4.17 that in excitation of a state $M_S = +1$ there is a reversal of the free-electron spin direction, whereas in excitation of $M_S = 0$ the spin directions of incident and scattered electrons are the same. Consequently, excitation by a totally polarized electron beam e↑ yields a scattered beam with two thirds e↓ and one third e↑, in other words, a polarization $P' = -\frac{1}{3}$ of the scattered beam.

Since the derivation of this result was straightforward, it does not seem very challenging to make such an experiment because one knows its result in advance. For helium, for example, where the underlying assumption (no explicit spin-

M_S −1   0   +1

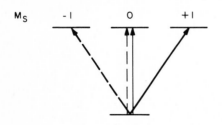

Fig. 4.18. Excitation of sublevels $M_s = 0, \pm 1$ (which may be part of sublevels M_j as in Figs. 4.10 and 11) by unpolarized electrons. (———) excitation by e↑; (– – – –) excitation by e↓

dependent forces) is well established, it seems quite obvious that a measurement of the ratio of the polarizations P' of the scattered beam and P of the incident beam would yield $P'/P = -\frac{1}{3}$. This assumes, of course, that there is no other influence on the polarization which has not been considered here (e.g., by compound states). The first polarization experiment of this kind has therefore been made with a target with which there was no question that one would learn something new.

For mercury it is well known that spin-dependent forces (spin-orbit forces) are no longer negligible. The excitation of a triplet state can in this case occur not only by an exchange of electrons but also by a direct process, in which the spin of one of the atomic electrons flips during excitation. This affects, of course, the value P'/P just derived so that a measurement of this ratio yields the extent to which the exchange processes discussed above still contribute to the excitation. This has been studied in a triple scattering experiment [4.42, 43].

Figure 4.19 is a schematic diagram of the apparatus. Scattering from a mercury-vapor beam, as described in Sect. 3.7.1, is used to produce a polarized-electron beam of 80 eV and $P = 0.22$. The polarized electrons are decelerated to energies between 5 and 15 eV and focused on a second mercury target. From the electrons scattered here, an energy analyzer selects those that have been scattered in the forward direction after excitation of the $6\,^3P$ states of the mercury atoms (energy loss ~ 5 eV). The polarization of these electrons is measured by a Mott detector.

There were two reasons for studying the electron scattering in the forward direction. First, there is the maximum of the scattered intensity; this helps in such a triple scattering experiment with its notorious lack of intensity. Second, the spins of the electrons scattered in the forward direction are not affected by spin-orbit coupling (see Sect. 3.6 where this has been illustrated by the fact that electrons scattered at small angles pass by the atom at a distance large enough that spin-orbit interaction is negligible). It was uncertain whether P' for electrons scattered in the forward direction would be different from the incident polarization P, because the theoretical treatment of exchange scattering in the

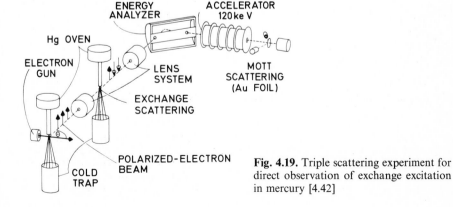

Fig. 4.19. Triple scattering experiment for direct observation of exchange excitation in mercury [4.42]

forward direction was particularly difficult [4.44, 45] and did not yield reliable results. Since it is the exchange processes that cause the change of the free-electron spin direction, P' would equal P if these processes did not play a role at small scattering angles.

The experimental results of Fig. 4.20 show that at incident energies below 8 eV there is a great number of processes which change the polarization, and at 6 eV the limiting value of $P'/P = -\frac{1}{3}$ is even observed (within the experimental error limits). That means that at this energy nearly all the excitation processes of the 6^3P levels occur by exchange scattering. On the other hand, the exchange excitation discussed above no longer plays an appreciable role at energies above

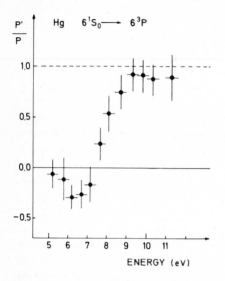

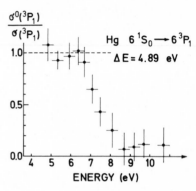

Fig. 4.20. Measured values [4.42] of depolarization vs. incident energy for $6^1S_0 \rightarrow 6^3P$ (forward direction)

Fig. 4.21. Contribution of the exchange processes discussed above to the excitation of the 6^3P_1 state [4.42]. σ^0 is the differential cross section for excitation by these exchange processes, σ is the complete differential cross section for excitation of 6^3P_1. All results in forward direction

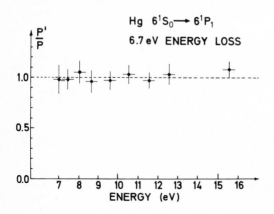

Fig. 4.22. Measured values [4.42] of depolarization vs. incident energy for $6^1S_0 \rightarrow 6^1P_1$ (forward direction)

10 eV. Figure 4.21 gives these facts directly for the $6\,^3P_1$ state. It is the evaluation of a measurement in which the fine structure of $6\,^3P$ has been resolved. A theoretical description of the results with the Bonham-Ochkur approximation reflects the strong energy dependence of the exchange contribution [4.46, 47].

Needless to say, experiments of this kind are rather delicate and need careful checks in order to ensure that the observed depolarization is not spurious. An essential check of the experiment discussed is shown in Fig. 4.22. Here the excitation of the 6^1P_1 level (energy loss 6.7 eV) has been studied in the same apparatus. In the excitation of a singlet state from a singlet ground state no change of spin directions can occur, no matter whether the excitation takes place by a direct or an exchange process.[7] (Remember that we are discussing cases where polarization effects due to spin-orbit interaction of the unbound electrons are negligible.) This is observed in the experiment, which shows that no spurious depolarization effects occur.

The derivation of the formulae expressing the polarization of the final beam in terms of the direct and exchange scattering amplitudes may be performed by the methods discussed in Sect. 4.5.2 and will therefore not be given here [4.30].

Another process which has been studied with two-electron atoms is the light polarization produced by polarized-electron impact. The picture given at the beginning of this section may be used to explain the basic mechanism responsible for the circular polarization of the light emitted in the direction of the quantization axis z. We have seen that the substates $M_S = -1$ are not populated by exchange excitation with e↑. In diagrams analogous to Figs. 4.10 and 11, where the M_J are expressed in terms of M_S and M_L $(M_S + M_L = M_J)$, the $M_S = -1$ substates are located preferentially on the left-hand side so that, in excitation by e↑, one has a significant disparity between the populations of the substates with positive and negative M_J. In other words, the spin orientation of the incident electrons produces an orientation of the excited atoms and can thus result in a preferential emission of σ^+ light. For e↓ the situation is reversed so that mainly σ^- light is emitted.

If the light is produced by electrons with transverse polarization P_y and is observed in the y direction, it may have not only circular but also linear polarization η_1 [cf. (4.72)] as discussed in the preceding section. The origin of the light polarization can be derived from the scheme of Fig. 4.18 as follows. According to Chap. 2, transverse spin states are a (coherent) superposition of longitudinal spin states parallel ↑ and antiparallel ↓ to the quantization axis z which we take as usual to be the direction of the incident electrons. Excitation by transversely polarized electrons is therefore described by coherent superposition of the transitions caused by e↑ and e↓ so that the three sublevels M_S of Fig. 4.18 are excited coherently. When an atom thus excited decays, one finds a coherent superposition of the radiation fields corresponding to transitions $\Delta M = \pm 1$ and

---

[7] The same is true, of course, for elastic scattering from atoms in a singlet ground state; there is no point in studying elastic exchange scattering from such atoms with polarized electrons.

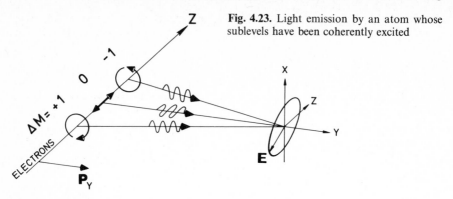

Fig. 4.23. Light emission by an atom whose sublevels have been coherently excited

$\Delta M = 0$, as illustrated in Fig. 4.23. Accordingly, the light observed in the y direction has, in general, both circular and linear components and is elliptically polarized.

An apparatus that has been used for extensive studies of the light polarization produced by polarized electrons is shown in Fig. 4.24. A GaAs source (cf. Sect. 8.2) produces electrons of a constant polarization whose value is between 0.35 and 0.45. By means of electrostatic deflection and magnetic rotation (cf. Sect. 8.1.1) the polarization is made transverse (in the y direction) or longitudinal before the electron beam hits the target. The results shown below were obtained with a mercury target whose resonance line $6\,^3P_1 \rightarrow 6\,^1S_0$ (254 nm) has been excited by a beam of 10 nA with an energy spread of 140 meV. For analysis of the light polarization, the advantages of a pile-of-plates analyzer forming the

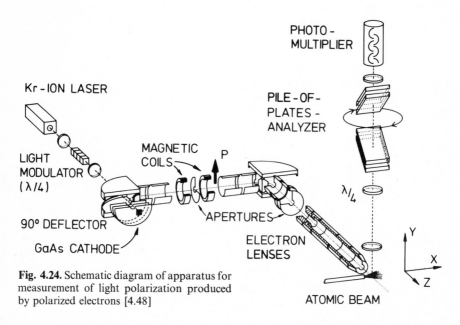

Fig. 4.24. Schematic diagram of apparatus for measurement of light polarization produced by polarized electrons [4.48]

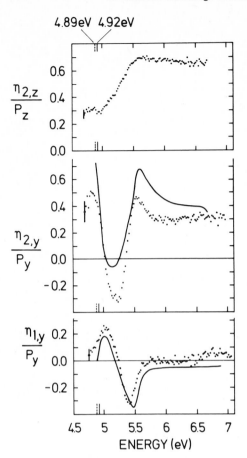

Fig. 4.25. Experimental values of light polarization [Stokes parameters defined by (4.72) and (4.73)] produced by polarized-electron impact. Transition $6\,^3P_1 \to 6\,^1S_0$ (254 nm) in mercury. All the Stokes parameters shown differ from zero only if the electrons are polarized. Linear and circular polarizations $\eta_{1,y}$ and $\eta_{2,y}$ are obtained with transverse electron polarization and photon detector in y direction; circular polarization $\eta_{2,z}$ is obtained with longitudinal electron polarization and detector in z direction. Light polarization is normalized to electron polarization. The threshold 4.89 eV for light excitation, the position of the 4.92 eV resonance, and typical error bars are indicated [4.48]. (——) theoretical curves [4.49]

Brewster angle with the direction of incidence have been exploited. For measurement of circular light polarization it was preceded by a quarter-wave plate (Pockels cell). The photons were detected by a photomultiplier in conjunction with a multichannel analyzer.

Figure 4.25 gives some results at electron energies between the excitation threshold and 7 eV for a few of the cases listed in Table 4.1: linear and circular light polarizations $\eta_{1,y}$ and $\eta_{2,y}$ detected in the y direction when the electrons have transverse polarization, and circular polarization $\eta_{2,z}$ detected in the z direction with longitudinal electron polarization. The light polarizations are normalized to the electron polarization by which they are produced. All of these light-polarization curves are strictly zero if produced by unpolarized electrons. This follows from the symmetry arguments discussed in the preceding section and has also been confirmed in the above experiment. With polarized electrons the light polarization becomes significant. The remarkable structure of $\eta_{1,y}$ and $\eta_{2,y}$ is caused by resonances due to the compound negative ion states of mercury which are found at these energies. The theoretical results depend sensitively on the total angular momenta assigned to the resonances, so that a classification of the resonances becomes possible [4.49] (see also Sect. 3.8). Bearing in mind the approximations which have to be made in these calculations and noting that other classifications of the resonances may even result in different signs of the calculated light polarization, the overall agreement between theoretical and experimental results in Fig. 4.25 can be regarded good.

The first experiment to determine electron polarization by measuring circular polarization of the emitted light [4.50, 51] was made with nonresonant radiation from zinc atoms [4.52]. Longitudinally polarized electrons excited the triplet state $5s\,^3S_1$ from the ground state $4s^2\,^1S_0$; the light emitted by the transitions $5s\,^3S_1 \rightarrow 4p\,^3P_J$ was observed in the z direction. From the measured circular light polarization of 14 % for the transition to $4\,^3P_0$ the electron polarization was derived to be (28 ± 2) %. Care was taken to avoid excitation of energy levels higher than the $5\,^3S_1$ state. Otherwise this state might have also been populated by decay from the higher excited levels ("cascading"). This would alter the population difference of its sublevels and thus change the polarization of the emitted light. The optical detection of electron polarization is therefore limited to electron energies in the small range between the excitation threshold of the level of interest and the first states that are able to cascade to this level.

More recently, the analogous transitions in mercury $7\,^3S_1 \rightarrow 6\,^3P_J$ have been studied with higher accuracy. Polarization curves $\eta_{2,y}/P_y$ of the type shown in Fig. 4.25 have been observed between the excitation threshold and 21 eV [4.53]. At threshold the polarization of 88 % predicted by theory [4.51] for the transition to $6\,^3P_0$ has been confirmed, while at higher energies the curve has a pronounced resonance structure dropping, for instance, from 88 % to 15 % within an energy range of 1 eV above threshold. Because of this rather involved dependence of the light polarization on electron energy, the original proposal [4.51] to use it without calibration for electron-polarization analysis cannot be recommended except for a very narrow region above threshold. The transitions discussed here have the advantage for polarization analysis that the light is not absorbed by ground-state Hg atoms so that problems of self absorption do not occur. Their intensity is, however, much lower than that of the 254 nm line (Fig. 4.25) ending in the ground state.

A more detailed analysis of the light-polarization curves would be beyond the scope of this book. Instead, we will explain in more general terms the physical import of such studies. We have emphasized before that the compatibility of certain observables with symmetry principles does not mean that these observables will definitely appear in experiment. Instead, they are found to be different from zero only if they are really produced by the dynamics of the process under consideration. Evidence of this fact can be found throughout this book, such as the electron polarization resulting from scattering of an initially unpolarized electron beam or the polarization of photoelectrons, both of which appear only if spin-orbit interaction plays a role. A more complicated example is the light polarization η_1. It has been shown that a nonvanishing value of η_1 is only possible if spin-orbit coupling cannot be neglected during the collision [4.39, 54]. These examples show that the mere presence or absence of certain observables in an experiment provides insight into the mechanisms effective in the processes studied. More quantitative information is, of course, obtained by the numerical values of these observables. If, for instance, exchange processes are the dominant mechanism, then in atomic excitation by electrons of transverse polarization P_y, the circular light polarization $\eta_{2,y}$ has the same sign as P_y. A change of sign of $\eta_{2,y}$ can only be caused by strong spin-orbit coupling during the scattering [4.39].

Measurements of η_1 and η_2 are therefore an important means for disentangling the various atomic interactions like Coulomb-, exchange-, and spin-orbit interaction, which are effective in the excitation process. When one finds $\eta_2 \neq 0$, one can conclude from the preceding discussions that exchange processes play a role. From differing signs of $\eta_{2,y}$ and P_y or from $\eta_1 \neq 0$ one can conclude that spin-orbit interaction is strong. In this way one can specifically test the assumptions made in theoretical models on the various mechanisms, whereas the usual comparison between experimental and theoretical results allows only a test of the theoretical model as a whole with all its approximations on the scattering dynamics and the atomic wave functions. While even such rough information as the sign of an observable, or whether it vanishes, gives insight into the dynamics of the process, more detailed data such as the energy-dependence curves of the observables discussed here put theoretical models to strong quantitative test. This is even more true if the scattered polarized electrons are observed in coincidence with the photons whose polarization is analyzed [4.55]. Such measurements have been done with the apparatus of Fig. 4.24 by detecting the electrons scattered in the forward direction in coincidence with the photons [4.56]. First theoretical attempts to calculate the light polarization in such coincidence experiments have been successful.

We have seen in the preceding sections how experiments with polarized electrons add new independent observables to the conventional ones. In addition to the customary quantities, like total or differential cross sections and linear light polarization η_3, they enable one to measure novel observables, such as the polarization P arising from scattering of an unpolarized beam, the scattering asymmetry of a polarized beam (which, in inelastic scattering, is in general different from P), the change of the electron polarization caused by scattering, and the light polarizations η_1 and η_2 which yield even more detailed information when analyzed in coincidence with the scattered electrons by which they are produced. The complexity of the inelastic processes discussed in this chapter does not, therefore, prevent their experimental analysis, since it results in a great number of measurable quantities. Studies of these observables have opened new dimensions in atomic physics with the promise of obtaining much better insight into atomic interactions.

Let us finally point out that some of the work on exchange scattering has been made with a view to a possible source of polarized electrons. It will be discussed in Sect. 8.2 [4.57–60].

Problem 4.5. Show that the cross section for excitation by $e\uparrow$ of a triplet state with $M_s = 1$ from a singlet ground state is twice as large as that for excitation of $M_s = 0$, provided that the other quantum numbers of these states are identical.

Solution. We evaluate the scattering amplitude (4.4) with the antisymmetric wave functions

$$\psi_i = \frac{1}{\sqrt{3}} [\exp(i\mathbf{k} \cdot \mathbf{r}_1)\eta(1)u(\mathbf{r}_2,\mathbf{r}_3)\chi_A(2,3) + \exp(i\mathbf{k} \cdot \mathbf{r}_2)\eta(2)u(\mathbf{r}_3,\mathbf{r}_1)\chi_A(3,1)$$

$$+ \exp(i\mathbf{k} \cdot \mathbf{r}_3)\eta(3)u(\mathbf{r}_1,\mathbf{r}_2)\chi_A(1,2)] \quad \text{and}$$

$$\psi_f = \frac{1}{\sqrt{3}} [\exp(i\mathbf{k}' \cdot \mathbf{r}_1)\eta'(1)u'(\mathbf{r}_2,\mathbf{r}_3)\chi_s(2,3) + \exp(i\mathbf{k}' \cdot \mathbf{r}_2)\eta'(2)u'(\mathbf{r}_3,\mathbf{r}_1)\chi_s(3,1)$$

$$+ \exp(i\mathbf{k}' \cdot \mathbf{r}_3)\eta'(3)u'(\mathbf{r}_1,\mathbf{r}_2)\chi_s(1,2)],$$

where the notation is the same as in Sect. 4.1; χ_s is given by (4.15a and b) for the final substates $M_S = +1$ and 0, respectively. Multiplication yields nine terms. The three terms corresponding to direct scattering (incoming and scattered electrons have the same label) disappear because $\chi_A(\lambda, \mu)$ is orthogonal to $\chi_S(\lambda, \mu)$. If we first consider the transitions to $M_S = +1$, the six nonvanishing terms have the form

$$\frac{g(\theta)}{3\sqrt{2}} \{\eta'(2)\alpha(3)\alpha(1)\eta(1)[\alpha(2)\beta(3) - \beta(2)\alpha(3)]\}, \quad \text{where}$$

$$g(\theta) = -\frac{m}{2\pi\hbar^2} \langle \exp(i\mathbf{k}' \cdot \mathbf{r}_2)u'(\mathbf{r}_3,\mathbf{r}_1)|T|\exp(i\mathbf{k} \cdot \mathbf{r}_1)u(\mathbf{r}_2,\mathbf{r}_3)\rangle, \tag{4.74}$$

etc., through permutation.[8] The spin function of the incident $e\uparrow$ is $\eta = \alpha$. If the spin function η' of the outgoing electron also equals α, the above product of the spin functions is zero. This is in agreement with Fig. 4.17 which shows a change of the spin direction of the free electron for excitation of $M_S = +1$. For $\eta' = \beta$ the product of the spin functions is -1. Since we have six terms of the kind given above, we find the scattering amplitude to be $-6[g(\theta)/3\sqrt{2}] = -\sqrt{2}g(\theta)$.

For $M_S = 0$ the six nonvanishing terms have the form

$$\frac{g(\theta)}{3 \cdot 2} \{\eta'(2)[\alpha(3)\beta(1) + \beta(3)\alpha(1)]\alpha(1)[\alpha(2)\beta(3) - \beta(2)\alpha(3)]\},$$

etc., through permutation. This time the spin product is zero for $\eta' = \beta$, in accordance with Fig. 4.17, which shows no change of the free-electron spin for excitation of $M_S = 0$. For $\eta' = \alpha$ the spin product is 1, so that the sum of the six nonvanishing terms yields the scattering amplitude $g(\theta)$.

The expressions (4.74) for $g(\theta)$ are identical for excitation of $M_S = 1$ and $M_S = 0$ if the space functions of the final states are identical. The ratio of the scattering amplitudes is therefore $-\sqrt{2}g(\theta)/g(\theta) = -\sqrt{2}$, and the ratio of the cross sections is 2.

---

[8] In permuting the electrons one has to take into account that the u are symmetric and the u' are antisymmetric.

4.8 Møller Scattering

Electron-electron scattering is spin dependent. This can be utilized for polarization analysis. Comparison with the Mott analyzer is made.

One of the few cases where calculations of the polarization effects in inelastic scattering are relatively easy is electron scattering from electrons whose binding energy is small compared to the energy transfer during the collision. One can then neglect the binding energy and consider the scattering process as an elastic collision between free electrons. This yields a simple relation between f and g which is generally derived in textbooks of quantum mechanics but can also be read off from (4.7). Since bonding is no longer considered, we can replace the wave functions u and u' by free-electron wave functions and obtain in the first Born approximation

$$f \propto \langle \exp{(i\mathbf{k}_1' \cdot \mathbf{r}_1)} \exp{(i\mathbf{k}_2' \cdot \mathbf{r}_2)} | \frac{e^2}{r_{12}} | \exp{(i\mathbf{k}_1 \cdot \mathbf{r}_1)} \exp{(i\mathbf{k}_2 \cdot \mathbf{r}_2)} \rangle$$

$$g \propto \langle \exp{(i\mathbf{k}_1' \cdot \mathbf{r}_2)} \exp{(i\mathbf{k}_2' \cdot \mathbf{r}_1)} | \frac{e^2}{r_{12}} | \exp{(i\mathbf{k}_1 \cdot \mathbf{r}_1)} \exp{(i\mathbf{k}_2 \cdot \mathbf{r}_2)} \rangle ,$$

(4.75)

where the scattering potential e^2/r_{12} has been introduced and the common factor has been omitted. In the center-of-mass system we have $\mathbf{k}_1 = -\mathbf{k}_2$, $\mathbf{k}_1' = -\mathbf{k}_2'$ so that, after separation of the motion of the center of mass, we obtain

$$f \propto \int \exp{[i(\mathbf{k}_1 - \mathbf{k}_1') \cdot \mathbf{r}]} \frac{e^2}{r} d^3r, \quad g \propto \int \exp{[i(\mathbf{k}_1 + \mathbf{k}_1') \cdot \mathbf{r}]} \frac{e^2}{r} d^3r, \quad (4.76)$$

where $\mathbf{r} = \mathbf{r}_1 - \mathbf{r}_2$. Since $|\mathbf{k}_1 - \mathbf{k}_1'| = 2k_1 \sin{\theta_c/2}$, $|\mathbf{k}_1 + \mathbf{k}_1'| = 2k_1 \cos{\theta_c/2}$, where θ_c is the scattering angle in the center-of-mass system, one easily sees, by taking the polar axis in the direction of the vectors $\mathbf{k}_1 - \mathbf{k}_1'$ and $\mathbf{k}_1 + \mathbf{k}_1'$, respectively, that

$$g(\theta_c) = f(\pi - \theta_c).$$

Let us assume that a totally polarized electron beam is scattered from a totally polarized target, the angle between the polarization of the target and that of the incident beam being ϑ. Then we find from (4.36) the cross section

$$\sigma(\theta_c) = \frac{1}{2}[(1 + \cos{\vartheta})|f(\theta_c) - f(\pi - \theta_c)|^2$$
$$+ (1 - \cos{\vartheta})(|f(\theta_c)|^2 + |f(\pi - \theta_c)|^2)]. \quad (4.77)$$

For the scattering angle $\theta_c = 90°$ we obtain

$$\sigma(90°) = (1 - \cos{\vartheta})|f(90°)|^2,$$

which is 0 for parallel spins ($\vartheta = 0$) and $2|f(90°)|^2$ for antiparallel spins ($\vartheta = 180°$).

The difference of the cross sections can be understood as follows. Electrons with parallel spins have symmetric spin functions but an antisymmetric space part of the wave function so that the probability of finding them close together tends to zero as the distance decreases. That is why they are, on average, further apart than electrons with antiparallel spins and are thus less likely to be scattered from each other.

If the binding energy of the target electron is negligible compared to the energy transfer, it can be considered to be at rest in the laboratory system. Since θ_c and the scattering angle θ in the laboratory system are then related by $\theta_c = 2\theta$ we have $\sigma_p/\sigma_a = 0$ for $\theta = 45°$, when σ_a and σ_p are the differential cross sections for parallel and antiparallel spins. This case where half of the incident energy is transferred to the target electron (see Fig. 4.26) is therefore most suitable for polarization analysis: polarized electrons may be scattered by a magnetic material whose spins are oriented first parallel and then antiparallel to the incident polarization. The relative difference of the scattering intensity yields the unknown polarization.

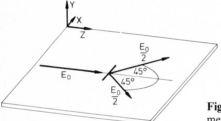

Fig. 4.26. Electron-electron scattering with symmetric energy transfer (nonrelativistic limit)

A relativistic treatment of electron-electron scattering does not, in principle, affect this scattering asymmetry, which is caused by electron exchange; the numerical results are, however, modified. Only at low velocities is the interaction of the electrons, at each instant of time, given by their static interaction. At higher electron velocities, there are retardation effects of the electromagnetic interaction due to the finite velocity of light. Exchange of photons between the two electrons then becomes important, so that one has no longer a problem of quantum mechanics but, instead, of quantum electrodynamics.

A solution to this problem under the aspect considered here has been given by *Bincer* [4.61] who calculated the spin dependence of the cross section for scattering of two Dirac electrons. The first treatment of relativistic electron-electron scattering by *Møller* [4.62] did not emphasize polarization phenomena. Bincer's results for longitudinally polarized electrons are shown in Fig. 4.27, where σ_p/σ_a is given as a function of the energy transfer $w = W/T$ [W is the kinetic energy lost in the collision by the incident electron and $T = mc^2 (\gamma - 1)$ is its kinetic energy before the collision]. As explained above, one has in the nonrelativistic limit $\sigma_p/\sigma_a = 0$ at $w = 0.5$. For higher energies σ_p/σ_a still has a minimum at $w = 0.5$, though its value increases from 0 to 1/8 in the extreme relativistic limit ($\gamma \rightarrow \infty$).

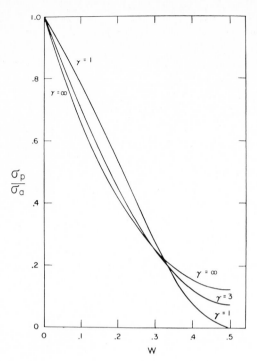

Fig. 4.27. Ratio of the differential cross sections for scattering of longitudinally polarized electrons with parallel (σ_p) and antiparallel (σ_a) spins [4.61]

The scattering angle θ belonging to $w=0.5$ decreases from $45°$ in the nonrelativistic limit $\gamma=1$ to $0°$ for $\gamma \rightarrow \infty$.

In our discussion of the spin dependence of the cross section in the nonrelativistic case no assumption has been made about the direction of the spin relative to the momentum. Accordingly, the results are valid both for longitudinal and for transverse polarization. This becomes different in the relativistic region, as Fig. 4.28 illustrates. The meaning of the asymmetry coefficients a_{ij} shown there can be seen if one writes the cross section in the form

$$\sigma(\theta) = I(\theta)(1 + a_{ij}) P_i P_j^{(t)} \tag{4.78}$$

where $I(\theta)$ is the Møller cross section for unpolarized electrons and P_i and $P_j^{(t)}$ $(i, j = x, y, z)$ are the polarization components of the incident electron and the target electron, respectively. The direction of the axes is as shown in Fig. 4.26.

The contours $a_{zz}=$ const. again illustrate the spin dependence of the cross section for scattering of longitudinally polarized electrons. In the nonrelativistic region, the spin dependence for transversely polarized electrons is the same, as can be seen from the contours $a_{xx}=$ const. and $a_{yy}=$ const. Between 100 keV and 1 MeV there is, however, a strong decrease in a_{xx} and a_{yy}, so that above 1 MeV the spin dependence of scattering for transversely polarized electrons is considerably lower than that for longitudinally polarized electrons.

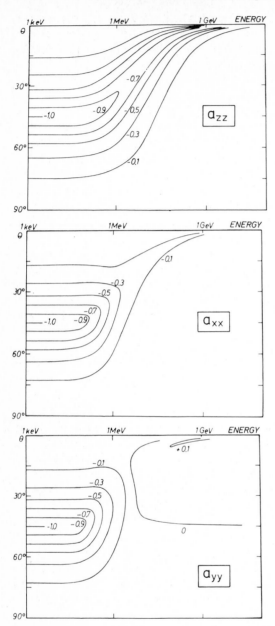

Fig. 4.28. Contours of the asymmetry coefficients for electron-electron scattering, cf. (4.78) [4.63]

The coefficients a_{ij} $(i \neq j)$ which describe scattering of transversely polarized by longitudinally polarized electrons or vice versa are, at all energies and angles, either smaller than 0.1 or vanish. This means that in those cases the spin dependence of scattering is negligible, particularly if one considers the experimental conditions under which these effects can be studied: even in an iron target that is magnetized to saturation, only approximately 8 % ($\approx 2/26$) of the electrons are orientated. A target polarization $P^{(t)} = \pm 8$ % yields, with the optimum value of $a_{ii} = -1$ in (4.78), a scattering asymmetry of only 8 %. With asymmetry coefficients that are much smaller, the observed effects are reduced below the 1 % limit, which makes experiments cumbersome.

The smallness of the observable effects is the reason why Møller scattering has not become the predominant method for measuring electron polarization although it certainly has several advantages. Many studies of spin polarization in β decay utilized this technique, taking advantage of the fact that the longitudinal polarization of the electrons need not be converted into transverse polarization as in Mott scattering, if the target is magnetized in the direction of the incident spins. The few electron-electron scattering events can be selected from the many background electrons, which are mainly due to scattering from the nuclear Coulomb field,[9] by detecting the two outgoing electrons in coincidence. This is done by two counters arranged under suitable angles (45° in the nonrelativistic range, as shown in Fig. 4.26; smaller angles at higher energies). Pulse-height analysis further distinguishes the two electrons sharing the incident energy from the electrons scattered by the Coulomb field. The technique described also reduces the influence of multiple scattering: electrons that undergo considerable scattering will not be recorded.

The experiments discussed in the various chapters of this book show however that Mott scattering is more widely used than Møller scattering for polarization analysis. The asymmetry effects are larger for the former and careful absolute measurements of the asymmetry function have been made, so that one does not have to rely on theory for the calibration of the polarization analyzer. Although careful measurements of the Møller cross section for unpolarized particles have been made [4.64–66], absolute measurements of the asymmetry coefficients are not known to the author, so that one has more or less to rely on theoretical results, such as those shown in Fig. 4.28, which are based on lowest order perturbation theory. On the other hand, Møller scattering does not have the disadvantage of Mott scattering that it is inapplicable in the extreme relativistic region. Since $\sigma_p/\sigma_a = 1/8$ even for $\gamma \rightarrow \infty$, a polarization analyzer based on this method works at all energies (see Sect. 8.4).

It is quite obvious from our discussion that the employment of Møller scattering would not lead to an efficient source of polarized electrons. Even with an iron target the polarization would never be larger than 8 %, and the necessity of discriminating against the large number of background electrons would make such a source even less attractive.

---

[9] The scattering intensity in electron-electron scattering is proportional to the number Z of electrons per atom, whereas the intensity scattered from the nuclear Coulomb field is, roughly speaking, proportional to Z^2.

5. Polarized Electrons by Ionization Processes

5.1 Photoionization of Polarized Atoms

Polarized electrons can be produced by photoionization of polarized atomic beams, which have high intensities when produced by six-pole magnets. The process has been investigated mainly with a view to building a source of polarized electrons.

In the first chapter it was shown that a Stern-Gerlach magnet cannot be used as a polarization filter to select free electrons with a certain spin direction. Electrons that are bound to atoms can, however, be polarized in this way. If the oriented atomic electrons are extracted from the atoms without affecting their spin directions, polarized free electrons are obtained. This can, for example, be achieved by photoionization.

Although such an experiment was suggested [5.1] as early as 1930, it was performed only much later [5.2–4]. Instead of using conventional Stern-Gerlach magnets, six-pole magnets (see Fig. 5.1) were used to polarize alkali atoms. The atoms with the required spin orientation, emerging divergently from the atomic-beam oven, can thereby be focused so that high intensities can be attained. The reason for this is as follows.

Let us assume a field that exerts a force proportional to $\mp \mu r$ on the magnetic dipoles, where $\mp$ refer to directions of the electron spin parallel and antiparallel to the magnetic field, and r is the distance from the axis. Such a field deflects away from the axis those atoms whose spins are antiparallel to its direction; it acts on them as a diverging lens. Atoms with the opposite spin directions are deflected towards the axis. They perform harmonic oscillations of uniform frequency about the axis and are therefore focused to one point if they have equal axial velocity. The field acts on these atoms as a converging lens.

It can be shown that a six-pole magnet as drawn in Fig. 5.1 possesses such lens properties: Near the axis, the magnitude $|B|$ of the magnetic field strength is to a good approximation proportional to r^2 (B itself is of course in no way axially symmetric, as one can see from Fig. 5.1). Thus the potential energy of the dipoles for the two spin directions is

$$V = \pm \mu |B| \propto \pm \mu r^2,$$

and the force is $-\boldsymbol{V} V \propto \mp \mu r$. In actual fact, the magnetic field and thus also the spins parallel to it have all possible directions. If, however, a magnetic field in the

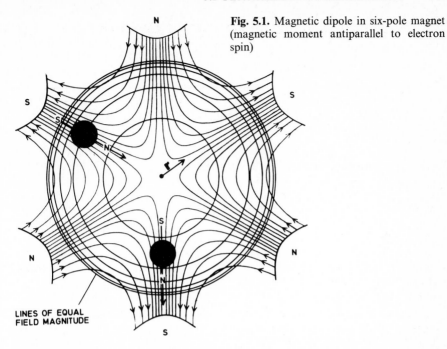

Fig. 5.1. Magnetic dipole in six-pole magnet (magnetic moment antiparallel to electron spin)

LINES OF EQUAL
FIELD MAGNITUDE

direction of the axis is attached to the six-pole magnet, the field as well as the spins lying parallel to the field gradually turn into the axial direction as one goes from the inside of the magnet to the outside. The change of the magnetic field direction as seen from the moving particles takes place slowly in comparison to the Larmor frequency so that the spins follow the change of the magnetic field adiabatically.

In this way the six-pole magnet produces a longitudinally polarized, well-focused beam of alkali atoms. The oriented valence electrons are then ejected by photoionization. This occurs within the axial magnetic field just mentioned which is made strong enough to decouple the atomic electron spin s from the nuclear spin I ($j=s=\frac{1}{2}$ in the ground state of alkali atoms). Without decoupling, the resultant angular momentum $F=s+I$ and not the electron spin would be oriented in the magnetic field.[1] The spin expectation value in the direction of orientation of the selected atoms would then decrease; in other words, the observed polarization would be diminished.

After extraction of the photoelectrons from the region of the magnetic decoupling field they were, in the aforementioned experiments, sent through a polarization transformer in order to convert their longitudinal polarization into transverse polarization, as required for analysis by a Mott detector. The

---

[1] This is analogous to the coupling of s and l to j which will be discussed in more detail in Sect. 5.2.

maximum polarization obtained was 85%. The experiments were not done primarily because of an interest in the underlying physical processes, but rather for the purpose of building an intense source of polarized electrons. A further discussion of this method will therefore be given in Sect. 8.2, so that we can refrain here from giving further experimental details.

For completeness, let us mention that polarized atoms have also been obtained by optical pumping with circularly polarized light, though the object of these experiments was not the production of polarized electrons by subsequent photoionization. In this method, of which an example is given in Sect. 5.5.2, the spin orientation of the photons in the circularly polarized pumping light is transferred to the atoms. The technique has been repeatedly used to produce polarized atoms for studies of spin-dependent electron-atom scattering of the type discussed in the preceding chapter [5.5–8].

5.2 The Fano Effect and Its Consequences

Polarized electrons can be produced by photoionization of unpolarized atoms with circularly, linearly, or even unpolarized light. The photoelectron polarization is caused by spin-orbit interaction in the continuum state or in the bound atomic states. Polarization measurements in conjunction with cross-section measurements may be utilized to determine completely the matrix elements characterizing the photoionization process.

5.2.1 Theory of the Fano Effect

The obvious idea that polarized electrons could be obtained by photoionization of polarized atoms was put forward long ago. That the same goal could be achieved with less effort by starting with unpolarized atoms and using circularly polarized light is more difficult to see and was first recognized by *Fano* [5.9] in 1969.

We describe the Fano effect with the aid of Fig. 5.2, which gives the relevant energy levels of an alkali atom.[2] The unpolarized atomic beam is a mixture of equal numbers of atoms A↑ and A↓ with spins parallel and antiparallel to the quantization axis, which we assume to be given by the direction of light propagation. This means that the levels $m_j = m_s = +\frac{1}{2}$ and $-\frac{1}{2}$ of the ground state $^2S_{1/2}$ are equally populated.

The transitions caused by the incident light lead to P states because of the selection rule $\Delta l = \pm 1$. For alkali atoms, the P states have the total angular momenta $j = \frac{1}{2}, \frac{3}{2}$. Radiation of a wavelength short enough for ionization leads to transitions into the continuous $P_{1/2}$ and $P_{3/2}$ states adjoining the bound $P_{1/2}$ and $P_{3/2}$ states at the ionization threshold. If the unpolarized atomic beam is ionized

---

[2]   It has been shown [5.9] that the influence of hyperfine interaction on the polarization is negligible.

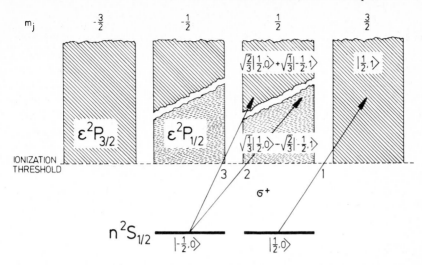

Fig. 5.2. Level diagram for the discussion of the photoionization of alkali atoms. The angular-momentum properties of the states are characterized by combinations of the kets $|m_s, m_l\rangle$

by circularly polarized σ^+ light one has the additional selection rule $\Delta m_j = +1$. One then obtains the transitions 1, 2, and 3 shown in Fig. 5.2.

The final-state angular momenta, which are our main interest, can be directly seen from the spin and angular parts of the wave functions. For the final state reached via transition 1 this part is

$$\binom{1}{0} Y_{1,1}(\theta, \phi) \tag{5.1}$$

(see Problem 5.1), if $Y_{lm}(\theta, \phi)$ denote the spherical harmonics. From the spin function $\binom{1}{0}$ it can be seen that the spin component in the z direction (quantization axis) is $+\frac{1}{2}$; the eigenvalues of $Y_{1,1}$ are given by $l = m_l = 1$. We therefore have abbreviated this state in Fig. 5.2 as $|m_s, m_l\rangle = |\frac{1}{2}, 1\rangle$. Its vector-model representation is given in Fig. 5.3a.

The spin directions of the states reached via transitions 2 and 3 can no longer be simply described by a single quantum number. The state $m_j = \frac{1}{2}$, for example, can be realized by $m_s = \frac{1}{2}$, $m_l = 0$ or by $m_s = -\frac{1}{2}$, $m_l = +1$. But neither of the corresponding eigenfunctions of j_z, $\binom{1}{0} Y_{1,0}(\theta, \phi)$ and $\binom{0}{1} Y_{1,1}(\theta, \phi)$, is simultaneously an eigenfunction of j^2, which, owing to the conservation of total angular momentum, is required for a realistic wave function. Using the Clebsch-Gordan coefficients, the correct eigenfunctions can, however, be constructed as linear combinations of these two parts. For $^2P_{3/2}(m_j = \frac{1}{2})$ one finds

$$\sqrt{\frac{2}{3}} \binom{1}{0} Y_{1,0}(\theta, \phi) + \sqrt{\frac{1}{3}} \binom{0}{1} Y_{1,1}(\theta, \phi). \tag{5.2}$$

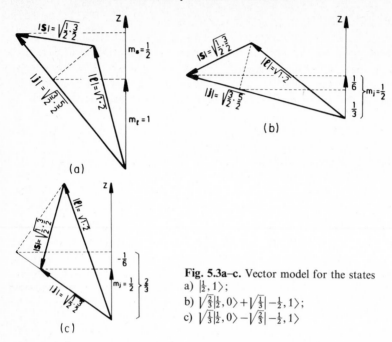

Fig. 5.3a–c. Vector model for the states
a) $|\frac{1}{2},1\rangle$;
b) $\sqrt{\frac{2}{3}}|\frac{1}{2},0\rangle+\sqrt{\frac{1}{3}}|-\frac{1}{2},1\rangle$;
c) $\sqrt{\frac{1}{3}}|\frac{1}{2},0\rangle-\sqrt{\frac{2}{3}}|-\frac{1}{2},1\rangle$

The reader who is not familiar with Clebsch-Gordan coefficients can easily check by direct calculation that (5.2) is an eigenfunction of j^2 and j_z with the eigenvalues $j(j+1)=\frac{3}{2}\cdot\frac{5}{2}$ and $m_j=\frac{1}{2}$ (see Problem 5.1).

According to the quantum mechanical interpretation of the expansion of a wave function, a measurement on the state (5.2) yields the spin $+\frac{1}{2}$ with probability $\frac{2}{3}|Y_{1,0}(\theta,\phi)|^2$ and the spin $-\frac{1}{2}$ with probability $\frac{1}{3}|Y_{1,1}(\theta,\phi)|^2$. If we are not concerned with the angular distribution of the photoelectrons (which is described by the spherical harmonics) but instead integrate over the solid angle, we obtain, since the Y_{lm} are normalized, $\frac{2}{3}\cdot\frac{1}{2}+\frac{1}{3}\cdot(-\frac{1}{2})=\frac{1}{6}$ for the expectation value of the spin component in the z direction. Correspondingly, the expectation value of the z component of the orbital angular momentum is $\frac{2}{3}\cdot0+\frac{1}{3}\cdot1=\frac{1}{3}$. These values can also be obtained from the vector model, as shown in Fig. 5.3b. (One must not, of course, directly form the projections of l and s in the z direction from the arbitrary positions shown. One must first form the projections in the fixed j direction. The time-averaged values of l and s thus obtained are then projected on the z axis.)

The situation in the state $^2P_{1/2}$ is quite analogous. For $m_j=\frac{1}{2}$ one has the eigenfunction[3]

$$\sqrt{\frac{1}{3}}\begin{pmatrix}1\\0\end{pmatrix}Y_{1,0}-\sqrt{\frac{2}{3}}\begin{pmatrix}0\\1\end{pmatrix}Y_{1,1}. \tag{5.3}$$

---

[3] cf. footnote 4 in Chap. 4.

Its representation in the vector model is given in Fig. 5.3c.

What can be learned about the polarization of the photoelectrons from the eigenfunctions just established? First of all we see that for transition 1 (Fig. 5.2) no change occurs in the spin direction since m_s is $+\frac{1}{2}$ in both the initial and final states. For transitions 2 and 3, however, which start from $m_s = -\frac{1}{2}$, there is a definite possibility of a spin flip to $m_s = +\frac{1}{2}$. To determine the fractions of the photoelectrons with $m_s = +\frac{1}{2}$ and $-\frac{1}{2}$ and thus the electron polarization, we must know the probabilities with which the various transitions occur. They are determined by the dipole matrix elements.

To calculate the matrix elements one needs the complete wave functions including their radial parts which are not known exactly. We denote the radial parts by $F(r)$, $F_1(r)$ and $F_3(r)$ which refer to the ground state, $^2P_{1/2}$ state, and $^2P_{3/2}$ state, respectively. $F_1(r)$ is generally different from $F_3(r)$ since the radial parts of the Hamiltonians that result in the $P_{1/2}$ or $P_{3/2}$ states differ in the sign of the spin-orbit coupling potential $(\frac{1}{2}m^{-2}c^{-2})(1/r)(dV/dr)(\mathbf{l}\cdot\mathbf{s})$, the scalar product $\mathbf{l}\cdot\mathbf{s}$ being negative for $j=\frac{1}{2}$ and positive for $j=\frac{3}{2}$ (see Figs. 5.3b and 5.3c, and Sect. 3.2). Since the radial parts of the Hamiltonians differ, one obtains different radial eigenfunctions. The difference between $F_1(r)$ and $F_3(r)$ increases with increasing spin-orbit coupling, i.e., with increasing atomic number.

Using the abbreviations $\alpha = \begin{pmatrix} 1 \\ 0 \end{pmatrix}$, $\beta = \begin{pmatrix} 0 \\ 1 \end{pmatrix}$, $R_{1,3} = \langle F_{1,3}(r)|r|F(r)\rangle$, and the dipole operator $x+iy$ of circularly polarized σ^+ light, the matrix elements for transitions 1, 2, 3 in Fig. 5.2 are

$$b_1 = \langle \varepsilon\,^2P_{3/2}, m_j=\tfrac{3}{2}|x+iy|\,^2S_{1/2}, m_j=\tfrac{1}{2}\rangle$$
$$= \langle F_3(r)\alpha Y_{1,1}|x+iy|F(r)\alpha Y_{0,0}\rangle = -\sqrt{\tfrac{2}{3}}\,R_3, \tag{5.4}$$

$$b_2 = \langle \varepsilon\,^2P_{1/2}, m_j=\tfrac{1}{2}|x+iy|\,^2S_{1/2}, m_j=-\tfrac{1}{2}\rangle$$
$$= \langle F_1(r)(\sqrt{\tfrac{1}{3}}\,\alpha Y_{1,0} - \sqrt{\tfrac{2}{3}}\,\beta Y_{1,1})|x+iy|F(r)\beta Y_{0,0}\rangle$$
$$= +\tfrac{2}{3}R_1, \tag{5.5}$$

$$b_3 = \langle \varepsilon\,^2P_{3/2}, m_j=\tfrac{1}{2}|x+iy|\,^2S_{1/2}, m_j=-\tfrac{1}{2}\rangle$$
$$= \langle F_3(r)(\sqrt{\tfrac{2}{3}}\,\alpha Y_{1,0} + \sqrt{\tfrac{1}{3}}\,\beta Y_{1,1})|x+iy|F(r)\beta Y_{0,0}\rangle$$
$$= -\tfrac{\sqrt{2}}{3}R_3. \tag{5.6}$$

The integrations carried out above are very simple as only the lowest spherical harmonics (which are merely sine and cosine functions) occur. In addition, the orthogonality of the spin functions α and β has been used (see Problem 5.2).

The wave function of the electrons that have made the transitions 2 and 3 starting from the same level is given by the coherent superposition

$$b_2 \cdot |\varepsilon\,^2P_{1/2}, m_j=\tfrac{1}{2}\rangle + b_3 \cdot |\varepsilon\,^2P_{3/2}, m_j=\tfrac{1}{2}\rangle \tag{5.7}$$

(common factors which cancel out in the calculation of the polarization, such as the intensity of the ionizing light, have been omitted). For transition 1, which describes the ionization of a different atom of our incoherent mixture of $A\uparrow$ and $A\downarrow$, the corresponding expression is

$$b_1 \cdot |\varepsilon\,^2P_{3/2}, m_j = \tfrac{3}{2}\rangle. \tag{5.8}$$

By substituting (5.1–6) into (5.7 and 8) and rearranging according to spin functions, we obtain

$$-\frac{1}{3}\sqrt{\frac{2}{3}}\,[\sqrt{2}(R_3 - R_1)\,Y_{1,0}\alpha + (2R_1 + R_3)\,Y_{1,1}\beta]$$

$$= -\frac{1}{3}\sqrt{\frac{2}{3}}\begin{pmatrix}\sqrt{2}(R_3 - R_1)\,Y_{1,0} \\ (2R_1 + R_3)\,Y_{1,1}\end{pmatrix} \tag{5.9}$$

in the case of transitions 2 and 3 for the part of the wave function which determines the angular momenta. In the case of transition 1 we obtain

$$-\sqrt{\frac{2}{3}}\begin{pmatrix}R_3\,Y_{1,1} \\ 0\end{pmatrix}. \tag{5.10}$$

According to Sect. 2.3, the density matrices of the final states are, from (5.9 and 10),

$$\varrho_{2+3} = C\begin{pmatrix}2(R_3 - R_1)^2|Y_{1,0}|^2 & \sqrt{2}(R_3 - R_1)(2R_1 + R_3)\,Y_{1,0}\,Y_{1,1}^* \\ \sqrt{2}(R_3 - R_1)(2R_1 + R_3)\,Y_{1,0}^*\,Y_{1,1} & (2R_1 + R_3)^2|Y_{1,1}|^2\end{pmatrix}, \tag{5.11}$$

$$\varrho_1 = C\begin{pmatrix}9\,R_3^2|Y_{1,1}|^2 & 0 \\ 0 & 0\end{pmatrix}, \tag{5.12}$$

where common factors that are not important in our considerations have been expressed by the constant C.

These density matrices can be used to calculate the electron polarization that arises in the photoionization of the $A\downarrow$ and $A\uparrow$ beams. To calculate the polarization of the electron mixture which arises from the photoionization of the unpolarized atomic beam, we form the density matrix of the mixed state, which according to (2.22) is the sum of the matrices (5.11 and 12):

$$\varrho = C\begin{pmatrix}9\,R_3^2|Y_{1,1}|^2 + 2(R_3 - R_1)^2|Y_{1,0}|^2 & \sqrt{2}(R_3 - R_1)(2R_1 + R_3)\,Y_{1,0}\,Y_{1,1}^* \\ \sqrt{2}(R_3 - R_1)(2R_1 + R_3)\,Y_{1,0}^*\,Y_{1,1} & (2R_1 + R_3)^2|Y_{1,1}|^2\end{pmatrix}. \tag{5.13}$$

Since from (2.21) one has $P_i = \mathrm{tr}\{\varrho\sigma_i\}/\mathrm{tr}\{\varrho\}$ for the components of the polarization vector, one obtains, if one chooses the z component as an example, the angle-dependent expression

$$P_z = \frac{9\,R_3^2\,|Y_{1,1}|^2 + 2(R_3 - R_1)^2\,|Y_{1,0}|^2 - (2\,R_1 + R_3)^2\,|Y_{1,1}|^2}{9\,R_3^2\,|Y_{1,1}|^2 + 2(R_3 - R_1)^2\,|Y_{1,0}|^2 + (2\,R_1 + R_3)^2\,|Y_{1,1}|^2}. \tag{5.14}$$

Using the rearrangement made in Problem 5.3, one obtains

$$\begin{aligned}
P_z &= \frac{\tfrac{9}{2}\,R_3^2 \sin^2\theta + 2(R_3 - R_1)^2 \cos^2\theta - \tfrac{1}{2}(2R_1 + R_3)^2 \sin^2\theta}{2(R_3 - R_1)^2 + (6\,R_1 R_3 + 3\,R_3^2)\sin^2\theta} \\[2mm]
&= \frac{2(R_3 - R_1)^2 + 2(R_3^2 + R_1 R_3 - 2\,R_1^2)\sin^2\theta}{2(R_3 - R_1)^2 + (6\,R_1 R_3 + 3\,R_3^2)\sin^2\theta}. \tag{5.15}
\end{aligned}$$

In this subsection we will deal with the case where all photoelectrons are collected regardless of their direction of emission. We are therefore interested in the polarization $\bar{P}$ averaged over all angles. In this averaging, the polarization values must be weighted with the corresponding intensities I. According to Problem 5.3, $\bar{P}_x$ and $\bar{P}_y$ vanish. Thus

$$\bar{P} = \bar{P}_z = \frac{\iint I P_z \sin\theta\, d\theta\, d\phi}{\iint I \sin\theta\, d\theta\, d\phi}. \tag{5.16}$$

From the form $P_z = (N_\uparrow - N_\downarrow)/(N_\uparrow + N_\downarrow) = \mathrm{tr}\{\varrho\sigma_z\}/\mathrm{tr}\{\varrho\}$ of the polarization formula it can be seen that $\mathrm{tr}\{\varrho\}$ is proportional to the intensity of the photoelectrons. Therefore, with (5.14 and 16), and using the fact that the spherical harmonics are normalized, one obtains

$$\bar{P}_z = \frac{9\,R_3^2 + 2(R_3 - R_1)^2 - (2\,R_1 + R_3)^2}{9\,R_3^2 + 2(R_3 - R_1)^2 + (2\,R_1 + R_3)^2} = \frac{1 + 2\,X}{2 + X^2} \quad \text{with} \tag{5.17}$$

$$X = \frac{2\,R_3 + R_1}{R_3 - R_1}. \tag{5.18}$$

Problem 5.1. Show that $\begin{pmatrix} 1 \\ 0 \end{pmatrix} Y_{1,1}$ is an eigenfunction of j^2 and j_z with the respective eigenvalues $j(j+1)\hbar^2 = 3/2 \cdot (5/2)\hbar^2$ and $m_j\hbar = (3/2)\hbar$. Show the analogous relations for the state $|{}^2 P_{3/2}, m_j = \tfrac{1}{2}\rangle$.

Solution.   Using the abbreviations $\alpha = \begin{pmatrix} 1 \\ 0 \end{pmatrix}$, $\beta = \begin{pmatrix} 0 \\ 1 \end{pmatrix}$ we have

$$j^2 \alpha Y_{1,1} = (l^2 + s^2 + 2l \cdot s)\alpha Y_{1,1} = [(2 + \tfrac{3}{4})\hbar^2 + 2l \cdot s]\alpha Y_{1,1}.$$

With the relations following from (2.2),

$$\sigma_x \alpha = \beta, \quad \sigma_y \alpha = i\beta, \quad \sigma_z \alpha = \alpha, \quad \sigma_x \beta = \alpha, \quad \sigma_y \beta = -i\alpha, \quad \sigma_z \beta = -\beta,$$

one has

$$2l \cdot s\alpha Y_{1,1} = \hbar(l_x \sigma_x + l_y \sigma_y + l_z \sigma_z)\alpha Y_{1,1} = \hbar(l_x \beta + il_y \beta + \hbar\alpha)\,Y_{1,1}.$$

From the following relations which can be found in textbooks on quantum mechanics,

$$(l_x+il_y)\,Y_{lm}=\hbar\sqrt{(l-m)(l+m+1)}\;Y_{lm+1}\quad{}^4$$
$$(l_x-il_y)\,Y_{lm}=\hbar\sqrt{(l+m)(l-m+1)}\;Y_{lm-1},$$

(5.19)

one has $(l_x+il_y)\,Y_{1,1}=0$. Thus it follows that

$$j^2\alpha Y_{1,1}=(2+\tfrac{3}{4}+1)\hbar^2\alpha Y_{1,1}=\tfrac{3}{2}\cdot\tfrac{5}{2}\,\hbar^2\alpha Y_{1,1}.$$

In addition one has

$$j_z\alpha Y_{1,1}=(l_z+s_z)\alpha Y_{1,1}=\left(\hbar+\frac{\hbar}{2}\right)\alpha Y_{1,1}=\tfrac{3}{2}\hbar\alpha Y_{1,1}.$$

For $|{}^2P_{3/2},m_j=\tfrac{1}{2}\rangle$ it follows that

$$j^2\sqrt{\tfrac{1}{3}}\,(\sqrt{2}\,\alpha Y_{1,0}+\beta Y_{1,1})$$
$$=[l^2+s^2+\hbar(l_x\sigma_x+l_y\sigma_y+l_z\sigma_z)]\sqrt{\tfrac{1}{3}}\,(\sqrt{2}\,\alpha Y_{1,0}+\beta Y_{1,1})$$
$$=\hbar^2(2+\tfrac{3}{4})\sqrt{\tfrac{1}{3}}\,(\sqrt{2}\,\alpha Y_{1,0}+\beta Y_{1,1})+\hbar(l_x\beta+il_y\beta)\sqrt{\tfrac{2}{3}}\,Y_{1,0}$$
$$+\hbar(l_x\alpha-il_y\alpha-\hbar\beta)\sqrt{\tfrac{1}{3}}\,Y_{1,1}.$$

From the last two terms, together with (5.19), one has

$$\frac{2}{\sqrt{3}}\,\hbar^2\beta Y_{1,1}+\sqrt{\tfrac{2}{3}}\,\hbar^2\alpha Y_{1,0}-\sqrt{\tfrac{1}{3}}\,\hbar^2\beta Y_{1,1}.$$

Thus one obtains

$$j^2\sqrt{\tfrac{1}{3}}\,(\sqrt{2}\,\alpha Y_{1,0}+\beta Y_{1,1})$$
$$=\hbar^2(2+\tfrac{3}{4})\sqrt{\tfrac{1}{3}}\,(\sqrt{2}\,\alpha Y_{1,0}+\beta Y_{1,1})+\sqrt{\tfrac{1}{3}}\,\hbar^2\beta Y_{1,1}+\sqrt{\tfrac{2}{3}}\,\hbar^2\alpha Y_{1,0}$$
$$=\tfrac{3}{2}\cdot\tfrac{5}{2}\hbar^2\sqrt{\tfrac{1}{3}}\,(\sqrt{2}\,\alpha Y_{1,0}+\beta Y_{1,1}).$$

In addition one has

$$j_z\sqrt{\tfrac{1}{3}}\,(\sqrt{2}\,\alpha Y_{1,0}+\beta Y_{1,1})=\sqrt{\tfrac{1}{3}}\left[\sqrt{2}\left(0+\frac{\hbar}{2}\right)\alpha Y_{1,0}+\left(\hbar-\frac{\hbar}{2}\right)\beta Y_{1,1}\right]$$

$$=\frac{\hbar}{2}\sqrt{\tfrac{1}{3}}\,(\sqrt{2}\,\alpha Y_{1,0}+\beta Y_{1,1}).$$

Problem 5.2.   Calculate the value of b_2 given in (5.5).

Solution.   Since $x=r\sin\theta\cos\phi$, $y=r\sin\theta\sin\phi$, and $\langle\alpha|\beta\rangle=0$, $\langle\beta|\beta\rangle=1$, it follows that

$$b_2=-\langle F_1(r)\sqrt{\tfrac{2}{3}}\,Y_{1,1}|r\sin\theta\,e^{i\phi}|F(r)\,Y_{0,0}\rangle$$
$$=\sqrt{\frac{2}{3}}\,R_1\left(\sqrt{\frac{3}{8\pi}}\right)\int_0^\pi\int_0^{2\pi}\sin\theta e^{-i\phi}\sin\theta e^{i\phi}\sqrt{\frac{1}{4\pi}}\sin\theta\,d\theta\,d\phi$$
$$=\frac{2\pi}{4\pi}\,R_1\int_0^\pi\sin^3\theta\,d\theta=\tfrac{2}{3}R_1.$$

---

[4]   Variations in the sign in various publications arise from differing definitions of Y_{lm}. We use the notation in which $Y_{l,-m}=(-1)^m Y_{lm}^*$.

Problem 5.3. Calculate P_x and show that the average values $\bar{P}_x$ and $\bar{P}_y$ vanish $\left(\text{use } Y_{1,0} = Y_{1,0}^*\right.$

$= \sqrt{\dfrac{3}{4\pi}}\, \cos\theta,\ Y_{1,1} = -\sqrt{\dfrac{3}{8\pi}}\, \sin\theta\, e^{i\phi}\bigg).$

Solution. With ϱ from (5.13) we obtain

$$P_x = \frac{\operatorname{tr}\left\{\varrho \cdot \begin{pmatrix} 0 & 1 \\ 1 & 0 \end{pmatrix}\right\}}{\operatorname{tr}\{\varrho\}}$$

$$= \frac{\sqrt{2}(R_3 - R_1)(2R_1 + R_3)(Y_{1,0}Y_{1,1}^* + Y_{1,0}^* Y_{1,1})}{9 R_3^2 |Y_{1,1}|^2 + 2(R_3 - R_1)^2 |Y_{1,0}|^2 + (2R_1 + R_3)^2 |Y_{1,1}|^2}$$

$$= \frac{\sqrt{2}(R_3 - R_1)(2R_1 + R_3) Y_{1,0}(Y_{1,1}^* + Y_{1,1})}{2(R_3 - R_1)^2 |Y_{1,0}|^2 + (4R_1^2 + 4R_1 R_3 + 10 R_3^2)|Y_{1,1}|^2}$$

$$= -\frac{(R_3 - R_1)(2R_1 + R_3)\cos\theta\,\sin\theta\, 2\cos\phi}{2(R_3 - R_1)^2(1 - \sin^2\theta) + (2R_1^2 + 2R_1 R_3 + 5R_3^2)\sin^2\theta}$$

$$= -\frac{2(R_3 - R_1)(2R_1 + R_3)\cos\theta\,\sin\theta\,\cos\phi}{2(R_3 - R_1)^2 + (6R_1 R_3 + 3R_3^2)\sin^2\theta}. \qquad (5.20)$$

It can immediately be seen from the first term that

$$\iint IP_x \sin\theta\, d\theta\, d\phi = \sqrt{2}(R_3 - R_1)(2R_1 + R_3) \iint (Y_{1,0}Y_{1,1}^* + Y_{1,0}^* Y_{1,1})\sin\theta\, d\theta\, d\phi$$

vanishes since the spherical harmonics are orthogonal. For $\bar{P}_y$ it is unnecessary to make the analogous calculation: Since in the arrangement discussed we have rotational symmetry, the x and y directions are equivalent. Because $\bar{P}_x$ is 0, $\bar{P}_y$ must also be 0.

5.2.2 Illustration of the Fano Effect. Experimental Results

According to the preceding formulae, the attainable polarization is determined by the radial matrix elements R_1 and R_3. It can be seen immediately that for $R_1 = R_3$ (vanishing spin-orbit coupling) the polarization vanishes. It is therefore not obvious that polarized photoelectrons arise from ionization with circularly polarized light (i.e., spin-oriented photons). The occurrence of electron polarization cannot simply be inferred from the balance of angular momentum: With vanishing spin-orbit coupling, the photon spin is transferred to the orbital angular momentum of the photoelectron, as described by the selection rule $\Delta m_l = \pm 1$ for circularly polarized light. Only if there is an interaction between spin and orbital angular momentum of the photoelectron can the preferential orientation of the photon spins lead to a preferential orientation of the electron spins.

To avoid getting lost in the calculations we shall try to illustrate how the Fano effect arises. First we shall consider the electron polarization which arises from photoionization of $A\downarrow$, i.e., from transitions 2 and 3 in Fig. 5.2. We must then start from ϱ_{2+3} [(5.11)] and instead of (5.17) which refers to photoionization of an unpolarized atomic beam we obtain

$$\bar{P}_z = \frac{2(R_3 - R_1)^2 - (2R_1 + R_3)^2}{2(R_3 - R_1)^2 + (2R_1 + R_3)^2} = \frac{N_\uparrow - N_\downarrow}{N_\uparrow + N_\downarrow} = \frac{Q_e^\uparrow - Q_e^\downarrow}{Q_e^\uparrow + Q_e^\downarrow}. \tag{5.21}$$

$Q_e^\uparrow$ and $Q_e^\downarrow$ are the cross sections for photoproduction of $e\uparrow$ and $e\downarrow$ for the target being considered. If the spin-orbit coupling vanishes $(R_1 = R_3)$ one obtains $\bar{P}_z = -1$. This is to be expected since the atomic beam is totally polarized in the $-z$ direction and the electron spin is not affected by the photoionization if there is no spin-orbit interaction. If $R_3 \neq R_1$ we see from (5.21) that $Q_e^\uparrow \neq 0$. This means that some of the spins will flip during photoionization and reduce the degree of polarization arising in these transitions.

The polarization of the photoelectrons produced by transition 1 can be calculated from (5.12):

$$\bar{P}_z = \frac{Q_e^\uparrow - Q_e^\downarrow}{Q_e^\uparrow + Q_e^\downarrow} = \frac{9R_3^2 - 0}{9R_3^2 + 0} = 1. \tag{5.22}$$

In this case no spin flips occur: $Q_e^\downarrow$ is always zero.

Thus we see that the spin-orbit interaction has two effects:
a) it may cause spin flips in transitions 2 and 3;
b) it leads to differing cross sections for the photoionization of $A\uparrow$ and $A\downarrow$:

$$\frac{Q(A\downarrow)}{Q(A\uparrow)} = \frac{2(R_3 - R_1)^2 + (2R_1 + R_3)^2}{9R_3^2} \tag{5.23}$$

$[Q(A\downarrow)/Q(A\uparrow)$ is the ratio of the denominators in (5.21) and (5.22), which refer to the photoionization of $A\downarrow$ and $A\uparrow$, respectively]. With vanishing spin-orbit coupling this ratio is 1.

The resulting polarization is determined by these two effects of spin-orbit coupling. It is quite easy to see that the average polarization $\bar{P}_z$ is not necessarily parallel to the spin direction of the incident photons. If, for example, $R_1 = 4R_3$ then from (5.23) $Q(A\downarrow)/Q(A\uparrow) = 11$ and from (5.21)

$$Q_e^\uparrow / Q_e^\downarrow = 2(R_3 - R_1)^2 / (2R_1 + R_3)^2 = 2/9.$$

This means that the probability for the $A\downarrow$ to be ionized is 11 times greater than that for the $A\uparrow$, and that most of the spin directions are retained. Therefore a negative polarization arises as also follows from (5.17): $\bar{P}_z = -\frac{1}{2}$.

If, on the other hand, $R_3 = -2R_1$, then (5.21) yields $Q_e^\downarrow = 0$. This means that in the ionization of the $A\downarrow$, all spins must flip into the $+z$ direction. We then obtain only photoelectrons with spins parallel to the z direction. Correspondingly, (5.17) yields $\bar{P}_z = 1$. This can also be explained as follows. In forming the expression (5.9) from $|\varepsilon^2 P_{1/2}, m_j = \frac{1}{2}\rangle$ and $|\varepsilon^2 P_{3/2}, m_j = \frac{1}{2}\rangle$, terms containing the spin-down functions cancel due to interference, if $R_3 = -2R_1$. In this exceptional case, $m_s = +\frac{1}{2}$ and $m_l = 0$ are good quantum numbers for the state resulting from transitions 2 and 3.

Since spin-orbit coupling is only a small interaction, such large polarization effects cannot usually be expected. Normally the polarization does not differ

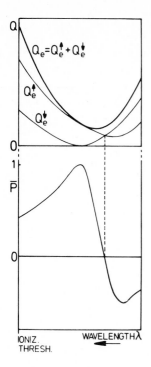

Fig. 5.4. Qualitative diagram of the photoionization cross section and the photoelectron polarization as a function of wavelength for alkali atoms ($Q_e^\downarrow = 0$ for $R_3 = -2R_1$)

significantly from 0, the value without spin-orbit coupling. The circumstances which favor high degrees of polarization can be illustrated by the following picture which in fact led to Fano's considerations. The photoionization cross sections, together with the transition matrix elements – and thus also the polarization of the photoelectrons – depend on the wavelength of the incident light. For most alkali atoms this dependence is particularly pronounced in an experimentally convenient region: the photoionization cross section $Q_e = Q_e^\uparrow + Q_e^\downarrow$ passes through a deep minimum near the threshold. Due to the spin-orbit interaction we obtain slightly different cross sections $Q_e^\uparrow$ and $Q_e^\downarrow$ for the photoproduction of e↑ and e↓ (see Fig. 5.4) since one obtains from (5.17)

$$\frac{Q_e^\uparrow}{Q_e^\downarrow} = \frac{9 R_3^2 + 2(R_3 - R_1)^2}{(2 R_1 + R_3)^2}. \tag{5.24}$$

Although the small spin-orbit interaction does not generally lead to large differences between the two cross sections, their ratios near the minima are quite considerable.[5] If one uses wavelengths near the minima, one obtains electrons in

---

[5] The strong deviation of the ratio R_1/R_3 from 1 in the numerical example above can only occur near zeros of R_1 and R_3 because the absolute difference between these quantities is very small due to the small spin-orbit coupling. Since the photoionization cross sections are determined by R_1 and R_3, the minima of the cross sections and the zeros of R_1 and R_3 lie in the same wavelength range.

predominantly one spin state, i.e., a high degree of polarization. The situation is therefore quite analogous to that in electron scattering, which was discussed in Sect. 3.4.2: the shape of the polarization curves is determined by the cross-section curves for producing e↑ and e↓.

To be able to quantitatively determine the curves given in Fig. 5.4, R_1 and R_3 must be known. It is, however, difficult to calculate the radial parts of the wave functions and thus the transition matrix elements with enough precision that even the difference $R_3 - R_1$ occurring in the formulae is still reliable. Fano therefore tried to evaluate the wavelength dependence $\bar{P}_z(\lambda)$ by making use of the fact that the spin-orbit interaction responsible for the polarization effect also has other consequences which had been studied earlier. Apart from the doublet splitting of the alkali energy levels, spin-orbit interaction determines the intensity ratio of the doublet lines and the exact shape of the photoionization cross section Q_e (even if the spin direction of the photoelectrons is not taken into account). From such data he estimated the parameter X – connected with R_1 and R_3 by (5.18) – as a function of the wavelength λ. He predicted approximately the results that were obtained by the measurements we will now discuss.

Figure 5.5 is a schematic diagram of an experimental setup for measuring the Fano effect. An unpolarized cesium-vapor beam is crossed by monochromatic circularly polarized uv light. The photoelectrons produced are collected by an extraction system, irrespective of their direction of emission. The subsequent measurement of their polarization which is carried out with a Mott detector (see Sect. 8.1.2) thus yields the average value $\bar{P} = \bar{P}_z$. The task of suppressing the numerous background electrons that come from the chamber walls or other parts of the apparatus is accomplished by suitable electron-optical potential barriers. In addition, it must be ensured, through the choice of a suitable oven temperature, that the portion of the Cs_2 molecules in the Cs beam is small; the high photoionization cross section of these dimers must not give rise to an appreciable number of unwanted photoelectrons.

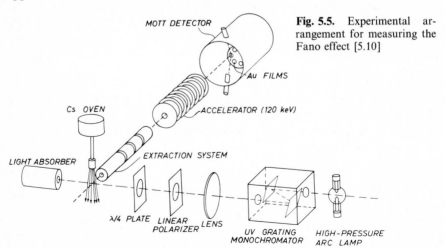

Fig. 5.5. Experimental arrangement for measuring the Fano effect [5.10]

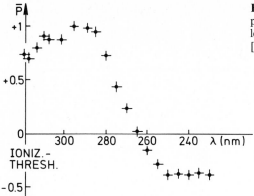

Fig. 5.6. Experimental values of the photoelectron polarization vs. wavelength for the Fano effect with cesium [5.10]

Figure 5.6 shows the results of these measurements which were made in the interesting wavelength region where the cross-section minimum occurs. It can be seen that at 290 nm total polarization is achieved within the limits of experimental accuracy. This makes the Fano effect of interest as a source of polarized electrons. In addition, such measurements yield information on the parameter $X(\lambda)$ from (5.18) and thus on the radial matrix elements. In this way one obtains more accurate information than was previously available on the influence of spin-orbit coupling on the aforementioned properties of the alkali atoms [5.11–15].

Although the wavelength dependence of the photoelectron polarization has not been measured for alkali atoms other than Cs (for solid alkalis, see Sect. 7.2), it can easily be drawn from an equivalent experiment which has been made with K, Rb, and Cs. A↑ and A↓ have been separately photoionized by circularly polarized light [5.11]. The ratio of the photoionization cross sections $Q(A\uparrow)/Q(A\downarrow)$ is given by (5.23). Comparison with (5.24) shows that the information obtained from a measurement of $Q(A\uparrow)/Q(A\downarrow)$ is equivalent to that obtained from measuring the polarization $\bar{P}=(Q_e^{\uparrow}-Q_e^{\downarrow})/(Q_e^{\uparrow}+Q_e^{\downarrow})=[(Q_e^{\uparrow}/Q_e^{\downarrow})-1]/[(Q_e^{\uparrow}/Q_e^{\downarrow})+1]$. We can therefore say that there is good quantitative knowledge of the Fano effect, not only for Cs, but also for Rb and K (see also [5.16]).

One of the consequences of the discovery of the Fano effect was a strong stimulation of further work on polarization effects in photoionization. Not long after the first surprise, Fano's discovery, came the second surprise, when it turned out that it is not even necessary to use circularly polarized light in order to eject polarized electrons from unpolarized atoms. This will be discussed in the following subsection.

Problem 5.4. Derive from the results of Sect. 5.2.1 the angular dependence of the photoionization cross section

$$\frac{d\sigma}{d\Omega}=\frac{Q}{4\pi}\left[1-\frac{\beta}{2}\left(\frac{3}{2}\cos^2\theta-\frac{1}{2}\right)\right]\tag{5.25}$$

for circularly polarized or unpolarized incident light. Q is the total cross section, β a parameter which has to be determined.

Solution. The intensity $I(\theta)$ of the photoelectrons produced by σ light is proportional to $\mathrm{tr}\{\varrho\}$, where ϱ is the density matrix (5.13) describing the photoelectrons. Using the expressions for the spherical harmonics given in Problem 5.3 we obtain

$$\frac{d\sigma}{d\Omega} \propto I(\theta) \propto \mathrm{tr}\{\varrho\} = R_1^2 + R_1 R_3 + \tfrac{5}{2} R_3^2 - \tfrac{3}{2}(2 R_1 R_3 + R_3^2)\cos^2\theta$$

$$= (R_1^2 + 2 R_3^2)\left[1 - \frac{1}{2}\frac{4 R_1 R_3 + 2 R_3^2}{R_1^2 + 2 R_3^2}\frac{3}{2}\cos^2\theta + \frac{R_1 R_3 + R_3^2/2}{R_1^2 + 2 R_3^2}\right].$$

Setting

$$\beta = \frac{4 R_1 R_3 + 2 R_3^2}{R_1^2 + 2 R_3^2} = 2\frac{X^2-1}{X^2+2}$$

with X from (5.18) and determining the constant of proportionality by the condition

$$2\pi \int_0^\pi \frac{d\sigma}{d\Omega} \sin\theta\, d\theta = Q,$$

we obtain (5.25). For reasons of mirror symmetry the result holds both for σ^+ and σ^- light. It is therefore valid also for unpolarized light.

5.2.3 Polarized Electrons Ejected from Unpolarized Atoms by Unpolarized Light

When unpolarized electrons are scattered from an unpolarized target one obtains a polarization of the scattered electrons which is perpendicular to the scattering plane and disappears when averaged over all azimuthal directions. This has been discussed in previous chapters and provokes the question whether the situation in photoionization might be analogous. It has in fact been predicted that photoelectrons produced by unpolarized light from unpolarized targets and ejected into a well-defined direction may have spin polarization [5.17–19]. From the symmetry arguments of Sect. 3.4.4 it follows that the polarization must be perpendicular to the reaction plane, which is the plane defined by the direction of the incident light and the direction of the photoelectrons observed. Axial symmetry then requires the polarization averaged over the directions of emission to be zero (see Fig. 5.7).

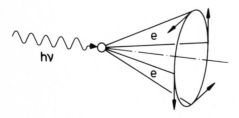

Fig. 5.7. Polarized photoelectrons produced by shining unpolarized light on unpolarized atoms

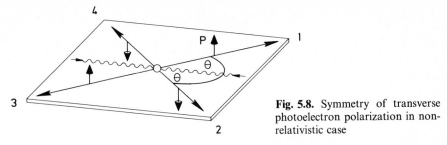

Fig. 5.8. Symmetry of transverse photoelectron polarization in non-relativistic case

Symmetry arguments also give some insight into the angular dependence of the polarization. Axial symmetry demands that, as in electron scattering, electrons emerging in directions 1 and 2 of Fig. 5.8 have opposite polarizations. In discussing the θ dependence of the polarization we consider the (non-relativistic) limit where the photon momentum can be neglected in comparison with the photoelectron momentum so that it does not play a role for the polarizations observed, for instance, in directions 1 and 2 whether the photons come from the right or from the left. It follows then from rotation of the right-hand side of Fig. 5.8 through 180° that the polarizations in directions 3 and 4 are opposite to those in directions 2 and 1, respectively. We have $P_n(\pi - \theta) = -P_n(\theta)$ where $\boldsymbol{P} \equiv P_n \hat{\boldsymbol{n}}$ with $\hat{\boldsymbol{n}} = \boldsymbol{k}_i \times \boldsymbol{k}/|\boldsymbol{k}_i \times \boldsymbol{k}|$ ($\boldsymbol{k}_i$ and $\boldsymbol{k}$ give the directions of the incident light and of the photoelectron which form an angle θ). This symmetry is reflected by the numerator in the polarization formula

$$P(\theta) = \frac{\xi \sin 2\theta}{1 - \frac{\beta}{2}\left(\frac{3}{2}\cos^2\theta - \frac{1}{2}\right)} \hat{\boldsymbol{n}}. \tag{5.26}$$

The denominator is, apart from a constant factor, the well-known expression for the total photoelectron intensity emerging into the direction of observation (cf. Problem 5.4). This is in accordance with the fact discussed previously [e. g., in connection with (5.16)] that the denominator of the polarization formula gives the total number of e↑ and e↓.

While the symmetry of the angular distribution follows from general principles, the parameters ξ and β are determined by the matrix elements of the particular photoionization process considered. In order to derive the polarization by means of the density matrix formalism, calculations along the line of Sect. 5.2.1 have to be repeated for σ^- light. Adding incoherently the density matrices obtained for σ^+ and σ^- light and using (2.18) one obtains the photoelectron polarization (5.26) for unpolarized incident light were ξ is proportional to

$$\frac{|R_1||R_2|\sin(\delta_1 - \delta_2)}{|R_1|^2 + |R_2|^2}. \tag{5.27}$$

$|R_i|$ and δ_i ($i = 1, 2$) are the moduli and phases of the (reduced) matrix elements describing the transitions to the two continua which we have assumed here to be allowed by the selection rules. In the case of alkali atoms, the phases of the two matrix elements differ only due to the influence of the spin-orbit interaction which is not strong enough to produce a significant deviation of $\sin(\delta_1 - \delta_2)$ from zero [5.14, 15], so that no appreciable polarization appears.[6] If, on the other hand, a state is photoionized where a p electron is ejected, the electron can reach S and D states. Consequently, we have in (5.27) the phase difference $\delta_S - \delta_D$ which is usually much larger since it is caused by the Coulomb force which is much stronger than the weak spin-orbit interaction in the continuum responsible for the polarization effects with alkalis.

Numerical values for ξ are shown in Fig. 5.9. The two curves hold for photoionization of the p shell of argon atoms when the residual argon ion is either in the $^2P_{1/2}$ or in the $^2P_{3/2}$ state. The electrons corresponding to these two states of the residual ion differ in energy by 180 meV and were separated experimentally by an electrostatic spectrometer. Such experiments could only be performed with highly developed experimental techniques since the photoelectrons have to be analyzed simultaneously for energy, angle of emission, and spin. The results of Fig. 5.9 have been obtained by observing the polarization of the photoelectrons ejected at the magic angle $\theta_m = 54°44'$. At this angle the denominator of (5.26) is 1 so that ξ follows directly from $P_n = \xi \sin 2\theta_m$. Measurements of this type have also been made for other noble gases and metal vapors like mercury [5.23, 24].

An arrangement which allows one to observe the complete angular dependence $P_n(\theta)$ is shown in Fig. 5.10. It takes advantage of the fact that there is a close connection between the polarizations of photoelectrons produced by unpolarized and by linearly polarized light. As shown in Problem 5.5, the photoelectron polarization $\boldsymbol{P}^{(l)}$ produced by linearly polarized light is consistent with the polarization (5.26) for unpolarized light if it is given by

$$\boldsymbol{P}^{(l)}(\theta) = \frac{2\xi \sin 2\theta}{1 + \beta(\tfrac{3}{2}\cos^2\theta - \tfrac{1}{2})}\,\hat{\boldsymbol{n}} = P_n^{(l)}(\theta)\hat{\boldsymbol{n}} \tag{5.28}$$

where θ is now the angle between the light polarization vector $\boldsymbol{E}$ and the photoelectron direction $\boldsymbol{k}$, and $\hat{\boldsymbol{n}}$ is given by $\hat{\boldsymbol{n}} = \boldsymbol{k} \times \boldsymbol{E}/\|\boldsymbol{k} \times \boldsymbol{E}\|$. For the measurement of $P_n^{(l)}(\theta)$ the light polarization vector $\boldsymbol{E}$ is rotated about the axis of the light beam by rotating the vuv polarizer as indicated in Fig. 5.10. Thus the angle θ is varied while the axis of the photoelectron polarization remains perpendicular to the scattering plane of the Mott detector used for polarization analysis.

The polarization of the vuv radiation is produced by reflecting the incident beam at three gold coatings, which produces a maximum polarization of 88%. The apparatus includes several parts which together reduce the electron intensity arriving at the counters of the Mott detector by a factor of 10^{-7}: light polarizer

---

[6] When discussing the Fano effect for alkali atoms we assumed both matrix elements to be real so that the phase difference was exactly 0 (or π if one of the matrix elements is negative), which turned out to be a very good approximation.

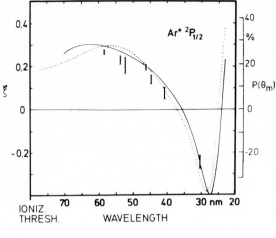

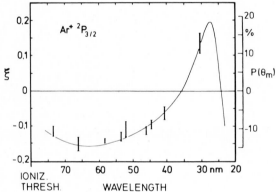

Fig. 5.9. Transverse polarization P_n of photoelectrons ejected from argon atoms by unpolarized radiation (in accordance with common practice we frequently drop the index of a polarization component when no other components exist). Angle of ejection $\theta_m = 54°44'$. The scales on the left hold for ξ defined by (5.26). Experimental data from [5.20]; theoretical values by *Cherepkov* [5.21] (– – – –) and *Huang* et al. [5.22] (———)

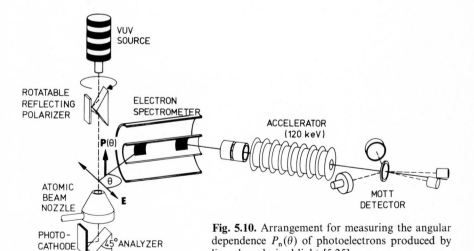

Fig. 5.10. Arrangement for measuring the angular dependence $P_n(\theta)$ of photoelectrons produced by linearly polarized light [5.25]

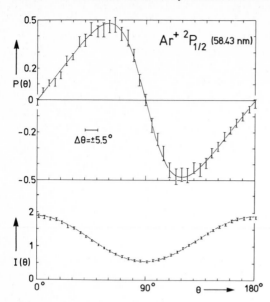

Fig. 5.11. Angular distribution of photoelectron polarization $P_n^{(l)}(\theta)$ (upper curve) and intensity (lower curve, arbitrary units) for the process $\mathrm{Ar} + h\nu(21.22\ \mathrm{eV}) \rightarrow \mathrm{Ar}^+(^2P_{1/2}) + \mathrm{e}$. (——) least squares fit yielding ξ and β of (5.28) [5.25]

(reduction of the light intensity to 4%), electron spectrometer (accepted angle $\Delta\theta = \pm 5.5°$, energy resolution 0.7%), Mott detector (reduction 1/1000). This is why the measurements were possible only with the strongest helium resonance line (58.43 nm $\hat{=}$ 21.22 eV), which was produced in a capillary discharge tube. With a primary intensity of 10^{13} photons/s typical count rates in the Mott detector were 2 to 30 counts per second. Instrumental asymmetries were eliminated by taking advantage of the reversal of the electron polarization which follows from mirror symmetry if the light polarizer is rotated from θ to $360° - \theta$.

The experimental results of Fig. 5.11 are found for argon when the residual ion is left in the $^2P_{1/2}$ state. The data are normalized to a light polarization of 100%. The parameters ξ and β of (5.28) yielding the best fit to the experimental results of Fig. 5.11 are $\xi = 0.25 \pm 0.01$, $\beta = 0.89 \pm 0.04$.

From the results so far presented in this chapter one can see that the polarization phenomena in photoionization may be produced by spin-orbit interaction in the bound state and/or in the continuum. For the polarization given in the last few figures, spin-orbit splitting of the argon p shell is essential. The opposite sign of the data for the $^2P_{1/2}$ and $^2P_{3/2}$ states in Fig. 5.9 results in a polarization $P \approx 0$ if the photoelectrons from these states are not separated: With the respective polarizations $P^{(1/2)}$ and $P^{(3/2)}$ from the $p_{1/2}$ and $p_{3/2}$ shells, one has

$$P = \frac{\sigma_{1/2} P^{(1/2)} + \sigma_{3/2} P^{(3/2)}}{\sigma_{1/2} + \sigma_{3/2}} = 0,$$

provided the cross-section ratio $\sigma_{3/2}/\sigma_{1/2}$ ("branching ratio") is equal to its statistical value 2 (4 electrons in $p_{3/2}$ shell, 2 electrons in $p_{1/2}$ shell; photo-ionization cross section of a shell proportional to number of electrons in this shell). Deviations from zero may be caused by spin-orbit interaction in the continuum, particularly in heavier elements [5.23]. In cases where fine-structure splitting does not play a role, such as photoionization of s shells, spin-orbit coupling in the continuum is the only mechanism for producing photoelectron polarization. Examples are the Fano effect in alkalis and the polarization observed in photoionization of the Hg $6s^2$ subshell by unpolarized light [5.24].

Let us conclude this section by pointing out the analogy between the results presented here and our findings in inelastic electron scattering, where we also saw spin-orbit interaction in bound atomic states and spin-orbit coupling in the continuum state of the unbound electron to be independent causes of electron polarization.

Problem 5.5. Show that the photoelectron polarization (5.26) produced by unpolarized light follows from the photoelectron polarization (5.28) produced by linearly polarized light.

Solution. As indicated in Fig. 5.12, the unpolarized light can be considered to be an incoherent mixture of linearly polarized light with the polarization vectors E_1 parallel and E_2 normal to the plane defined by k_i and k. Denoting the intensities and the polarizations of the photoelectrons produced by these two components by I_1, I_2 and $P^{(l1)}$, $P^{(l2)}$, respectively, we have for the resulting polarization of the photoelectrons produced by the unpolarized light according to (2.16)

$$P = \frac{I_1}{I_1 + I_2} P^{(l1)} + \frac{I_2}{I_1 + I_2} P^{(l2)}.$$

The last term disappears according to (5.28) since in this case the angle θ between the light polarization and the direction of the photoelectrons is $90°$ so that $P^{(l2)} \propto \sin 2\theta = 0$. As mentioned repeatedly, the denominators of the polarization formulae are proportional to the numbers of electrons observed, so that from (5.28) we have $I_1(\theta_1) = 1 + \beta(3/2 \cos^2 \theta_1 - 1/2)$, $I_2(90°) = 1 - \beta/2$. Consequently,

$$P = \frac{2 \xi \sin 2\theta_1}{2 + \beta(\frac{3}{2} \cos^2 \theta_1 - 1)} \hat{n} \quad \text{with}$$

$$\hat{n} = \frac{k \times E_1}{|k \times E_1|} = \frac{k_i \times k}{|k_i \times k|}.$$

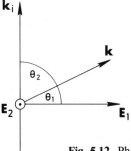

Fig. 5.12. Photoelectrons produced by unpolarized light (k_i and k refer to incident photon and photoelectron, respectively)

Introducing $\theta_2 = \pi/2 - \theta_1$, the angle between the directions of the light and the photoelectrons, we have $\sin 2\theta_1 = \sin 2\theta_2$ and $\cos^2 \theta_1 = 1 - \cos^2 \theta_2$.

Thus

$$P = \frac{\xi \sin 2\theta_2}{1 - \frac{\beta}{2}\left(\frac{3}{2}\cos^2 \theta_2 - \frac{1}{2}\right)} \hat{n}.$$

Dropping the index of the angle we obtain (5.26). In Fig. 5.12 we have arbitrarily drawn the light polarization vector E_1 to the right. The result is the same if E_1 is reversed, since both $\hat{n}$ and the sine function change sign then.

5.2.4 The "Perfect" Photoionization Experiment

When dealing with Mott scattering or with exchange scattering of electrons, we have discussed the measurements necessary for a "perfect" or complete experiment (cf. Sects. 3.3.3 and 4.3). Such an experiment was defined to yield the complete set of parameters (moduli and phases of the complex amplitudes) describing the process under consideration. Let us now discuss the measurements necessary for that purpose in photoionization.

It can easily be seen that, as a result of the selection rules for dipole radiation, photoionization cannot lead to more than three continuum states. If, for instance, a state with quantum numbers l and $j = l + \frac{1}{2}$ is photoionized, the selection rules $\Delta l = \pm 1$, $\Delta j = 0, \pm 1$ allow only the three final states with $l' = l - 1$, $j' = l - \frac{1}{2}$, and $l' = l + 1$, $j' = l + \frac{1}{2}$, $l + \frac{3}{2}$ to be reached. The process is therefore described by three matrix elements whose radial parts will be denoted by $R_k = |R_k| e^{i\delta_k}$ with three values for k. Unlike the angular parts, which are determined by the spherical harmonics, the radial parts are not well known. Since, according to the principles of quantum mechanics, an absolute phase determination cannot be achieved, one of the phases can be chosen arbitrarily. There are thus five independent parameters, three moduli and two phases, which must be determined in order to describe the photoionization process completely.

Which are the observable quantities from which these five parameters can be obtained? The extension of our previous density matrix calculations for two continuum states to the general case of three continuum states is straightforward. It shows that the following quantities may be observed when unpolarized atoms are photoionized and that they depend, in general, on the matrix elements indicated:

$Q(|R_k|^2)$, total cross section,

$\beta(|R_k|^2, |R_k||R_l| \cos(\delta_k - \delta_l))$, asymmetry parameter of angular distribution,

$\bar{P}(|R_k|^2, |R_{1/2}||R_{3/2}| \cos(\delta_{1/2} - \delta_{3/2}))$, average photoelectron polarization for incident σ light; the indices $\frac{1}{2}$ and $\frac{3}{2}$ refer to the spin-orbit-split continua $j' = l + \frac{1}{2}$, $l + \frac{3}{2}$ (interference terms of $l' = l - 1$ and $l' = l + 1$ disappear by averaging in view of the orthogonality of the wave functions)[7],

---

[7] The reason why, in the discussion of the Fano effect with alkali atoms, phase-dependent terms do not appear is given in the preceding footnote.

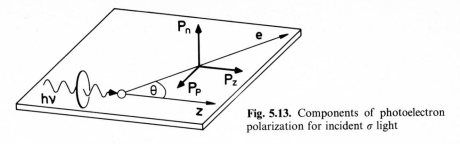

Fig. 5.13. Components of photoelectron polarization for incident σ light

$\xi(|R_k|^2, |R_k||R_l| \sin(\delta_k - \delta_l))$, which specifies the angular distribution (5.26) of the polarization normal to the reaction plane, and

$\gamma(|R_k|^2, |R_k||R_l| \cos(\delta_k - \delta_l))$, which specifies, for incident σ light, the angular distribution of the polarization component P_z ($z =$ direction of photon spins) according to

$$P_z(\theta) = \frac{\bar{P} - \gamma \cdot \left(\frac{3}{2} \cos^2 \theta - \frac{1}{2}\right)}{1 - \frac{\beta}{2}\left(\frac{3}{2} \cos^2 \theta - \frac{1}{2}\right)}. \tag{5.29}$$

For unpolarized target atoms, γ also specifies the transverse photoelectron polarization parallel to the reaction plane for incident σ light (see Fig. 5.13):

$$P_p(\theta) = \frac{3}{2} \frac{\gamma \sin \theta \cos \theta}{1 - \frac{\beta}{2}\left(\frac{3}{2} \cos^2 \theta - \frac{1}{2}\right)} \hat{e}_p, \tag{5.30}$$

with $\hat{e}_p = e_z \times \hat{n}$, where $\hat{e}_z$ is the unit vector along the z direction and $\hat{n}$ has been defined in the preceding subsection.

At first glance, it looks quite reasonable to have the five independent measurable quantities Q, β, $\bar{P}$, ξ, and γ for determining the aforementioned five independent parameters (moduli and phases) of the matrix elements. But the measurements do not yield the phase differences directly. Instead, it is $\sin(\delta_k - \delta_l)$ and $\cos(\delta_k - \delta_l)$ which derive from the measurable quantities. For an unambiguous determination of a phase difference neither the sine nor the cosine alone suffices; both are required. This means that, for one phase difference, two measurements are required. They are not independent, but they are necessary. The situation is as in electron scattering (cf. Sects. 3.3.3 and 4.3) where we found that, for the same reason, four measurements were needed in order to determine three independent parameters. Since in photoionization two independent phases have to be determined, four measurements are required for ascertaining the phases. Consequently, a total of seven instead of five measurable quantities are needed to determine the five parameters of the matrix elements describing the photoionization process.

From the results obtained hitherto, it might seem doubtful whether there exist additional quantities that could be measured. So far, however, we have only discussed the photoionization of unpolarized atoms. The additional information required may be obtained by photoionization of polarized atoms [5.26–28]. Let us briefly indicate the reasons for this.

First, $P_z(\theta)$ and $P_p(\theta)$ – (5.29, 30) – are no longer described by one and the same parameter γ if polarized targets are used. Instead, $P_p(\theta)$ is characterized by another quantity ξ_p, a function of $|R_k|^2$ and $|R_k||R_l|\cos(\delta_k - \delta_l)$ which depends on the argument in a different way than γ. Furthermore, the four angle-dependent quantities – (5.26, 29, 30), plus the intensity distribution – do not merely depend on $(3/2)\cos^2\theta - 1/2 = P_2(\cos\theta)$ or $\sin 2\theta = (2/3)P_2^{(1)}$ as in the case of unpolarized targets, but contain additional Legendre polynomials $P_{2l}(\cos\theta)$, $P_{2l}^{(1)}(\cos\theta)$, with factors which again depend on the matrix elements we want to determine. Besides, the ratio $Q(\pi)/Q(\sigma)$ of the cross sections for photoionization by σ and by π light yields another independent relation for evaluating the $|R_k|^2$. We will return to the latter point in Sect. 5.4.

This shows that there are even more measurable quantities than is necessary for a complete determination of the complex matrix elements if photoionization of polarized atoms is taken into consideration. The redundancy desirable for such difficult measurements is thus granted. The various experimental possibilities differ, however, in intricacy. In particular, experiments requiring polarized vuv radiation are difficult to perform since linear polarizers in the vuv are inefficient and vuv circular polarizers are even worse. It has been the construction of intense sources of (polarized) synchrotron radiation which now makes possible the measurement of even such unhandy parameters as γ [5.29].

Experimental determination of the matrix elements has so far been focused on cases where less than the entire set of observables is needed. This occurs [5.30, 31] if the number of matrix elements is reduced either by making simplifying theoretical assumptions or by studying photoionization processes that need only two matrix elements for their description; for example, transitions from a $P_{1/2}$ initial state lead, due to the selection rules, only to $S_{1/2}$ and $D_{3/2}$ continuum states. Examples will also be found in the following sections [5.32, 33]. In addition to the theoretical data mentioned earlier, numerical evaluations of the polarization parameters have been made for some noble gases [5.34], metal vapors [5.35], and cesium [5.15].

Let us conclude with a summary of the various possibilities of producing photoelectron polarization. It is given in Table 5.1. For unpolarized and for linearly polarized light the electron polarization has only a component perpendicular to the reaction plane. The other components and the polarization averaged over the directions of the photoelectrons disappear because otherwise parity conservation or axial symmetry would be violated (cf. Sect. 3.4.4). Only for circularly polarized incident light is it compatible with parity conservation that all the electron polarization components are different from zero. The average polarization has, for reasons of axial symmetry, a component only along the direction of the light propagation.

Table 5.1. Polarization of photoelectrons ejected from unpolarized atoms for different polarizations of incident light. The parameters given characterize the different components

Light polarization	Electron polarization			
	P_z	P_n	P_p	$\bar{P}$
unpol.	—	ζ	—	—
π	—	ζ	—	—
σ	γ	ζ	γ	$\bar{P}_z$

The polarization mechanisms which are effective in photoionization of atoms work also in molecules though the calculation of quantitative theoretical results is more difficult [5.36]. Experimental results which have so far been obtained with unoriented molecules yield a photoelectron polarization much smaller than with atoms [5.37].

The discussion of this chapter, as briefly summarized in Table 5.1, reveals clearly that spin polarization of photoelectrons is not at all exceptional, though it has been believed for a long time that it occurs only as an effect of higher order in forbidden transitions [5.38] or in the relativistic region where the photon energy is comparable with the electron rest mass (cf. Sect. 6.1). Theoretical and experimental developments of the past few years have shown that spin polarization of photoelectrons is quite a common phenomenon. No matter whether circularly, linearly, or unpolarized light is used for photoionization, the photoelectrons are generally polarized.

5.3 Ionizing Transitions of Excited Atomic States

The polarization of photoelectrons may be substantially influenced by autoionization resonances. The resonance behavior of the spin polarization is discussed using the example of thallium atoms. Significant resonance structure in the polarization curves has been found for several atoms and has been utilized for classification of the resonances. Auger electrons are polarized not only if ejected from polarized atomic shells or produced by polarized projectiles, but also if the initial hole state is aligned and the final-state angular momentum is nonzero.

5.3.1 Autoionizing Transitions

The discussion of the preceding section has been restricted to direct photo-ionization processes. At certain wavelengths, autoionizing transitions play a dominant part by causing resonances in the photoionization cross section. Such autoionizing resonances are also found in the polarization curve as will be seen in the following.

Let us first briefly recall the mechanism of autoionization. When an outer electron of the atom is excited beyond the ionization limit, its energy levels are no

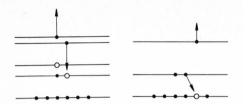

Fig. 5.14. Possibilities of autoionizing transitions. (*Left part*) after excitation of two outer electrons; (*right part*) after excitation of one inner electron

longer discrete, but it reaches the energy continuum as illustrated in Fig. 5.2. It is, however, equally possible to excite the atom to *discrete* energies exceeding this ionization limit. This can occur, for instance, if two outer electrons or one inner electron are excited, as shown schematically in Fig. 5.14. One of the electrons in the outer orbitals is likely to return to the lowest unoccupied energy state while its excess energy, instead of being radiated away, is concentrated on an excited electron. This electron obtains then more than enough energy to leave the atom, so that we have a spontaneous ionization or autoionization. We can also say that there exists a degeneracy between the singly excited continuum and the discrete atomic states above the ionization threshold. If selection rules permit, an atom in such a discrete state can decay into the continuum state of the same energy.

Figure 5.15 illustrates the situation for the thallium atom, where autoionizing resonances of the polarization curve were first studied. Due to the selection rules $\Delta l = \pm 1$, $\Delta j = 0$, ± 1, and $\Delta m_j = +1$ (σ^+ light), the states $\varepsilon^2 S_{1/2}$, $\varepsilon^2 D_{3/2}(m_j = \frac{1}{2})$, and $\varepsilon^2 D_{3/2}(m_j = \frac{3}{2})$ are accessible from the $6\,^2P_{1/2}$ ground states of the unpolarized thallium atoms.

The spin polarization of the photoelectrons in a specific final state follows immediately from the coupling coefficients of the wave functions $|m_s, m_l\rangle$ given in Fig. 5.15. For example, we find for the state $\varepsilon^2 D_{3/2}(m_j = \frac{3}{2})$ the polarization $P = (\frac{1}{5} - \frac{4}{5})/(\frac{1}{5} + \frac{4}{5}) = -0.6$ (see Problem 5.6). Superposition of the spin polarizations of the various final states weighted with the corresponding transition probabilities yields the photoelectron polarization. Evaluation of the transition probabilities similar to Problem 5.2 shows that the two final $\varepsilon^2 D_{3/2}$ states

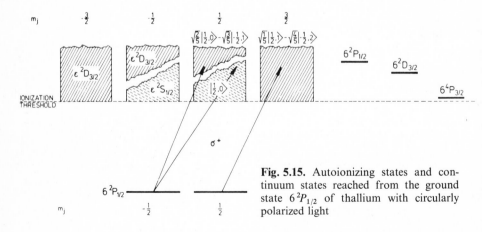

Fig. 5.15. Autoionizing states and continuum states reached from the ground state $6\,^2P_{1/2}$ of thallium with circularly polarized light

together contribute a polarization of -0.5, whereas the $\varepsilon\,^2S_{1/2}$ state has a polarization of $+1$. Because of the different signs of these values the resulting polarization

$$\bar{P}=(1\cdot Q_S-0.5\cdot Q_D)/(Q_S+Q_D) \tag{5.31}$$

would normally not be very large, the exact value depending on the ratio of the cross sections for transitions to the S and D states, Q_S/Q_D, which depends on the wavelength λ.

So far we have not taken into account, however, the autoionizing states of thallium which are also shown in Fig. 5.15. They result from excitation of a $6s$ electron and decay after a short lifetime, so that eventually the same final states are reached via these indirect transitions. The autoionizing state $6\,^2P_{1/2}$ decays into the $\varepsilon\,^2S_{1/2}$ continuum, and $6\,^2D_{3/2}$ and $6\,^4P_{3/2}$ into $\varepsilon\,^2D_{3/2}$, since, due to the conservation of angular momentum, J must be the same before and after the decay. At the excitation wavelengths of the autoionizing states one has a resonance behavior of the photoionization cross section Q as shown in Fig. 5.16 which is based on results of *Marr* and *Heppinstall* [5.39] and of *Berkowitz* and *Chupka* [5.40].

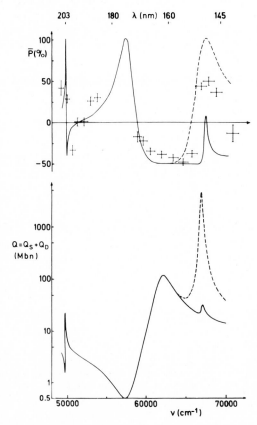

Fig. 5.16. (*Lower part*) photoionization cross section of thallium according to [5.39] (———) and [5.40] (– – – –). (*Upper part*) spin polarization of photoelectrons; experimental results and values calculated using the cross sections from the lower part

Whenever one of the cross sections Q_S and Q_D dominates due to the autoionization resonances, the polarization tends to $+100\%$ or -50%, respectively, according to (5.31). As an example, let us consider the wavelengths where Q_S dominates. The analysis of the cross-section resonances shows that Q_D disappears at approximately 49730 and 57239 cm^{-1}, leaving Q_S alone to contribute to P; in these cases the polarization should be exactly 100%. On the other hand, Q_S dominates near 67137 cm^{-1}. A tendency of the polarization towards 100% is then to be expected. Similarly, the negative peaks of $\bar{P}$ occur at those wavelengths, where Q_D dominates.

Since the cross-section resonances produce polarization resonances one frequently has polarization peaks located at those wavelengths where the photoionization cross section has a maximum. In the other polarization phenomena discussed so far we often found that the polarization peaks were located near cross-section minima, a fact which renders experimental studies difficult.

Our considerations show that the polarization of the photoelectrons can be predicted on the basis of (5.31) if the relative contributions of Q_S and Q_D to the photoionization cross section can be evaluated. The result of a semi-empirical calculation is seen in the upper part of Fig. 5.16, where the solid and broken lines are the polarization curves based on two different experimental cross-section curves. The figure also shows that experimental studies of this polarization phenomenon have to use wavelengths where it was cumbersome before the advent of synchrotron radiation to produce circularly polarized light of sufficient intensity.

In the experiment [5.41] sketched in Fig. 5.17 the uv light produced by a H_2 discharge lamp was circularly polarized by a MgF_2 prism in conjunction with a MgF_2 quarter-wave plate and focused onto a thallium-vapor beam emerging

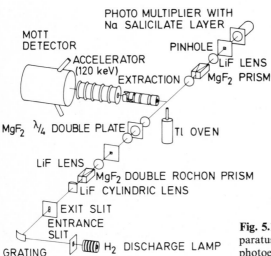

Fig. 5.17. Schematic diagram of the apparatus for measuring the polarization of photoelectrons from thallium [5.41]

from an oven at 1100 °C. The photoelectrons were produced in the middle of an asymmetric quadrupole field (not shown) which ensured that all electrons were extracted regardless of their direction of emission and that electrons produced at the walls were not detected. The electron polarization was measured by a highly efficient Mott detector whose I/I_0 ratio was better than 10^{-3} (see Sect. 8.1.2). It is worth pointing out that for the measurement of $\bar{P}(\lambda)$ the intensity distribution $I(\lambda)$ of the light need not be known. This is due to the fact that measuring the electron polarization means measuring the *ratio* of the intensities scattered into the two detectors in the Mott chamber, so that knowledge of incident intensities is unnecessary.

The deviation of the measured from the predicted polarization near 150 nm emphasized the necessity of remeasuring the photoionization cross section [5.42]. The new cross-section data are in accordance with the polarization given in Fig. 5.16. The observed value $\bar{P} = -50\%$ at 64400 ± 600 cm^{-1} means that at this wave number, Q_S disappears. This special case illustrates the general fact that at each wavelength the polarization measurement determines the ratio of the individual cross sections Q_S and Q_D. On the other hand, measurement of the photoionization cross section yields the sum $Q = Q_S + Q_D$, so that one has two independent experimental results for determining Q_S and Q_D separately. This is another example of the application of spin-polarization studies (for a further discussion see [5.43, 44]). They provide additional information on the autoionization process that is not obtainable from cross-section measurements alone; it can help one to classify and understand the autoionization resonances in cases more complicated than we have discussed here [5.36].

That is why polarization studies in autoionization have also been made for other metals and for noble gases. A most appropriate vuv source for that purpose is synchrotron radiation because it has a continuous spectrum. With the help of a monochromator, the wavelength dependence of the resonance can therefore be scanned. Figure 5.18 gives as an example the rich resonance structure which was found in the polarization curve obtained by photoionization of xenon atoms

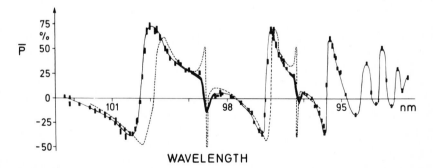

Fig. 5.18. Polarization $\bar{P}$ of photoelectrons ejected by circularly polarized vuv radiation from xenon atoms: (——) experimental results [5.45], (– – – –) theoretical results [5.19]. Further theoretical results in [5.46]

with circularly polarized radiation. All the photoelectrons were extracted regardless of their direction of emission. In these measurements advantage has been taken of another useful property of synchrotron radiation: the high degree of circular polarization of the vuv light emitted along a certain direction. For details of the evaluation of these results, and for the polarization phenomena observed in autoionization of other elements, we refer to some of the original papers [5.30, 32, 47]. It should, however, be emphasized that such resonance features occur not only in the angle-averaged polarization $\bar{P}$, but also in the other polarization parameters presented in Table 5.1. Measurements of a complete set of parameters as discussed in Sect. 5.2.4 have been used to determine the resonance behavior of the complex matrix elements for autoionization of the outer $6s^2$ shell of mercury [5.32].

Problem 5.6. Calculate the spin polarization of the photoelectrons in the state $|\varepsilon\,^2D_{3/2}, m_j = 3/2\rangle$ using the eigenfunctions given in Fig. 5.15.

Solution.   We are interested in all the photoelectrons regardless of their direction of emission. Since the z axis (propagation direction of the incident light) is the only preferential direction, P_x and P_y vanish when averaged over all angles. Hence we find from the definition of P_z as the expectation value of the spin operator σ_z

$$\bar{P}_z = \left\langle \sqrt{\frac{1}{5}}\begin{pmatrix}1\\0\end{pmatrix}Y_{2,1} - \sqrt{\frac{4}{5}}\begin{pmatrix}0\\1\end{pmatrix}Y_{2,2}\middle|\sigma_z\middle|\sqrt{\frac{1}{5}}\begin{pmatrix}1\\0\end{pmatrix}Y_{2,1} - \sqrt{\frac{4}{5}}\begin{pmatrix}0\\1\end{pmatrix}Y_{2,2}\right\rangle$$

$$= \left\langle \sqrt{\frac{1}{5}}\begin{pmatrix}1\\0\end{pmatrix}Y_{2,1} - \sqrt{\frac{4}{5}}\begin{pmatrix}0\\1\end{pmatrix}Y_{2,2}\middle|\sqrt{\frac{1}{5}}\begin{pmatrix}1\\0\end{pmatrix}Y_{2,1} + \sqrt{\frac{4}{5}}\begin{pmatrix}0\\1\end{pmatrix}Y_{2,2}\right\rangle.$$

Since the eigenfunctions are orthonormal, one has

$$\bar{P}_z = \frac{1}{5} - \frac{4}{5} = -0.6.$$

5.3.2 Auger Transitions

Autoionization is found in neutral atoms which have an internal energy high enough to decay by electron emission. If such states of high internal energy are produced not by excitation but by ionization of an inner shell, the subsequent decay is called Auger transition.

It is obvious that Auger transitions from polarized states may result in polarized free electrons. First attempts to utilize such processes for studying magnetic solids by polarized-Auger-electron spectroscopy are promising [5.48–51]. Auger electrons may also become polarized if the incident ionizing projectiles are polarized. This is plausible from the numerous processes treated where polarization of the incident particles is transferred to the decay products: from photons to electrons in photo- and autoionization; from electrons to photons in bremsstrahlung, and in atomic light emission after electron-impact excitation, etc.

We will focus our attention on the less obvious phenomenon in which polarization of Auger electrons can be produced although neither the atom nor the ionizing projectile is polarized [5.52–55]. From the symmetry arguments used in Sect. 5.2.3 we find again that the polarization must be normal to the reaction plane, have a forward-backward asymmetry $P(\pi-\theta)=-P(\theta)$, and disappear if averaged over the directions of emission. The analogies to photoionization go even further. Let us discuss this for the specific case of an initial hole in the L_3 shell ($J=3/2$), assuming that two decay channels (two possible angular momenta for the Auger electrons) exist. The angular dependence of the Auger-electron polarization is then given by (5.26), where $\hat{\boldsymbol{n}}$ is the normal to the reaction plane and ξ is again proportional to (5.27):

$$\xi \propto \frac{|R_1||R_2|\,\sin\,(\delta_1-\delta_2)}{|R_1|^2+|R_2|^2}\,A. \tag{5.32}$$

$|R_k|$ and δ_k are now the moduli and phases of the matrix elements describing the Auger transitions from the initial hole state to the final states where the L_3 hole is filled and an additional shell has been ionized. The parameter A will be discussed later. Again, polarization can only occur if there is interference of (at least) two transitions as described by the factor $\sin\,(\delta_1-\delta_2)$, i.e., more than one partial wave must be allowed by the selection rules for the Auger electrons.

This brings us to the differences between photoelectron and Auger-electron polarization. Auger emission is not governed by the well-known selection rules for dipole transitions since its cause is the Coulomb interaction between the electrons which participate directly in the transition. One of the consequences is that their polarization disappears if the final ion state has zero angular momentum $J_f=0$: since conservation of angular momentum and parity holds, the angular momentum J of the initial inner-shell hole is transferred to the Auger electron and one obtains only a single partial wave, a situation which we have just found to yield zero polarization.

Another difference from photoionization is the appearance of an additional factor A in the parameter ξ. It describes the alignment of the initial state. The atoms which are ionized in an inner shell may be aligned, i.e., the $2J+1$ sublevels $|JM\rangle$ may have unequal populations such that the states $|JM\rangle$ and $|J-M\rangle$ are equally populated. Figure 5.19a illustrates such a situation; the lengths of the vectors representing the states $|JM\rangle$ are assumed to be proportional to the numbers of particles in these states. It is seen that the average angular momentum of the system is zero. For comparison, the figure shows an isotropic system and a polarized system which possesses a net angular-momentum component along the z axis. If $Q(|M|)$ are the excitation cross sections for the magnetic substates, one has for the above-mentioned case of an initial L_3 hole ($J=3/2$)

$$A=\frac{Q(|\tfrac{3}{2}|)-Q(|\tfrac{1}{2}|)}{Q(|\tfrac{3}{2}|)+Q(|\tfrac{1}{2}|)}.$$

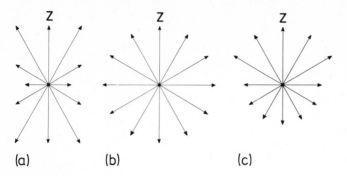

Fig. 5.19a–c. Axially symmetric systems (**a**) aligned, (**b**) isotropic, (**c**) polarized. The arrows represent the angular momentum J precessing around the axis of quantization. The length of an arrow indicates the number of particles in this state

It was explained at the end of Sect. 4.5.3 why, for quantization along the direction of the incident particle, the excitation cross sections for different sublevels differ from each other.

The appearence of the alignment parameter in the polarization formula implies a further restriction on the Auger-electron polarization. It is reduced by this factor and vanishes if $A = 0$. Alignment can only exist if $J > \frac{1}{2}$; otherwise one has only a single value of $|M|$ so that unequal populations of different $|M|$ cannot occur. As a consequence, polarized Auger electrons will not be emitted by decay of a vacancy in the K, L_1, L_2, M_1, M_2, etc., shells where J is too small.

The fact that the alignment appears in the polarization formula can be visualized as follows. We have seen that parity arguments like those of Sect. 3.4.4 demand the polarization to be normal to the reaction plane, which is spanned by the direction of the incident projectile and the direction of the Auger electron. Unlike photoionization, the Auger effect can be considered as a two-step process where, in the first step, the incident projectiles may produce an alignment of the vacancies along the direction of incidence. If such an alignment exists, a memory of the direction of the incident particles is retained in the decay of the vacancies so that there is a well-defined reaction plane in this second step. If there is no alignment, one has no such plane and definition of a polarization normal to a reaction plane is impossible. The decaying system is then isotropic, the incident particles might have come from any direction, and the polarization averaged over all those directions vanishes.

Let us point out again that the results obtained are valid for the case where neither the atoms nor the incident particles are prepolarized. Polarized incident particles may, of course, engender polarization of the vacancy states, (cf. Fig. 5.19c) which may result in Auger-electron polarization in any direction, depending on the incident polarization.

Having emphasized the similarities and differences between photoionization and Auger effect let us finally remark that in both cases the electron polarization is, of course, a result of spin-orbit interaction. In either case significant polarization may occur due to spin-orbit coupling in the continuum. In cases where this interaction is negligible, electrons coming from well-defined fine-structure levels may still have polarization due to spin-orbit interaction within the atoms.

Measurements of Auger-electron polarization should therefore resolve the fine structrue of the Auger lines or they should work with heavy elements. Observance of the conditions $J > \frac{1}{2}$ and $J_f \neq 0$ for the angular momenta of vacancy and final state, respectively, is mandatory provided that neither atoms nor incident projectiles are polarized. As in photoionization, such measurements yield information on the transition matrix elements which complements that obtained from the usual studies of intensity and angular distribution. For completeness we add that (5.26), with ξ from (5.32), gives the angular dependence of Auger-electron polarization only in the common case where the alignment can be described by a single parameter. For a vacancy in an inner shell with $J \geq 2$, $|M|$ has more than two values so that one needs more parameters to describe the alignment. They appear in the polarization formula as factors of the higher order Legendre polynomials [5.52, 53].

Since the polarization of Auger electrons ejected from unpolarized atoms is a new topic, experimental studies are as yet rare. In a first attempt, studies of the transitions $M_4 N_1 N_{2,3}(^3P_2)$, $M_5 N_1 N_{2,3}(^3P_1)$ in Kr and $M_5 N_{4,5} N_{4,5}(^3F_4)$ in Xe showed no significant polarization within the experimental error limits [5.56].

5.4 Multiphoton Ionization

An atom can also be photoionized by absorption of several photons whose total energy is greater than the ionization energy. Cases are discussed where the photoionization takes place via an intermediate state of the atom, i.e., by absorption of a resonance frequency. Numerous possibilities arise for producing polarized electrons. Studies of photoelectron polarization in conjunction with cross-section measurements allow comprehensive knowledge about multiphoton ionization to be obtained.

Until now we have been discussing photoionization that takes place by absorption of one photon. It can, however, also occur by absorption of several photons of lower energy, as long as the sum of the photon energies is greater than the ionization energy. In this case one can work with longer wavelengths and have the advantage of being outside the uv region which, below ca. 180 nm, presents particular experimental difficulties.

Such multiphoton ionization processes may occur by excitation of intermediate resonance levels which are subsequently ionized by photon absorption. However, when the photon frequencies do not match the energies of any intermediate states the processes will also take place, though with lower cross sections [5.57–60]. We will discuss here the first case of resonant multiphoton processes which is the easiest to realize experimentally and which can be described as photoionization of excited states.

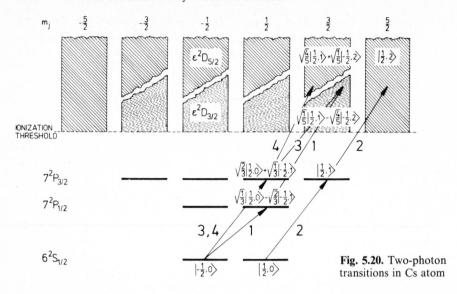

Fig. 5.20. Two-photon transitions in Cs atom

The polarization effects in multiphoton ionization are analogous to those discussed in Sect. 5.2. The first processes that were discussed are those where polarized photoelectrons are produced by circularly polarized light [5.57, 61–63]. To be specific, let us consider two-photon ionization of cesium. A photon of resonance frequency excites an intermediate state of the atom which is then ionized by absorption of a second photon. If we illuminate with the wavelength 459.3 nm, the intermediate state $7\,^2P_{1/2}$ is reached (see Fig. 5.20). If σ^+ light is used, the photoionization can only follow route 1 because of the selection rules $\Delta l = \pm 1$, $\Delta j = 0, \pm 1$, and $\Delta m_j = +1$. There are no transitions starting from the ground state with $m_j = +\frac{1}{2}$, nor are there transitions from the excited state $7\,^2P_{1/2}(m_j=\frac{1}{2})$ into the S states (since $m_j=\frac{3}{2}$ is not possible there) or into the P states of the continuum.

As in Sect. 5.2.1, by using Clebsch-Gordan coefficients or by direct calculation, it can be seen that the final state $\varepsilon\,^2D_{3/2}(m_j=3/2)$ has the angular momentum eigenfunction

$$\sqrt{\frac{1}{5}}\binom{1}{0}Y_{2,1} - \sqrt{\frac{4}{5}}\binom{0}{1}Y_{2,2}.$$

Since we are still interested in all photoelectrons regardless of their direction of emission we calculate their average polarization $\bar{P}_z$ in the direction of the light propagation and obtain $\bar{P}_z = \frac{1}{5} - \frac{4}{5} = -0.6$ (see Problem 5.6). Thus with σ^+ light of wavelength 459.3 nm one obtains, by absorption of two photons, a photoelectron polarization of -60%. We have here a different situation than in the case of the Fano effect. There it is essential that, owing to an appreciable spin-orbit coupling in the continuum, the radial matrix elements R_1 and R_3 depend on *j*. In the present discussion, however, use is made of the energy splitting of the bound states which is likewise caused by spin-orbit coupling. A polarized final state is reached via a polarized intermediate state and dependence of the radial matrix elements on *j* is not necessary for producing polarization of the photoelectrons.

If one uses σ^+ light of the shorter wavelength 455.5 nm one first reaches, by routes 2, 3, and 4, the sublevels $m_j = \frac{1}{2}$ and $\frac{3}{2}$ of the $7\,^2P_{3/2}$ state and finally the continuum states shown. Superposition of the polarizations of the final states, weighted with the corresponding transition probabilities, yields the resulting photoelectron polarization. Accordingly, the polarization obtained depends in this case on the transition probabilities into the various continuum states. If one assumes that the radial matrix elements to the continuum are equal for $j = \frac{3}{2}$ and $\frac{5}{2}$ (thus excluding the circumstances which cause the Fano effect), a simple calculation similar to that in Sect. 5.2.1 yields $\bar{P}_z = \frac{9}{11} \approx 82\,\%$. If the radial matrix elements differ, the polarization is somewhat changed.

Only if spin-orbit coupling were to disappear both in the continuum and in the discrete region (radial matrix elements to the continuum not *j* dependent, vanishing energy splitting in the discrete region) would the polarization vanish, since in that case the photon spins would not be coupled into the electron spin system.

Multiphoton ionization can only be achieved with light sources of high intensities. The ionization of short-lived intermediate states, discussed above, only yields measurable intensities if there are enough atoms in these states. This explains why such experiments have been stimulated by the advent of the laser.

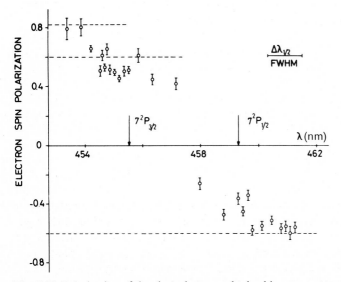

Fig. 5.21. Polarization of the photoelectrons obtained in resonant two-photon ionization of cesium via the $7\,^2P_{1/2,\,3/2}$ intermediate states [5.64]

The result of a polarization experiment [5.64] in two-photon ionization is shown in Fig. 5.21. The polarization has been measured as a function of the wavelength of a tunable dye laser which has been used for excitation as well as for ionization. The bandwidth of the radiation was approximately 1.2 nm, the pulse duration was 1.4 ns with a repetition rate of the pulses up to 100 Hz, and the maximum peak power was roughly 500 W.

The dashed lines in Fig. 5.21 indicate the polarizations -60% and $+82\%$ calculated above for the $^2P_{1/2}$ and $^2P_{3/2}$ intermediate states, respectively. The latter value is, however, to be expected only if the intensity of the exciting light is not too large, so that the populations of the intermediate sublevels are proportional to the transition probabilities into these states. If, at the other extreme, the transition to $7\,^2P_{3/2}$ is saturated (equal rates for excitation and induced emission) the excited sublevels are equally populated, since their populations equal those of the ground-state sublevels. Since the numbers of photoelectrons in the various continuum substates change when the populations of the intermediate states are changed, one obtains a different photoelectron polarization for the case of saturation. The polarization is then $+60\%$ which is also given by a dashed line in Fig. 5.21. When the wavelength of the laser was tuned on the resonance line 455.5 nm the observed polarization was, indeed, almost 60%, the laser power being more than sufficient to saturate the transition. By tuning the laser away from resonance, it was possible to use only the smaller power in the tail of the wavelength distribution for excitation while the full power of the pulse remained available for ionization. In that way a setting could be found where saturation no longer played a role. Under this condition 82% polarization was observed. The polarization obtained by two-photon ionization via $7\,^2P_{1/2}$ is not affected by saturation effects: in this case it is only one intermediate sublevel that is populated so that the final polarization does not depend on the relative populations of two competing sublevels as in the $^2P_{3/2}$ case. The experimental result was close to the theoretical value of -60% for the $7\,^2P_{1/2}$ resonance. When the laser was tuned in between the two resonances, the tails of the wavelength distribution produced photoelectrons with different signs of the polarization $\bar{P}$. This explains the small values for $\bar{P}$ between the resonance lines.

Multiphoton ionization is another example of the fact that it does not suffice to discuss the polarization phenomena "in principle". Instead, the results depend very much on the specific conditions of the experiment. We have seen that, owing to saturation effects, the polarization obtained can depend on the intensity of the exciting light. Another drastic change of the electron polarization may be brought about by the fact that the circular polarization of the light used does not reach the ideal value of 100%. The effect of a small admixture of σ^- light in high-intensity σ^+ light is quite obvious: If the exciting light intensity is much higher than is necessary to saturate the transitions $\Delta m_j = +1$, electrons originating from transitions $\Delta m_j = -1$ may contribute appreciably and reduce the polarization. This is why, near the resonances, the measured values in Fig. 5.21 do not quite reach the theoretical saturation values of $\pm 60\%$. At very high exciting light intensities there will be enough σ^- light available to saturate even the transitions $\Delta m_j = -1$. The subsequent photoionization of the states reached by these unwanted transitions results in a strong reduction of the overall polarization observed. There are still other reasons why the polarization may change in strong radiation fields: As the light intensity is increased beyond a certain limit, the width of the intermediate resonance level is changed as well as its energy [5.57]. The polarization of the photoelectrons is then affected due to the influence of neighboring levels with different polarizations; this has been theoretically discussed and experimentally verified by preliminary results obtained with sodium [5.65]. Besides the intensity, the other parameters of the laser pulse such as rise time and line shape may also affect the photoelectron polarization as has been shown by several calculations both for resonant and for nonresonant multiphoton processes [5.66].

A further reason for observing polarization values other than calculated above can be optical pumping processes of the kind discussed in Sect. 5.5.2. Optical pumping, a consequence of successive absorption and spontaneous emission processes, changes the relative population of the various sublevels. With σ^+ light, the population of the high-m_j sublevels is increased at the expense of the other sublevels which results in a higher photoelectron polarization. In the experiment discussed, optical pumping did not occur because the laser pulse length was much shorter than the natural lifetime of the excited states.

The length of the interaction time between the light and the atom affects the electron polarization in still another way. We have so far treated multiphoton ionization in the fine-structure scheme, not taking into account the coupling between the electron angular momentum and the nuclear spin. The relaxation time τ (or, if one prefers, the precession time) of this hyperfine interaction is usually of the order 10^{-9} to 10^{-8} s. If the ionization process takes place within a time interval $\Delta t \ll \tau$ after excitation, the fine-structure scheme is a proper description because the excited state has no time to relax into the hyperfine states. One has this situation if, e.g., the ionization pulse is very intense or if the same short pulse is used for excitation as well as for ionization. If, on the other hand, one uses a long ionization pulse with a moderate intensity so that it takes on average a time $t \gg \tau$ to ionize the intermediate state, the hyperfine-structure scheme gives a proper description of this state. The precession of its angular momentum caused by the hyperfine interaction results in a lower average orientation of the intermediate state. Its average spin polarization along the quantization axis is thus reduced, so that one also obtains a smaller polarization of the photoelectrons. For a detailed discussion see [5.67, 68].

Polarization experiments in multiphoton ionization allow determination of key parameters governing photoionization of excited states. In particular, the bound-free matrix elements involved in the processes can be found. This is similar to the situation in ground-state photoionization discussed in the preceding sections. Since the matrix elements (and the polarization depending on them) are very sensitive to the choice of the wave functions and the theoretical approximations [5.58], polarization measurements provide a refined test of

Fig. 5.22. Apparatus for measurement of polarization of photoelectrons produced by linearly polarized light in two-photon ionization of cesium. (Cs) cesium beam; (E) electric field direction; (θ) angle between directions of light polarization and observation [5.33]

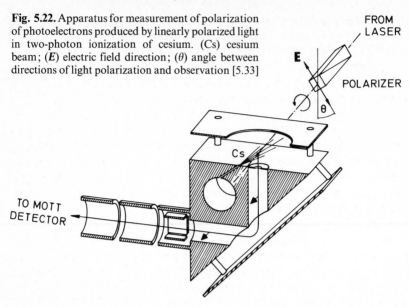

calculational models. Such studies of matrix elements have been made not only for photoionization of cesium ^2P states, but also for cesium ^2D states which have been excited by quadrupole transitions [5.69]. Two-photon ionization which includes such "forbidden" steps can be performed without difficulty by suitable lasers.

Up to now we have discussed photoionization of excited states by means of circularly polarized light. Polarized photoelectrons can, however, also be produced by linearly polarized or even unpolarized light, as we have seen in Sect. 5.2.3, where we considered photoionization of the ground state. We can now transfer these results to photoionization of excited states. Let us pick out the case of two-photon ionization by linearly polarized light, since here the electron polarization has been studied experimentally. From the results of Sect. 5.2.3 we expect the photoelectrons to have a polarization which is perpendicular to the polarization vector of the light and to the direction of observation of the photoelectrons. The polarization was measured with the apparatus shown in Fig. 5.22. An atomic beam of cesium was crossed with the light of a dye laser which produced pulses of 400 ns duration, focused into 1 mm^2. The photoelectrons ejected into a fixed direction were sent into a Mott detector for polarization analysis. The light polarizer was rotated in order to vary the angle θ between the directions of light polarization and of photoelectron observation. Being perpendicular to these two directions, the electron polarization in the arrangement of Fig. 5.22 did not change its direction when θ was varied, so that one always had maximum scattering asymmetry in the Mott detector. From the measured angular dependence of the electron polarization the characteristic parameter ξ of (5.28) was determined.

Applying the discussion of Sect. 5.2.4 to photoionization of the $7\,^2P_{1/2}$ intermediate state of cesium we see that one needs four measurements to completely evaluate the two matrix elements R_S and R_D for the transitions into the $\varepsilon\,^2S_{1/2}$ and the $\varepsilon\,^2D_{3/2}$ continuum state. For determination of the matrix elements advantage has been taken of the fact that one is not restricted to measurements of the quantities Q, β, $\bar{P}$, ξ, and γ of Sect. 5.2.4 if *polarized* atoms are photoionized: With the help of Fig. 5.20 it can be seen that two-photon ionization by σ light via $7\,^2P_{1/2}$ produces a polarized intermediate state from which only a D continuum state can be reached. (The S continuum is not shown in the figure.) The photoionization cross section $Q(\sigma)$ is therefore determined by $|R_D|^2$ alone. Since both S and D continuum states can be reached by two-photon transitions with linearly polarized light (selection rule $\Delta m_j = 0$, z direction = direction of light polarization), $Q(\pi)$ is determined by $|R_S|^2 + |R_D|^2$. From a measurement of the ratio $Q(\pi)/Q(\sigma)$, together with an absolute cross section measurement yielding $|R_S|^2 + |R_D|^2$, one therefore obtains $|R_S|^2$ and $|R_D|^2$ separately. From the measured electron polarization one obtains ξ and thus the expression (5.27) (in which R_1 and R_2 are now the matrix elements R_S and R_D). From the angular dependence of the intensity, which has also been measured in the experiment described, one obtains the asymmetry parameter $\beta(|R_S|^2, |R_D|^2, |R_S||R_D| \cos(\delta_S - \delta_D))$ (cf. Sect. 5.2.4). Having determined $|R_S|$ and $|R_D|$ from the cross sections, one can evaluate the phase difference $\delta_S - \delta_D$ from ξ and β and accomplish the complete determination of the matrix elements. For numerical results, see [5.33].

We have dealt with only a few of the simplest cases for producing polarized electrons by multiphoton ionization. There are abundant possibilities if one varies the number of photons participating and the kinds of atoms used. One can, for example, work [5.61] with atoms which have the ground state $n\,^2P_{1/2}$ (B, Al, Ga, In, Tl) and illuminate with σ^+ light of the wavelength that excites the intermediate state $(n+1)\,^2S_{1/2}$ (see Fig. 5.23). Due to the aforementioned selection rules for σ^+ light, the two-photon transition then follows a specific path into the continuum state $\varepsilon\,^2P_{3/2}(m_j = \tfrac{3}{2})$ in which all electrons have the spin quantum number $m_s = +\tfrac{1}{2}$. The photoelectrons are thus totally polarized.

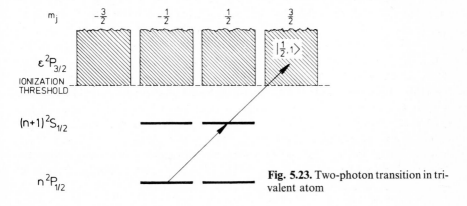

Fig. 5.23. Two-photon transition in trivalent atom

By playing around with energy levels of various atoms, selection rules, the light polarization, and the number of photons participating, the reader can easily find innumerable other transitions which yield polarized electrons in multiphoton ionization.

5.5 Collisional Ionization of Polarized Atoms

By collisional ionization of polarized atoms, polarized free electrons can be produced. The processes discussed in detail are the collision of polarized deuterium atoms with H_2 or He and Penning ionization of polarized helium atoms in a helium discharge. Measurement of the electron polarization can be used as a diagnostic tool for the analysis of collision processes.

5.5.1 Collisional Ionization of Polarized Metastable Deuterium Atoms

If one has polarized atoms one can eject their oriented electrons by ionization and so produce polarized free electrons. This can be done by photoionization, as discussed in Sect. 5.1, but it can also be done by collisional ionization. Useful degrees of polarization are obtained only if it is mainly the oriented electrons that are ejected in the collisions; ionization of states with nonoriented electrons must be avoided.

In the method [5.70] now to be discussed, polarized deuterium atoms in the state $2\,{}^2S_{1/2}$ are used. This state is metastable because radiative transitions into the ground state $1\,{}^2S_{1/2}$ are forbidden by the selection rule $\Delta l = \pm 1$. The state lies only 3.4 eV below the ionization threshold and is much easier to ionize than the ground state whose ionization requires about 13.6 eV. When the metastable deuterium atoms collide with H_2 molecules or He atoms, which themselves have high ionization energies, the metastable deuterium atoms are preferentially ionized.

The details of the experiment are given in Fig. 5.24. A deuteron beam which passes through Cs vapor picks up electrons by charge exchange. At deuteron energies from 500 to 1000 eV, about 25 % of the neutral D atoms emerging from D^+-Cs collisions are in the metastable state $2\,{}^2S_{1/2}$. The rest are in states that decay immediately to the deuterium ground state. The remaining charged particles that leave the Cs cell are removed from the beam by a weak electric deflector field.

To polarize the metastable deuterium atoms the "level-crossing" technique is used which takes advantage of the Zeeman splitting in an external magnetic field [5.71]. The $m_j = -\frac{1}{2}$ sublevel of the $2\,{}^2S_{1/2}$ state and the $m_j = \frac{1}{2}$ sublevel of the $2\,{}^2P_{1/2}$ state cross in a magnetic field of 57.5 mT (see Fig. 5.25). If the states overlap, a weak electrical perturbation field causes strong mixing of their wave functions. Consequently, the $2\,{}^2S_{1/2}$ state obtains a large admixture of the $2\,{}^2P_{1/2}$ wave function so that the $2\,{}^2S_{1/2}(m_j = -1/2)$ atoms can rapidly decay into the ground state by an optically allowed transition. Accordingly, the arrangement

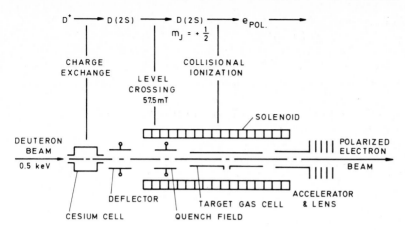

Fig. 5.24. Production and collisional ionization of polarized metastable deuterium atoms [5.70]

shown in Fig. 5.24 yields polarized metastable D atoms. (During the short time needed by the metastables to traverse the field, the weak perturbation field does not suffice to depopulate the $2^2S_{1/2}(m_j= +1/2)$ state.) The static magnetic field that gives rise to the Zeeman splitting and simultaneously decouples the electron spins from the nuclear spins may, in principle, be used as the perturbation to bring about the mixture between the S and P wave functions, since it causes a motional electric field $E=(v/c)\times B$. In the present experiment an additional weak electrostatic field ("quench field") is used to ensure that the unwanted states are completely depopulated before the atoms reach the ionization cell.

The collisional ionization of the metastable D atoms whose spins are oriented parallel to the magnetic field takes place in a gas cell filled with H_2 or He. (To photoionize the atoms would be ineffective here. The photoionization cross section is small, and the D atoms of a few hundred eV have high velocities so that their density in the ionization area is low.) The electrons ejected by collisional ionization are collected by an extraction field. After conversion of their

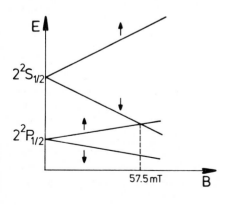

Fig. 5.25. $2^2S_{1/2}$ and $2^2P_{1/2}$ energy levels of atomic deuterium as a function of the magnetic field

longitudinal polarization into transverse polarization (see Sect. 8.1.1) the polarization is measured with a Mott detector. The maximum value found was 33 %.

Since the ionization cross section of the metastable state is about 15 times larger than that of the ground state and since the ratio of polarized metastable atoms to atoms in the ground state is about 1/8, the polarization ought to be considerably higher than 33 % (about 60 %) if the polarized bound electrons were directly knocked out during the collision. The fact that the measurement gave P ≤ 33 % indicates that the collisional ionization predominantly proceeds as follows: When the polarized D atoms pass through the target gas cell, first D^- ions are formed; they subsequently eject one of their electrons by autoionization. The spins of the electrons that are picked up by the polarized atoms from the unpolarized target gas can be parallel or antiparallel to those of the polarized atoms. Since in the autoionization process there is no difference between the attached electrons and the previously oriented atomic electrons, the observed electron polarization is reduced.

The experimental result thus leads to the conclusion that the collisional ionization of the metastable D atoms does not take place directly, but instead occurs predominantly by autoionization. This is an example of how polarization measurements can yield essential information on atomic collision processes.

5.5.2 Penning Ionization

We will now give another example of the diagnostic possibilities arising from electron-polarization studies. It has been shown by such investigations [5.72] that the free electrons in a weak He discharge are mainly produced by collisions between metastable He atoms (He*) and not by ionization of the atoms through electron collisions, as one might suppose.

Figure 5.26 is a schematic diagram of the experiment. A weak rf discharge produces a steady-state population of the metastable helium states. Because of their long lifetime (fraction of a millisecond) the metastable states are populated to a much higher extent than the other excited states. The metastable 2^3S_1 atoms

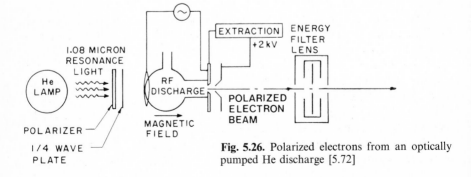

Fig. 5.26. Polarized electrons from an optically pumped He discharge [5.72]

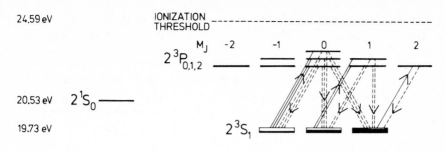

Fig. 5.27. Optical pumping in helium. Not drawn to scale. The $2^3P_{0,1,2}$ levels lie so close together that they are all excited by the 1.08-μm line

are polarized by optical pumping with circularly polarized light. The polarized He* atoms thus obtained give rise to polarized free electrons in the discharge. An exit canal in the discharge cell allows extraction of electrons and analysis of their polarization by a Mott detector.

Figure 5.27 explains how the metastable atoms are polarized by optical pumping. By irradiating with infrared circularly polarized σ^+ light of wavelength 1.08 μm one obtains transitions into the 2^3P states with the selection rule $\Delta M_J = +1$. The subsequent spontaneous emission is governed by the selection rules $\Delta M_J = 0, \pm 1$, so that only some of these transitions go back to the initial states; the rest go to higher M_J of the state 2^3S_1. By continuous pumping, the population of the state $2^3S_1(M_J = +1)$ is therefore increased at the expense of the two other sublevels of 2^3S_1. Since in the $2^3S_1(M_J = +1)$ state both electron spins are parallel to the incident direction of the light, one obtains a partial polarization of the He*(2^3S_1) atoms in this direction. These metastables are thus labeled by their polarization. If there are polarized free electrons in the discharge they could only have originated from these atoms.

On the other hand, there are numerous processes in the discharge that can produce unpolarized electrons, for example, collisions of ions with the walls of the exit canal and other secondary processes. The fact that, in spite of this, the extracted electron beam was still found to be 10 % polarized, shows that a large fraction of the free electrons in the discharge originates from the polarized He* atoms (an estimate showed that – since the metastable atoms are only partially polarized – in the ideal case where all extracted electrons are produced by collisions between metastable atoms, 30–40 % polarization should be measured).

This proves that collisional ionization of unpolarized He atoms (which are at least a factor 10^5 more numerous than the polarized He* atoms) by electrons is not the main source of the free electrons. Ionization of the polarized He* atoms by electron collisions (ionization energy 4.7 eV) can, however, also be excluded: Since in the discharge there are not enough electrons available with energies >4.7 eV (estimated on the basis of the electron temperature of 2.5 eV), this process cannot possibly balance the large electron loss rate due to diffusion to the walls, as would be necessary for maintaining a steady discharge. Exchange scattering of slow electrons by the polarized He* atoms does not release enough polarized electrons either, as another estimate by the authors shows. Thus Penning ionization, that is, ionization by collision of two metastable atoms, remains as the predominant electron production process.

Let us consider this reaction in more detail. When a He*$(2^3S_1, M_J=+1)$ atom collides with a He*$(2^3S_1, M_J=0)$ atom, the reaction can be written

$$\text{He*}(2\,^3S_1) + \text{He*}(2\,^3S_1) \;\rightarrow\; \text{He}(1\,^1S_0) + \text{He}^+ + e,$$
$$\uparrow\uparrow \qquad \uparrow\downarrow + \downarrow\uparrow \qquad\qquad \uparrow\downarrow - \downarrow\uparrow \quad \uparrow \quad \uparrow$$

where the spin eigenfunctions of the various particles are given symbolically. Conservation of total spin, which is assumed in the above reaction, has been separately proved by the authors [5.73]. They showed that with strong polarization of the metastable atoms, the production rate of free electrons decreases. This has the following explanation: with increasing polarization, the Penning ionization must increasingly occur with metastable atoms in the state $M_J=1$. Since

$$\uparrow\uparrow + \uparrow\uparrow \;\rightarrow\; \uparrow\downarrow - \downarrow\uparrow + \uparrow + \uparrow$$

is incompatible with spin conservation, the decrease of the free electrons with increasing polarization shows that spin conservation holds. Furthermore, the observation of the decrease of the free electrons provides additional evidence for the fact that Penning ionization is the predominant source of ionization in the discharge.

Polarized electrons are also produced by the reaction

$$\text{He*}(2^1S_0) + \text{He*}(2^3S_1) \;\rightarrow\; \text{He}(1\,^1S_0) + \text{He}^+ + e$$
$$\uparrow\downarrow - \downarrow\uparrow \qquad \uparrow\uparrow \qquad\qquad \uparrow\downarrow - \downarrow\uparrow \quad \uparrow \quad \uparrow$$

in which metastable atoms in the singlet state are involved. The electrons produced by Penning ionization of two H*(2^3S_1) atoms with $M_J=0$ are, however, unpolarized.

The conclusions drawn here were confirmed by measuring the electron polarization during the afterglow. When the discharge voltage is terminated the electrons rapidly come to thermal equilibrium with the gas atoms so that electron-impact ionization can no longer take place. Nevertheless, for 10^{-3} to 10^{-4} s (during the so-called afterglow) charged particles are still produced. It is known that the Penning ionization is responsible for this. There are no competitive electron-production processes in the afterglow, so that an increase of the polarization is to be expected. Indeed, the polarization measurements on electrons which were extracted during the afterglow yielded polarizations of up to 17%.

These measurements have demonstrated that while collisions between metastable atoms are the only source of free electrons in the afterglow, they also produce most of the electrons in the active discharge. The process discussed here is but one of many (e.g., chemi-ionization [5.74] or atom-surface collisions [5.75]) in which electron polarization can be used as a diagnostic tool for the analysis of collision processes [5.76–78]. Such processes can also be considered with a view to building a source of polarized electrons. We will return to this point in Sect. 8.2.

6. Further Relativistic Processes Involving Polarized Electrons

6.1 Bremsstrahlung and Other Relativistic Electron-Photon Processes

Few of the theoretical results on polarization phenomena in relativistic electron-photon interaction have been put to a quantitative experimental test. Among those which have are the emission asymmetry of the bremsstrahlung produced by transversely polarized electrons and the transverse polarization of photoelectrons produced by unpolarized radiation. Some of the numerous polarization correlations between electrons and photons that were previously inaccessible to measurement may be checked with modern electron-polarization techniques. Compton scattering may be used for polarization analysis of high-energy electrons.

The reader will have noticed that our presentation does not follow the historical development of polarized–electron physics. Historically, the electron spin was explained from first principles by the Dirac theory. As a consequence, spin polarization was first discussed in connection with solutions of Dirac's relativistic equation. Polarization effects in electron scattering, for instance, were assumed to be significant only if the electron velocity were comparable to the velocity of light. The early polarization experiments in scattering were therefore generally made with fast electrons even though slow electrons could be handled much more conveniently and successfully [6.1]. It was not until the 1960s that the large polarization effects in low-energy electron scattering were ascertained.

Similarly, polarization phenomena in relativistic electron–photon interaction were discussed soon after the development of the Dirac equation. It thus occurred that the higher–order effects of photoelectron polarization in the relativistic region were predicted much earlier than the pronounced polarization phenomena in the visible and the uv which follow from the dipole approximation (cf. Sect. 5.2) and which were long overlooked.

The attention of experimentalists was therefore focused on relativistic polarization effects at a time when techniques for producing and analyzing polarized electrons were at an early stage. This is why the experimental verification of the processes treated in the following section often does not come up to the standard of present–day experiments or is even missing. Perhaps the following brief summary of relativistic processes involving polarized electrons will stimulate further experiments with modern techniques: in a few cases they will be seen to have even begun.

A process that is well known because it has frequently been used for measuring the polarization of β particles is the production of circularly polarized x–rays (bremsstrahlung) by longitudinally polarized electrons [6.2]. The polarization of the photons can be detected by Compton scattering or absorption in magnetized iron because the Compton cross section depends on the orientation of the photon spins with respect to the spin direction of the scattering electrons. The circular polarization has its maximum for photons emitted in the forward and backward directions. It increases with the atomic number of the target and the energy of the incident electrons. At a fixed electron energy, hard quanta have larger polarization than soft quanta. In the high–energy limit the circular–polarization transfer is complete at the tip of the photon spectrum.

A circular bremsstrahlung component is also produced by transversely polarized electrons if their polarization vector lies in the reaction plane (cf. Fig. 6.1). If the latter restriction had not been made, one would have a violation of parity: reflection at the reaction plane would not change the initial state (electron polarization normal to the reaction plane) whereas the helicity of the photons would be reversed!

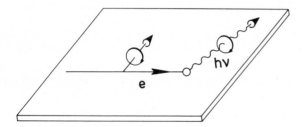

Fig. 6.1. Circular bremsstrahlung produced by electrons whose transverse polarization vector lies in the reaction plane spanned by electron and photon momentum

The latest quantitative theoretical results on the polarization of bremsstrahlung have been computed by *Tseng* and *Pratt* [6.3] in a partial–wave expansion taking screening of the atomic Coulomb field by the outer electrons into account. The authors also give quantitative results for the following polarization correlations between electrons and bremsstrahlung.

In addition to circular photon polarization, electrons with polarization components in the reaction plane produce some linear photon polarization with respect to an axis forming an angle of $45°$ with the reaction plane. This bremsstrahlung polarization, which is described by the Stokes parameter η_1 defined in Sect. 4.6, is independent of the linear polarization with respect to an axis normal to the reaction plane. Bremsstrahlung polarization of the latter kind which is described by η_3 is quite common since it is produced even by unpolarized electrons and has therefore been studied for a long time; theoretical and experimental work by Sommerfeld and Kulenkampff dates back to the 1930s.

The only bremsstrahlung effect that has been analyzed with the help of modern polarized-electron sources is one produced by electrons with polarization normal to the reaction plane. Apart from linear polarization characterized by η_3, such electrons engender a left-right asymmetry of bremsstrahlung emission. Analogous to (3.70), which describes the asymmetry of polarized-electron scattering, the intensity distribution of the bremsstrahlung is given by the differential cross section

$$\sigma = \sigma_u (1 + C_{20} \boldsymbol{P} \cdot \hat{\boldsymbol{n}}), \tag{6.1}$$

where σ_u is the cross section if the incident electrons are unpolarized, $\boldsymbol{P}$ is the electron polarization, and $\hat{\boldsymbol{n}}$ is the unit vector normal to the reaction plane defined by the momenta of the incident electron and the emitted photon. C_{20} depends, like the Sherman function in (3.70), on the scattering angle θ, the atomic number Z of the target atom, the electron energy E, and – in addition – on the photon energy k. The subscript 0 indicates that all photons are considered regardless of their polarization; the 2 is to indicate that only the electron-polarization component normal to the reaction plane contributes to the asymmetry.

Two experiments of this kind have been published in this decade, one using polarized electrons produced by Mott scattering [6.4], while the other uses a Fano-effect source of polarized electrons [6.5]. According to (6.1), the asymmetry can be determined either by measuring the intensity of the bremsstrahlung at symmetric angles to the right and to the left or by observing the intensity at a fixed emission angle for incident e↑ and e↓. The latter method has been used in [6.5], where the bremsstrahlung produced by 128-keV ($= mc^2/4$) electrons in a gold foil of 200 μg/cm^2 thickness has been observed at the emission angles of 60°, 100°, and 145°. The x-ray intensity was recorded with a spectrometer of 750 eV energy resolution (Ge detector) so that the dependence of C_{20} upon the photon energy k was obtained. In order to avoid instrumental asymmetries the characteristic K_α radiation was simultaneously recorded. Since K_α emission is isotropic, independent of the spin direction of the incident electrons, the measurements of bremsstrahlung spectra with e↑ and e↓ were normalized to the K_α intensity.

The asymmetry function $C_{20}(k)$, found as the relative difference of the numbers of photons of energy k produced by e↑ and e↓, is shown in Fig. 6.2 for the emission angle $\theta = 60°$. The experimental data are compared with theoretical values obtained from a partial-wave analysis of bremsstrahlung production in the screened Coulomb field (Hartree-Fock-Slater potential) of the target atom. The agreement with theory is good in Fig. 6.2 but at $\theta = 145°$ discrepancies were found.

None of the other above-mentioned correlations between electron polarization and bremsstrahlung has yet been checked by quantitative experiments. The experimental work on circular bremsstrahlung produced by longitudinally polarized electrons was carried out with a different goal, namely, detecting the longitudinal polarization of β particles predicted in 1956 as a consequence of parity violation in weak interactions [6.2]. For an exact analysis of the electron

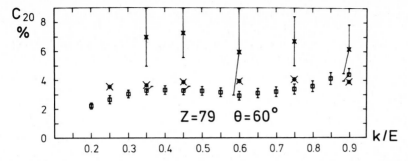

Fig. 6.2. Bremsstrahlung asymmetry function C_{20} vs. ratio of photon energy k and electron energy $E = 128$ keV for gold at emission angle $\theta = 60°$. (*Small error bars*) data from [6.5]; (*large error bars*) data from [6.4]; ($\times$) theoretical values from partial-wave analysis for screened Coulomb potential [6.3]

polarization one needs, of course, the coefficient describing the transfer of the electron helicity to the bremsstrahlung helicity. Like C_{20}, it depends on Z, θ, E, and k. The experiments on β decay had to rely on theoretical data for this coefficient, though different approximations (Born approximation, partial-wave analysis for Coulomb field and screened Coulomb field) have been shown to yield somewhat different results [6.3] which have not been seriously checked by experiment yet.

Now that polarized-electron currents which are almost ten orders of magnitude larger than those obtained in the earlier bremsstrahlung experiments with β particles are becoming available, the techniques for studying the correlations between the polarization of electrons and their bremsstrahlung are accessible. (A quite different aspect, namely, the intensity asymmetry of the bremsstrahlung emitted by slow polarized electrons in magnetic substances, will be discussed in Sect. 7.1.1.).

Similar polarization correlations occur when, instead of producing photons by electrons, one produces free electrons by photons, as in the photoelectric effect, the Compton effect, or in pair production. We have discussed photo-ionization in some detail and have found that the polarization phenomena which have been theoretically predicted in the visible and the uv have recently been studied by many experiments. But experimental verification of the theoretical work on photoelectron polarization in the relativistic region is extremely meager. The reason for this becomes clear when we look at the first experiment of this kind, which has been performed only recently. It is a measurement of the transverse polarization normal to the reaction plane of K-shell photoelectrons produced in a gold target by unpolarized γ-rays [6.6].

A strong (87 mCi) source of ^{137}Cs was used for producing the photons. Its 662-keV γ-rays generate 581-keV photoelectrons from the K shell of gold. A target of high Z was imperative since the effect to be observed is predicted to be approximately of order αZ (α = fine-structure constant) and because the K-shell photoelectric cross section is proportional to Z^5. The 581-keV photoelectrons produced in a gold foil of 23.2 mg/cm² thickness were separated from the higher-shell photoelectrons, and from the Compton continuum which has a maximum energy of 477 keV, by a double focusing 90°-sector magnetic spectrometer. By rotation of the γ source, photoelectrons of emission angles $0° \leq \theta \leq 60°$ (angular uncertainty $\pm 8°$) could be detected. Their polarization normal to the reaction plane was analyzed by a Mott detector. Measurements could be made for symmetrical positions of the γ source with respect to $\theta = 0°$. They yield, for reasons of axial symmetry, opposite polarizations. Instrumental asymmetries could be eliminated by rotation of the detection system about the symmetry axis.

The mean counting rates per detector varied from 5 to 3 counts/min in the signal-plus-background mode with a mean background rate of about 2 counts/min. The highly unfavorable signal-to-background ratio explains the error bars shown in Fig. 6.3 which range from 9 % at $\theta = 10°$ to 32 % at $\theta = 50°$, even though the total time required for data collection was about 5 months. One can clearly see that, compared with the experimental situation in the vuv described in Sect. 5.2.3, the intensity problem is severely aggravated at the photon energies considered here as long as γ-ray sources are employed.

Within the error limits given, Fig. 6.3 shows agreement with theoretical expectations. The same holds for a later experiment using 334- and 439-keV photons and a thorium target [6.8]. For a summary of theoretical work on the photoeffect in the relativistic region we refer to [6.9].

In Compton scattering and pair production knowledge of the electron-polarization phenomena has not improved much. There are calculations on some of the polarization correlations between photons and electrons, such as the production of longitudinally and transversely polarized Compton electrons, and

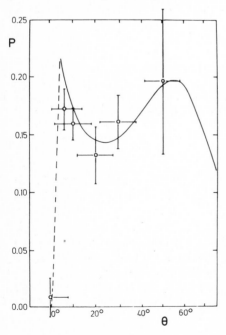

Fig. 6.3. Transverse polarization of K-shell photoelectrons vs. angle of emission produced in a gold target by unpolarized 662-keV γ-rays. Experimental points [6.6] compared with a theoretical prediction (*full curve*) [6.7]

of polarized electron-positron pairs by polarized photons [6.10]. But since the difficulties of measurement exceed those of the experiment just described, experimental and – as a consequence – theoretical interest has been limited. Present knowledge is therefore still well described by older reviews [6.2, 11] and need not be repeated here. We will, however, mention an aspect which is of practical importance.

As will be discussed in Sect. 8.1, there are only a few methods applicable to polarization analysis at high electron energies. One of the techniques which turned out to be practical at the highest energies exploits the asymmetry in Compton scattering of circularly polarized photons from the electrons whose polarization is to be determined [6.12]. The high photon intensities required are produced by a laser in conjunction with a Pockels cell for circular polarization. Energy and momentum transfer in a collision of a low-energy photon with a high-energy electron traveling in the opposite direction shifts the photon to shorter wavelengths. This situation is different from conventional Compton scattering where high-energy photons collide with slow electrons. To have a numerical example, let 15-GeV electrons in a storage ring make head-on collisions with photons of 3 eV which corresponds to a photon energy of 180 keV in the rest frame of the electrons. After the collision the photons have an energy of about 6 GeV. For polarization analysis the intensity asymmetry which arises by scattering of left- and right-handed photons from the polarized electrons can be measured with a shower detector. Absolute polarization measurements with this method still have to rely on the underlying theory.

We have emphasized that many of the polarization effects treated in this section lay unexplored because the experimental techniques of polarized-electron studies were underdeveloped. Now that the situation has changed the field may revive.

6.2 Spin-Flip Synchrotron Radiation

Electrons (or positrons) which circulate at high energy in a magnetic field become polarized through the emission of spin-flip synchrotron radiation. The polarization arises owing to unequal transition rates for the two states of spin orientation with respect to the magnetic field. An elementary description of the polarization mechanism and a brief discussion of results obtained with storage rings is given.

It is one of the fundamental facts of electrodynamics that an accelerated charged particle emits electromagnetic radiation. A conspicuous example is the "synchrotron radiation" emitted by the electrons circulating in the magnetic field of a storage ring or a synchrotron.

When discussing synchrotron radiation, one usually does not take account of the electron spin because it has practically no influence on the emitted intensity. If, however, the attention is not focused solely on the intensity alone one finds an interesting contribution of the electron spin: there is a certain probability that the magnetic moment of the electron orbiting at high energy in the magnetic field

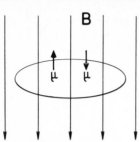

Fig. 6.4. Magnetic dipole transition

will change its direction (see Fig. 6.4). The change of magnetic energy which such a spin flip implies results in emission of radiation. Even though the intensity of this magnetic dipole radiation is negligible compared to the ordinary synchrotron radiation, the spin-flip synchrotron radiation has a remarkable consequence. When one starts with an unpolarized electron beam one obtains a gradual buildup of transverse polarization because the spin flips lead to a preferred population of the spin states with magnetic moment parallel to the magnetic field of the synchrotron or the storage ring (*Sokolov-Ternov* effect [6.13]). Consequently, one need not necessarily inject polarized electrons into an accelerator in order to perform polarized–electron experiments at high energies, but one may rely on the inherent mechanism of a suitable machine.

Unfortunately the buildup of the polarization would need hours (or at least minutes) under the operating conditions of present synchrotrons. This is due to the small probability of magnetic dipole transitions and means that an appreciable polarization cannot be attained in a synchrotron; it can, however, be attained in a storage ring where the electrons are stored for a long period. The polarization limit that can be reached in this way is 92.4 %.

We will now give an elementary description of the process leading to electron polarization through emission of spin-flip synchrotron radiation. For a thorough though highly didactic treatment see [6.14].

Our main interest is focused on the probability for spin-flip transitions, since this is the quantity that determines the buildup time for the polarization. Let us assume that the behavior of the spin can be calculated nonrelativistically in an inertial frame where the electron is at rest for a moment. In such a frame, moving with the constant velocity v, the electron orbiting with velocity v' is at rest when the direction of v' coincides with that of the frame velocity. The energy difference between the two possible spin orientations in the magnetic field B' which the electron experiences in this instantaneous rest frame is

$$2\mu B' = \frac{e\hbar B'}{mc},$$

since $\mu = e\hbar/2\,mc$ for the electron. Accordingly, one has for the frequency of the spin-flip transition

$$\omega' = \frac{eB'}{mc} = \frac{e\gamma B}{mc} = \gamma^2 \omega_c, \tag{6.2}$$

where use has been made of the expression (8.2) for the cyclotron frequency, and of the relation $B' = \gamma B$ between B' and the magnetic field B in the laboratory frame.

Let us mention in passing that one must not conclude from (6.2) that there is only a single frequency to be observed in the laboratory frame. The radiation is emitted in all directions, and transformation from the instantaneously comoving frame into the laboratory frame gives different frequency shifts for different directions. That alone can cause a broad frequency spectrum.

The transition probability per unit time for the spin flip, which is a spontaneous magnetic dipole transition, is given by the general expression

$$w' = \frac{4}{3\hbar} \left(\frac{\omega'}{c}\right)^3 |\langle\psi_f|\boldsymbol{\mu}|\psi_i\rangle|^2. \tag{6.3}$$

The formula differs from the well-known expression for electric dipole transitions as derived in textbooks of quantum mechanics (see, e.g., [6.15]) merely by a substitution of the magnetic dipole operator $\boldsymbol{\mu}$ for the electric dipole operator $e\boldsymbol{r}$. Evaluation of the matrix element (cf. Problem 6.1) yields, after substitution of (6.2) into (6.3),

$$w' = \frac{2}{3} \frac{e^2\hbar}{m^2c^5} \gamma^6\omega_c^3. \tag{6.4}$$

Assuming that orbital and spin motion are completely decoupled, and that we have only transitions from the upper to the lower energy state of the spin in the magnetic field, we have the rate equations

$$dN_\downarrow = -w'N_\downarrow dt'$$
$$dN_\uparrow = -dN_\downarrow = w'N_\downarrow dt' = w'(N_0 - N_\uparrow)dt', \tag{6.5}$$

where N_0 is the total number of electrons and $N_\uparrow$ is the number of electrons at lower energy (recall that spin and magnetic moment have opposite directions!). When the beam was unpolarized at $t' = 0$, $N_\uparrow = N_\downarrow = N_0/2$, the rate equations have the solutions

$$N_\downarrow = \frac{N_0}{2} e^{-w't'}, \qquad N_\uparrow = N_0 \left(1 - \frac{1}{2} e^{-w't'}\right),$$

which yields for the time dependence of the polarization

$$P(t') = \frac{N_\uparrow - N_\downarrow}{N_\uparrow + N_\downarrow} = 1 - e^{-w't'}. \tag{6.6}$$

We see that in the inertial frame considered, the time characteristic for the buildup of the polarization is $\tau' = 1/w'$. The characteristic time τ in the laboratory frame is

$$\tau = \tau'\gamma \approx \frac{m^2 c^5}{e^2 \hbar \gamma^5 \omega_c^3} \approx \frac{m^2 c^2 \varrho^3}{e^2 \hbar \gamma^5}, \tag{6.7}$$

since one has for the orbital radius ϱ of the relativistic particle $\varrho = c/\omega_c$. It has been taken into account here that, owing to our simplified argumentation, (6.4) holds only to within a factor of order unity.

Physically the simplifications are more serious than they may seem at first sight, even though the quantitative results are almost correct. A piece of evidence for the failure of our elementary description is the aforementioned polarization limit of 92.4 %. Nothing in the discussion so far indicates why all the electron spins should not eventually be found in the state of lowest magnetic energy, so that one could expect a polarization limit of 100 %.

Some understanding of the failure of the simplified description can be found by considering the exchange of energy between spin and orbital degrees of freedom. As a result of the recoil caused by the photon emission, the orbital energy of the electron is changed. The change of energy depends on the direction of emission and is comparable in size to the transition energy of the magnetic dipole. Owing to this interplay of orbital and spin motion it is not appropriate to take account of the magnetic dipole energy alone when talking about a transition from an upper to a lower energy level; instead, one has to consider the total of orbital plus magnetic energy. For this reason, emission of spin-flip synchrotron radiation can also occur when the magnetic moment flips into the upper energy level as long as the orbital energy takes care of the energy balance. It is these processes that prevent the polarization from reaching the limit of 100 %. A quantitative treatment shows that the transitions to lower magnetic energy which have been discussed at the beginning are 25.25 times as probable as the transitions into the opposite direction. This leads to a modification of the rate equations (6.5) into

$$dN_\downarrow = -dN_\uparrow = -w'N_\downarrow dt' + \frac{w'}{25.25} N_\uparrow dt'$$

and thus to a modification of (6.6) into

$$P(t) = 0.924 \, (1 - e^{-t/\tau}) \tag{6.8}$$

in the laboratory frame.

From a practical point of view, it is not only the degree of polarization that is of interest, but also the time needed for the buildup of the polarization. In (6.7) we found an approximate expression for the time τ needed to build up 63 % of the limiting polarization. Let us apply this formula to the storage ring SPEAR at the Stanford Linear Accelerator Center. The bending radius ϱ of the magnets is 12.7 m. The electron orbit is, however, not strictly circular, the storage ring consisting of bending magnets and straight sections combining to an orbit of circumference $2\pi R$, with $R = 37.3$ m. This gives rise to a correction factor R/ϱ in (6.7) [6.14]. For an electron energy of 2 GeV one thus obtains a characteristic buildup time of about 5 h, whereas at 4 GeV one obtains only 10 min due to the strong energy dependence of τ. If a storage ring like PETRA at Hamburg is operated at an electron energy of 15 GeV, because of the larger dimensions of this machine ($2\pi R = 2.3$ km, $\varrho = 197$ m) one obtains $\tau \approx 30$ min.

The buildup of the polarization has been experimentally verified at various storage rings. In the first observations the fact that colliding beams of polarized electrons and positrons (which are polarized by the same mechanism) produce azimuthal asymmetries in the intensity distribution of the reaction products has been exploited. From the observed asymmetries for the processes $e^- e^+ \rightarrow e^- e^+$ and $e^- e^+ \rightarrow \mu^- \mu^+$ the buildup time τ and the polarization limit could be determined, since the reactions can be reliably calculated from quantum electrodynamics [6.16]. More recent experiments employ the asymmetry in Compton scattering of circularly polarized photons from the polarized electrons [6.12] (cf. Sect. 6.1). The observed values agree approximately with the results expected from (6.7 and 8) which do not, of course, consider depolarization effects in the storage ring. Electron polarization up to 80 % was found in PETRA [6.17, 18].

The polarization of the electrons has been applied for measurement of the azimuthal cross-section asymmetries in reactions of the type $e^- e^+ \rightarrow$ hadrons showing that hadrons (mostly pions) are preferentially produced perpendicular to the polarization direction [6.19]. From these measurements information on the dynamics of hadron production in $e^- e^+$ annihilation could be obtained with much greater accuracy than from measurements of the polar angular distributions with unpolarized beams. For a recent review of the topics presented in this section see [6.20].

Problem 6.1. Evaluate $|\langle\psi_f|\boldsymbol{\mu}|\psi_i\rangle|^2$ in (6.3).

Solution.

$$|\langle\psi_f|\boldsymbol{\mu}|\psi_i\rangle|^2 = \frac{e^2\hbar^2}{4\,m^2c^2}\,|\langle\psi_f|\boldsymbol{\sigma}|\psi_i\rangle|^2. \tag{6.9}$$

Since the discussion leading to (6.3) concerns the simple case where initial and final state differ merely by their spin-dependent parts we have, since the space part is normalized,

$$|\langle\psi_f|\boldsymbol{\sigma}|\psi_i\rangle|^2 = |(0\ 1)\,(\sigma_x\hat{e}_x + \sigma_y\hat{e}_y + \sigma_z\hat{e}_z)\begin{pmatrix}1\\0\end{pmatrix}|^2. \tag{6.10}$$

Using (2.6) we obtain for the last expression

$$|1 + i + 0|^2 = 2$$

so that we have

$$|\langle\psi_f|\boldsymbol{\mu}|\psi_i\rangle|^2 = \frac{e^2\hbar^2}{2\,m^2c^2}.$$

7. Polarized Electrons from Solids and Surfaces

7.1 Polarized Electrons from Magnetic Materials

Polarized free electrons can be obtained by photoemission, field emission, and secondary emission of spin-oriented electrons from magnetized materials. Such measurements provide information on the structure of magnetic solids. The results can be interpreted within the framework of current models of magnetism, but necessitate quantitative refinement of the theory. Inverse photoelectron spectroscopy with polarized electrons yields information on empty magnetic states above the Fermi level.

In magnetized solids there are electrons whose spins have a preferential orientation. An obvious idea is to try to somehow extract these polarized electrons and thus to obtain polarized free electrons. This could, for example, be done by the photoeffect or by field emission. However, for a long time all attempts of this kind failed. This was probably due to inadequate vacuum conditions and failure to minimize stray magnetic fields which may cause precession of the spins. One must also take into account that even a magnetizing field B along the direction of extraction, as frequently used, has substantial effects on the shape and the position of the electron beam. This may be a strong source of error when scattering asymmetries are measured as a function of B, or when instrumental asymmetries are to be eliminated by reversal of B, because large spurious asymmetries can be caused by the magnetic field.

Polarized electrons were not successfully extracted from magnetic materials before the end of the 1960s. Since that time experiments have been made on ferromagnetic and antiferromagnetic substances, metals as well as semiconductors and insulators [7.1–4].

In the classical ferromagnets Fe, Co, and Ni the magnetism is caused by the $3d$ electrons. For the description of the magnetism in these materials the band model is widely used. In this model, the exchange interaction which is responsible for the ferromagnetism causes an energy shift of the e↑ and e↓ in the $3d$ band. The shift is determined by the strength of the exchange interaction and has different directions for e↑ and e↓, as shown schematically in Fig. 7.1. Since both subbands are filled up to the Fermi energy, the $3d_\uparrow$ band holds more electrons than the $3d_\downarrow$ band. For nickel there are 5 electrons per atom in the $3d_\uparrow$ band and 4.46 electrons per atom in the $3d_\downarrow$ band. Accordingly, there is a resulting magnetic moment of about $0.54\,\mu_B$ per atom.

Other materials that have been studied are systems with localized magnetic moments, such as rare-earth elements and their compounds which are ferro-

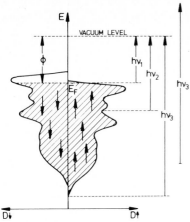

Fig. 7.1. Schematic diagram of the $3d$-band splitting below the Curie temperature. (E_F) Fermi energy, (ϕ) work function, (D) density of states. The $3d$ band is overlapped by a $4s$ band (not shown here) which contains 0.54 electrons per atom

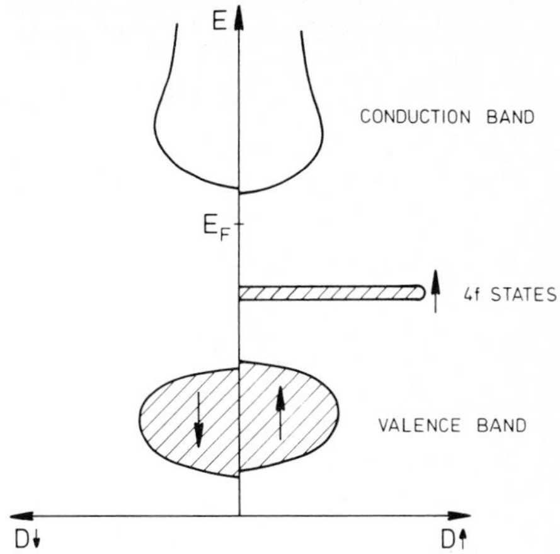

Fig. 7.2. Schematic diagram of the energy states in europium chalcogenides

magnetic or antiferromagnetic at low temperatures. In such materials it is mainly the $4f$ electrons of the rare-earth atoms that are responsible for the magnetism, as is illustrated by the schematic energy-level diagram of Fig. 7.2. There are 7 electrons per europium atom in the localized $4f$ states. Since $l = 3$ for f states, this means that half of the $2(2l+1) = 14$ states available are occupied. The Pauli principle still allows all the electrons to have the same spin direction. In accordance with Hund's rule, this state, which has the spin quantum number $\frac{7}{2}$, occurs in europium. It is evident that extraction of the electrons from the $4f$ states should yield highly polarized electrons.

7.1.1 Photoemission

If nickel is irradiated with short-wave light of frequency $v \geq v_1$ (see Fig. 7.1), $3d$ electrons are photoemitted. With the frequency v_3 the $3d$ electrons can be knocked out irrespective of their energetic position. If one uses the numbers from page 196 (and ignores complications which have yet to be mentioned) one would therefore expect a polarization $P \approx (5 - 4.46)/(5 + 4.46) \approx 5$ to 6% if the sample is magnetically saturated. Let us make a remark here on the sign of the polarization. Since it is the magnetic moment of the electron that gives rise to the magnetization, we follow common usage and define the polarization of the electrons obtained from magnetic materials to be $P = (N_\uparrow - N_\downarrow)/(N_\uparrow + N_\downarrow)$, where $N_\uparrow$ and $N_\downarrow$ are the numbers of electrons with their magnetic moments (not their spins) parallel and antiparallel to the magnetization of the crystal. With this definition P is positive for electrons with their magnetic moments parallel and,

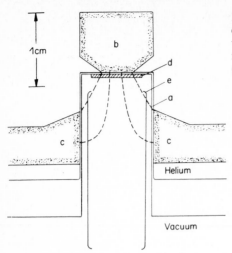

Fig. 7.3. Magnetic photocathode for the emission of polarized electrons [7.6]. (*a*) vacuum envelope, (*b, c*) poles of the electromagnet, (*d*) sample, (*e*) extraction electrode, (————) magnetic field

due to the negative gyromagnetic ratio, their spins antiparallel to the magnetization. (One might better call this "magnetic-moment polarization" instead of spin polarization.)

For nickel the Fermi energy is located just above the majority d band so that only the $\downarrow$ band in Fig. 7.1 is intersected by E_F. Consequently, if one irradiates with light of a small frequency band around ν_1, so that only electrons close to the Fermi level can be emitted, one should observe much more e$\downarrow$ than e$\uparrow$, which results in a negative polarization (recall that the arrows in this discussion represent the magnetic moment). We will come back to this point later.

These examples and Fig. 7.1 show that the polarization of the photoelectrons is determined by the photon energy, the shape of the density-of-states curve $D(E)$, the energy shift of the $\uparrow$ and $\downarrow$ bands, and the position of the Fermi level. Thus we see that measurements of the photoelectron polarization as a function of the photon wavelength yield detailed information on the electronic structure of magnetic materials. More can be learned if, with the use of electron spectrometers, the dependence of the polarization on the photoelectron energy is measured. Experiments which analyze simultaneously energy, angle of emission, and polarization of the outgoing photoelectrons even yield information on the dependence $E(\mathbf{k})$ of the e$\uparrow$ and e$\downarrow$ energies on the electron wave vector [7.5].

Let us now pick out a few examples of the extensive experimental studies on the polarization of photoelectrons from magnetized materials [7.1–4]. In order to allow the study of clean surfaces, the samples must be prepared in ultrahigh vacuum. In the arrangement of Fig. 7.3, a strong external field is applied to magnetize the samples in order to reach magnetic saturation. The magnetic field, like the electric extraction field, is perpendicular to the photoemitting surface of the sample so that the extraction of the electrons is not hindered by the Lorentz force. The photoelectrons released by uv light of variable wavelength are deflected through 90° by a cylindrical condenser so that they leave the region of

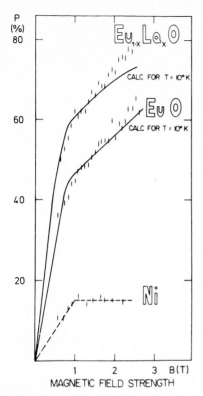

Fig. 7.4. Dependence on the magnetic field strength of the polarization of photoelectrons from the neighborhood of the Fermi level in Ni films and from the $4f$ levels of EuO and $Eu_{1-x}La_xO$, where $x = 1$ at. % [7.7, 8]

the light beam and their longitudinal polarization becomes transverse (see Sect. 8.1.1). The entire arrangement can be placed in a He cryostat. The polarization is measured with a Mott detector after the electrons have been accelerated to 100 keV.

An example of an experimental result is given in Fig. 7.4 which shows the dependence of the photoelectron polarization on the strength of the external magnetic field. With nickel, the polarization (which is proportional to the magnetization) clearly reflects the well-known saturation behavior of the magnetization curve. Whereas in the Ni curve the photoelectrons originated not far from the Fermi level, with pure or lanthanum-doped EuO crystals the photon energy was chosen so that most of the photoelectrons came from the $4f$ levels. The high polarization which is expected in emission from these levels was actually found, as shown in Fig. 7.4. The fact that in this case no saturation occurs is explained by the influence of the surface layer: The measured photoelectric magnetization curves represent the magnetic behavior of a thin sheet of material at the surface. Its thickness, which is determined by the mean free path of the photoexcited electrons in the material, amounts to a few lattice constants only. The solid curves for EuO and $Eu_{1-x}La_xO$ have been calculated under the assumption that the bulk material saturates at the kinks of the curves, whereas the magnetization of an unsaturated surface layer increases further as the field increases. This layer reduces by exchange scattering the polarization of electrons emerging from layers deeper in the bulk. Quantitative information on such exchange processes in paramagnetic surface layers has been obtained more recently by measuring the depolarizing power of various surface-layer materials [7.9, 10]. Striking differences have been found with layers made of K, Au, Ni, Gd, and Ce.

Another type of measurement deals with the dependence of the polarization on the photon energy. When one studies Eu and Gd compounds one expects, according to Fig. 7.2, valence states, $4f$ states, and conduction states to emit electrons of different polarization. The contribution of the $4f$ states should be particularly significant since they contain electrons of only one spin direction. This was confirmed by measurements of the photoelectron polarization as a function of photon energy [7.1]. The sign of the polarization and its dependence on the photon energy was as expected or could be easily interpreted within the framework of current models.

Among the magnetic materials where such studies have been made are chalcogenides of rare earths and actinides [7.3], ferromagnetic gadolinium [7.11], the $3d$ ferromagnetic transition metals Ni, Co, and Fe (where $P > 50\%$ was found [7.12–14]), ferromagnetic alloys [7.15], and materials of great technical importance like magnetite and other simple ferrites [7.2, 3]. We will not list here all substances that have been studied but rather give an example showing the power of such investigations and a few of the problems inherent in them: we will discuss the wavelength dependence of the polarization obtained with nickel [7.16–18]. Results of measurements on the Ni (100) face are shown in Fig. 7.5. They were made on a clean single crystal in a magnetic field large enough to produce magnetic saturation, at a temperature of 273 K, and at a pressure lower than $5 \cdot 10^{-10}$ mbar. Since in experiments of this type all the photoelectrons are extracted regardless of their energy, it is the integrated electron polarization that is observed.

At photon energies near the threshold of about 5.2 eV a large negative polarization is found, which is in accordance with the band model as explained before: there are many more $e\downarrow$ than $e\uparrow$ near the Fermi level so that near threshold one has mainly photoemission of minority-spin electrons. In a similar experiment near threshold with incident radiation of smaller bandwidth it was exclusively the $e\downarrow$ that were photoemitted so that a polarization of -100% from Ni (110) was indeed observed [7.18]. As the photon energy is increased, the polarization changes sign and reaches values of about $+30\%$ because more and more $e\uparrow$ are released. At a certain energy which depends on the details of the band structure, the polarization reaches a maximum and then decreases because the deeper regions of the d band tend to reduce the polarization to the mean value of the whole d band. The maximum energy $\hbar\omega \approx 11$ eV used in the experiment suffices to release all the d electrons. According to the calculation at the beginning of this subsection one would then expect $P \approx 5\%$ provided that all electrons are emitted with equal probability. Roughly speaking, this is consistent with the experimental result. The wavelength dependence of the polarization curve

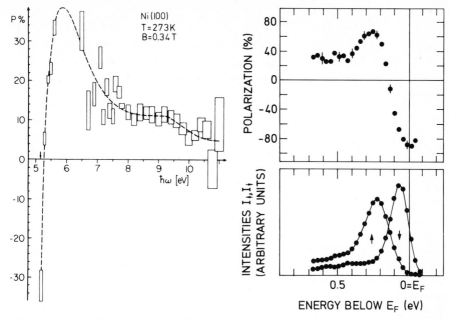

Fig. 7.5. Dependence of photoelectron polarization on photon energy $h\nu$ for Ni(100). Photothreshold indicated by arrow [7.16]

Fig. 7.6. Spin-resolved energy distribution curves of photoelectrons from Ni(110). $I_\uparrow$ and $I_\downarrow$ follow from the measured polarization P. Photon energy 16.85 eV. (E_F) Fermi energy [7.21]

near threshold and the energy of the crossover point are of particular interest since they give information on important details of the band structure such as the exchange splitting of the d bands and the energy difference between the top of the majority-spin d band and the Fermi level. This explains the interest in such studies and the great number of measurements which have already been made with various materials.

External magnetizing fields of the kind shown in Fig. 7.3 interfere with the analysis of photoelectron energy and angle of emission in experiments aiming at more detailed information [7.19, 20]. More recent experiments have therefore been made with spontaneously magnetized single crystals without application of an external field during the polarization measurements [7.3, 14, 18]. The stray magnetic field of the thin, transversely magnetized sample was so small that it did not disturb the measurement. The results of an angle- and energy-resolved polarization measurement are shown in Fig. 7.6, which displays the intensities of e↑ and e↓ following from the measured polarization. The curves can be understood from what has been said before about the density of states of majority and minority electrons. While such experiments so far have had to be made with vuv resonance radiation (see also [7.22]), the most recent measurements take advantage of high-intensity synchrotron radiation, which is tunable over a wide range of wavelengths [7.23].

There are quite a few problems in the measurement and the interpretation of polarized photoemission which we do not want to gloss over. Whereas the attempts to detect the polarization of electrons extracted from magnetic solids failed altogether in the first decade of such experiments, later more successful measurements failed in some cases to obtain the correct sign of the polarization near threshold [7.1, 2]. This did not imply, however, a failure of the band model of ferromagnetism, as suggested by the authors, but was mainly due to the rapid change in sign of the polarization near threshold of which we have given an example. The later results [7.14–18, 21] are compatible with band theory [7.24, 25]. One has also to be careful when drawing from the experiments detailed conclusions concerning the electronic structure of the magnetic substances. We have assumed that the polarization of the photoelectrons is identical to or at least proportional to the polarization in the bulk material. This assumption would certainly not be valid if – in analogy to the processes discussed in Sect. 5.2 – the initial polarization were changed by the photoabsorption process, or if it were highly probable that the spins of the electrons would flip en route to, or when emerging from, the surface. Indeed, an exact treatment of all these processes is difficult; estimates [7.6] and additional measurements [7.26] have been made to prove that in the experiments made so far, no appreciable errors have been caused by these processes. The fact that a small fraction of the electrons is in the $4s$ band (in our numerical example 0.54 of the total 10 outer electrons per atom) and may have some polarization there likewise is not expected to have an appreciable effect. A detailed discussion of controversial problems such as the validity of the various assumptions made in the interpretation of the experimental data, the influence of surface layers, and cross checks with other experimental results can be found in review articles by experimentalists [7.1, 3, 4] and by theoreticians [7.27]. In particular the latter article establishes the need for careful analyses of such experiments.

An interesting experimental feature of electron-spin spectroscopy results from the fact that the photoelectric probing depth is determined by the small mean free path of the photoelectrons. The method is therefore appropriate for studying surface magnetism or for measuring magnetization in very thin films [7.28]. Assuming a light spot of 1 mm^2 and a mean free path of the order of 10 Å, we see that the volume of material required is $\approx 10^{-9}$ cm^3. A sample of that size would not give a detectable signal in a conventional magnetometer. Apart from the cases discussed here, other intriguing applications can be found where polarized photoemission has been used as a surface magnetometer to monitor surface segregation and chemical reactions at the surface [7.3].

The inverse process to photoemission is the transition, by emission of bremsstrahlung, of free electrons into an energy band of the sample. If, for instance, the sample is a magnetized nickel crystal, the incident electrons cannot end up in the occupied $3d_\uparrow$ band. They can, however, fill the empty states above the Fermi level in the $3d_\downarrow$ band. Accordingly, if the crystal is bombarded by polarized electrons one expects an asymmetry of the bremsstrahlung intensity; more photons should be emitted, if the polarization is parallel to the $3d_\downarrow$

orientation than if it is antiparallel. An important feature of this technique is that it permits investigation of the empty states above the Fermi level, a range inaccessible in ordinary photoemission spectroscopy. The method of polarized inverse photoelectron spectroscopy has been tested on a nickel crystal and the results could be well explained in terms of its band structure [7.29]. First applications to studies of exchange splitting of empty energy bands in iron have been reported [7.4, 30]. The technique is still in its infancy, but is presently being further developed for exploring the electronic structure of magnetic surfaces.

7.1.2 Field Emission

Another possibility for extracting electrons from solids is field emission in a high electric field. If the field strength is E near the surface of the sample, the potential energy of an electron outside the surface is roughly given by $-eEz$. In this expression the electrostatic charges induced in the sample by the electron are not yet taken into account. They reduce the potential energy so that the potential has the shape of the solid curve in Fig. 7.7. By tunneling through this potential barrier, the electrons of the sample can reach the free states. The tunneling probability decreases sharply with increasing width and height of the potential barrier so that virtually only states within a few tens of millielectronvolts below the Fermi level contribute to the field emission.

   Figure 7.7 shows schematically the case of a ferromagnetic metal in which the valence band is split due to exchange interaction. For the $3d$ band of nickel, the splitting is illustrated in Fig. 7.1 where the superposition of the $4s$ band mentioned in the figure caption must also be taken into account.

   For the polarization of the field-emitted electrons one might expect approximately $P = (N_\uparrow - N_\downarrow)/(N_\uparrow + N_\downarrow)$, where $N_\uparrow$ and $N_\downarrow$ are the numbers of e↑ and e↓ in the neighborhood of the Fermi level. The extent to which $N_\uparrow$ and $N_\downarrow$ differ from each other depends on the density-of-states curves near the Fermi level.

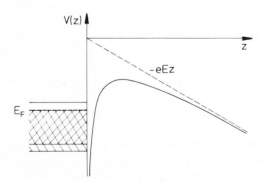

Fig. 7.7. Model for the field emission from a ferromagnetic metal

The assumption made here that the polarization of the emitted electrons is the same as the polarization within the ferromagnetic metal cannot, however, be justified in field emission. The exchange interaction which shifts the position of the energy bands for e↑ and e↓ also causes somewhat differing shapes of the potential barrier for each of the two spin directions [7.31]. Thus the e↑ and e↓ will have different tunneling probabilities so that the polarization measured externally need not be the same as the internal polarization. Another even more important point is that, as a detailed analysis shows, the tunneling probability is not only energy dependent but also a sensitive function of the symmetry type of the electronic state [7.27, 32]. As a consequence, the small fraction of electrons near the Fermi level which are in the 4s band and have only slight, if any, polarization has a considerably larger tunneling probability than the 3d electrons. The polarization of the field-emitted electrons is therefore strongly diminished and is by no means the same as the polarization within the solid. The situation is here quite different from that in photoemission. The photoelectrons, due to their higher energy, have a much greater probability of leaving the metal so that the relative differences in escape probabilities for the different states are small and affect the observed polarization only slightly.

Measurements that have been carried out on ferromagnetic metals have mainly concentrated on nickel and gadolinium and yielded polarizations of a few percent [7.33–35]. In order to obtain reproducible results in such experiments, the change of the polarization vector caused by the magnetic und electric fields in front of the field-emitting tip must be small, which requires good centering of the tip in the magnetic field [7.36]. Experimental studies with single-crystal Ni tips showed that the polarization is strongly direction dependent. This is due to the fact that the barrier transmission probability and the local density of tunneling states are sensitive to the orientation of the emitting surface. In addition, the magnetization is anisotropic in k space since the exchange splitting is k dependent. For Ni (100) a polarization of -3% has been found [7.35]. The sign is in agreement with the predictions of the band model and with the results obtained from photoemission. The small magnitude of P is due to the aforementioned predominance of s-band tunneling. The fact that the polarization observed in both field emission and photoemission near threshold can be understood by the same model shows that the spin-dependent density of states near the Fermi energy is similar in the bulk and on the (100) surface of Ni. This conclusion can be drawn because, owing to the rather large escape depth of the slow threshold electrons, photoemission tests primarily bulk properties whereas field emission probes the surface properties. Field emission from single-crystal iron tips resulted in positive polarization up to 25% from the (100) face [7.37]. This is anticipated qualitatively from the band model in Fig. 7.1 if the Fermi level is located below the top of the majority band at an energy where the density of states is larger for e↑ than for e↓.

An example of an experimental setup for measuring polarization by field emission is shown in Fig. 7.8. The investigations were made on ferromagnetic EuS, a thin film of which was evaporated in situ onto a tungsten tip of 50 to

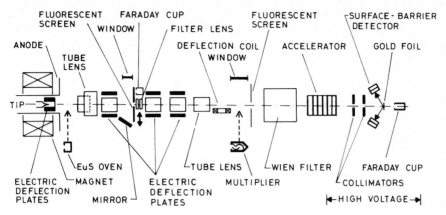

Fig. 7.8. Arrangement for measuring the polarization of field-emitted electrons [7.38]

100 nm radius. The tip was mounted on a cold finger which could be cooled below the Curie temperature of 16.6 K. A helium-flow system allowed temperature variation between 9 and 300 K. The tip could also be heated to produce an ordered structure in the EuS film by annealing. During field-emission studies a pressure of better than 10^{-10} mbar was maintained at the tip in order to avoid contamination of the emitting surface. The electric field strength of more than 10^7 V/cm required for emission from the tip was produced by a voltage of -2 keV with respect to ground. The emission pattern could be moved by electrostatic deflection, and for electron-polarization analysis one point of the pattern was selected by steering it onto the probe hole of the fluorescent screen. In the Mott analyzer two pairs of detectors were positioned at a scattering angle of $\theta = 120°$ and azimuthal directions of $0°$, $180°$ and $90°$, $270°$, respectively, for measuring both transverse components of the polarization simultaneously. The longitudinal component was measured by converting it to transverse with a Wien filter.

The measurements showed how the direction of the polarization depends on the magnetic field applied to the tip. In a weak longitudinal magnetic field (produced by a solenoid surrounding the emitter), or in zero field, the EuS layer was spontaneously magnetized in a direction tangential to the surface, which leads to largely transversely polarized electrons (the actual direction of magnetization was determined by stray fields). With a magnetic field of several kilogauss a longitudinal polarization was obtained. Most of the measurements were made with a weak field which was applied to the tip in order to produce an image of the field-emission pattern at the fluorescent screen. Spots of the emission pattern corresponding in size to a single Weiss domain were spin analyzed.

Figure 7.9 shows the temperature dependence of the polarization $P = |P|$ and the emission current as observed with an emitter annealed under well-defined conditions. The longitudinal magnetic field was 20 mT and the total emission current at $T = 10$ K was 20 nA. Polarizations up to 90 % were found at the lowest temperatures applied.

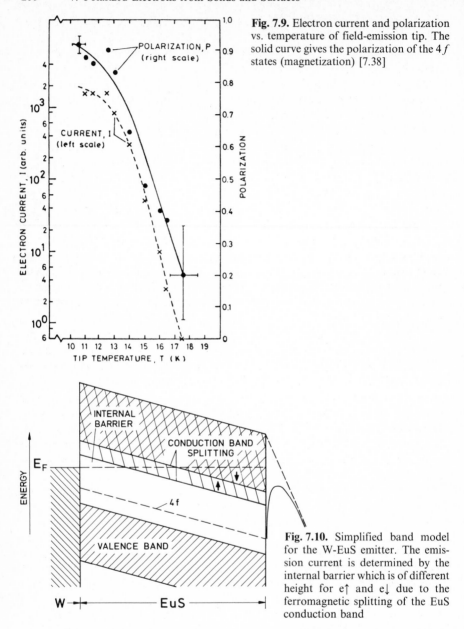

Fig. 7.9. Electron current and polarization vs. temperature of field-emission tip. The solid curve gives the polarization of the $4f$ states (magnetization) [7.38]

Fig. 7.10. Simplified band model for the W-EuS emitter. The emission current is determined by the internal barrier which is of different height for e↑ and e↓ due to the ferromagnetic splitting of the EuS conduction band

The experimental results can be roughly explained by the simplified band model shown in Fig. 7.10. Below the Curie temperature T_c one has a spontaneous magnetization of europium sulfide due to a polarization of the europium $4f$ states as represented by the solid line in Fig. 7.9. The exchange interaction of the conduction electrons with the polarized $4f$ states causes an energy shift of the e↑

and e↓ conduction electrons. The EuS conduction band splits into two subbands differing by their spin directions. Since europium sulfide is a good insulator at low temperatures the electric field applied to the tip penetrates into the EuS layer and causes a drop $-(1/\varepsilon)eEx$ of the potential energy (ε = dielectric constant of EuS). The larger the field E, the thinner is the barrier which the electrons have to overcome in order to tunnel from the tungsten Fermi level into the empty EuS conduction band and from there into vacuum. The exchange splitting is known to be proportional to the spontaneous magnetization of the $4f$ states, which increases as the temperature decreases (solid curve in Fig. 7.9). Consequently, the width of the tunnel barrier decreases for e↑ and increases for e↓ as the temperature decreases. One therefore expects an exponential current increase for e↑ and an exponential decrease for e↓. Because of the exponential dependence of the tunneling currents on barrier height and width the e↓ current can, however, be neglected.

The model discussed explains the exponential curve for the emission current of polarized electrons. But it predicts 100% polarization below T_c since practically only e↑ can tunnel into the vacuum. Various T-dependent spin-flip mechanisms (tunneling through "spin-flip bands" [7.39], magnon excitation [7.40]) have been suggested in order to explain the experimental temperature dependence $P(T)$.

We may interpret the field-emission experiment by saying that EuS has been used as a polarization filter for the electrons emitted by the W tip into the vacuum. The physical mechanism of the spin filter is based on the spin-dependent interaction of these electrons with the $4f$ states which gives rise to internal barriers of different heights for the e↑ and e↓. Although the discussion of Sect. 1.2 showed that "conventional" spin filters do not work with electrons, we see now that "unconventional" spin filters for electrons do exist. Discussions of the kind presented in Sect. 1.2 should not discourage us from searching for electron polarization filters which are just as effective as those for light and for atoms!

7.1.3 Secondary Electron Emission

Secondary electrons emitted from magnetic substances have also been found to have significant polarization. This has first been demonstrated with the experimental arrangement of Fig. 7.11. The secondary emitter, which is a small tantalum tube coated with europium oxide, is hit by the primary electrons coming from a tungsten field-emission tip. The sample is magnetized by a field between 0.5 and 7 T along the beam axis. The sample temperature is kept between 10 and 30 K, the cryovacuum is better than $3 \cdot 10^{-10}$ mbar.

The free electrons pass through a Wien filter which transforms the polarization from longitudinal to transverse for analysis by a Mott detector. The energy spread of the electrons leaving the Wien filter is less than 500 eV. Since the voltage between the field emitter and the secondary emitter is 1 to 1.5 kV, the

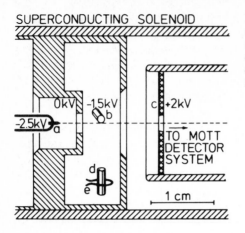

SUPERCONDUCTING SOLENOID

0kV -1.5kV   c +2kV
-2.5kV a b
d
e
1 cm

TO MOTT
DETECTOR
SYSTEM

Fig. 7.11. Production of polarized secondary electrons. (*a*) Tungsten field emitter (primary electron source), (*b*) secondary electron emitter, (*c*) fluorescent screen, (*d*) europium oxide oven, (*e*) resistance heater loop [7.41]

more highly energetic primary electrons and the low-energy secondary electrons are easily separable, and one or the other can be permitted to pass into the Mott detector. The primary electrons were found to be unpolarized ($P = 0 \pm 1.5\%$), whereas the secondary electrons from the EuO layer turned out to have a polarization up to 32% depending on the temperature of the sample and on the magnetic field. The polarization was stable over at least one hour.

More recently, studies of polarized secondary electron emission from the classical ferromagnets [7.42, 43] and from the ferromagnetic glass $Fe_{81.5}B_{14.5}Si_4$ [7.44] have been made with more elaborate techniques. One of the results is depicted in Fig. 7.12 showing the polarization P of secondary electrons as a function of their kinetic energy. They have been produced by a primary electron beam of 400 eV impinging on a magnetized Ni (110) single-crystal surface, the angle of incidence being 30° with respect to the surface normal. At the lowest

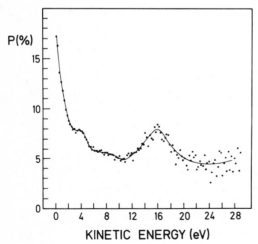

P(%)

15

10

5

0

0   4   8   12   16   20   24   28

KINETIC ENERGY (eV)

Fig. 7.12. Polarization vs. kinetic energy of secondary electrons from Ni(110) [7.43]

kinetic energy, P reaches a maximum of $(17 \pm 2)\%$. That is much higher than the mean $3d$-band polarization which we have seen to be about 5.5%. Such a behavior of the polarization curve has also been found for the other substances where energy-resolved measurements have been made. The maxima which are seen in the polarization curve correspond to structures in the energy distribution curve. They have been attributed to maxima in the density of states.

Various models have been proposed in order to explain the results in terms of the inelastic scattering processes by which the secondary electrons are produced. The spin dependence of scattering resulting in different scattering probabilities for e↑ and e↓ in the magnetic solid certainly plays an important part, but the details of this mechanism have not yet been quantitatively understood [7.45]. This is not too surprising since even the corresponding processes in electron scattering from free atoms are an object of present research as we have seen in Chap. 4. Still, a recent theoretical approach describes approximately the overall behavior of the polarization curve [7.46].

Systematic and comprehensive investigations of polarized secondary electrons and spin-dependent electron energy losses [7.3, 4, 47–49] would be an important contribution to the development of a quantitative theory of spin-dependent electron collisions in solids. This holds true also for polarization studies of secondary and backscattered electrons from nonmagnetic materials [7.3, 50–52].

7.2 Polarized Electrons from Nonmagnetic Materials

Polarized photoelectrons have also been obtained from nonmagnetic solids, such as alkali metals, tungsten, platinum, and GaAs. The polarization is a result of spin-orbit interaction which splits the energy bands. This is quantitatively confirmed by experiments made with GaAs crystals for which the polarization curve obtained can be easily explained by means of the well-known band structure. Studies of the polarization in photoemission from nonmagnetic solids allow one to determine the spin-orbit splitting of energy bands, a quantity which is presently unknown for many materials.

As we saw in Chap. 5, one does not need materials with oriented spins in order to produce polarized photoelectrons. Photoionization of unpolarized atoms with circularly polarized light also yields polarized electrons. Even linearly polarized and unpolarized light will do if only photoelectrons with definite directions of emission are observed.

An obvious question is whether this works only with free atoms or whether solids could also be used. The first experimental answer to this question was affirmative [7.53]. The measurements were carried out with the apparatus shown in Fig. 5.5, the only major change being that the target was no longer an atomic beam. Instead, the alkali atoms were evaporated on a substrate in normal high vacuum at a rate of about 50 atomic layers per second.

Comparison of the measured curve in Fig. 7.13 with that for free cesium atoms (Fig. 5.6) shows some characteristic differences: The polarization

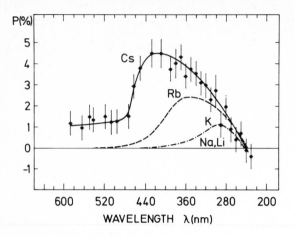

Fig. 7.13. Comparison of the photoelectron polarization obtained with different alkali films. Measured points with error bars are presented here for cesium only; the curves denote the average values [7.53]

maximum is shifted to longer wavelengths and no longer lies in the uv, but rather in the easily accessible visible wavelength region. The degree of polarization is much smaller, but the photocurrents obtained from solid samples are about 1000 times larger than those from atomic beams, which are targets of much lower density.

Films of other solid alkalis also yielded polarized electrons when photo-ionized with circularly polarized light, as shown in Fig. 7.13. It can be seen that the polarization decreases with decreasing atomic number. This is to be expected if, as in Sect. 5.2, spin-orbit coupling is responsible for the polarization. Within the error limits indicated in Fig. 7.13, no polarization was found for Na and Li with the aforementioned experimental arrangement.

What has already been said in other sections of this chapter applies to an even greater extent to the results now being discussed: The polarization phenomena obtained with solids are less well understood, quantitatively, than those obtained with free atoms. Until now, the above results have been explained only in principle. A model calculation has shown that the spin polarization of the photoelectrons can be explained by the energy-band splitting in solids caused by spin-orbit interaction [7.54].

The energy-band splitting, which is analogous to the spin-orbit splitting of the energy levels in free atoms, is schematically illustrated in Fig. 7.14. The left-hand side represents two energy states for a given magnitude and the [100] direction of the reduced wave vector k. The lower state Δ_1 belongs to the partly filled valence band,[1] whereas the upper state Δ_5 lies above the vacuum level E_∞. Electrons with energies $E \geq E_\infty$ can escape from the solid into the vacuum. When the magnitude of our specific wave vector is changed, the position of each energy state is shifted within the band of the allowed energies (Δ band for the specific

---

[1] The subscripts of the band symbols are related to the crystal symmetry and need not be explained for the qualitative description given here. For an introduction to the connection between crystal symmetry and band structure see [7.55].

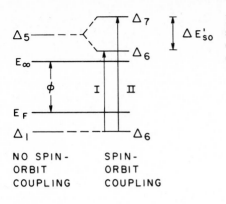

Fig. **7.14.** Allowed transitions between Δ states induced by circularly polarized light. ($\Delta E'_{so}$) spin-orbit splitting, (E_F) Fermi energy, (E_∞) vacuum level, (ϕ) work function

example chosen, where k has the direction [100]. For other directions of k the energy bands are labeled differently). The right-hand side of Fig. 7.14 shows, for the same k as chosen on the left, the splitting of the energy states due to spin-orbit interaction. Variation of the magnitude of k yields the split Δ band. We see from the figure that, in the example discussed, it is only the upper band that splits. We now consider direct interband transitions from states below the Fermi level E_F to states above the vacuum level E_∞ and look into the polarization of the photoelectrons thus produced.

Since neither quantitative calculations of the band splitting in Cs nor complete wave functions are available, it is impossible at the moment to calculate the curve $P(\lambda)$ quantitatively. Salient features of the polarization curve can, however, be predicted, knowing only the spin and angular parts of the wave functions. We have already seen this in Chap. 5. Spin and angular parts of a wave function are determined by the symmetry properties of the system considered and can therefore be derived with the help of group theory. This is much less complicated than the numerical calculation of the radial part. Evaluation of the matrix elements with the basic functions which contain the angular and spin parts and which are analogous to the coupled wave functions used in Sects. 5.2–4 yields the selection rules for the various interband transitions.

An essential difference between the situation in free atoms and solids is the following: In free atoms the sublevels of a given angular momentum all have the same energy owing to the high symmetry of the atomic field. In solids this symmetry is disturbed by the crystal fields which act as perturbations and, in general, remove the degeneracy. Consequently, we may have a more favorable situation here than in the atomic case discussed earlier, where irradiation with a given wavelength usually produced several simultaneous transitions starting from degenerate sublevels and ending in states of different polarizations so that frequently the overall polarization was not very high. In a solid where the degeneracy is removed we may, at a suitable photon energy, have only one transition to one final state so that the polarization becomes higher.

This makes plausible the theoretical result that in transition I of Fig. 7.14 totally polarized e↓ are produced with σ^+ light, whereas in transition II only e↑

are produced. Usually the situation is, however, less favorable and a given wavelength can produce e↑ and e↓. Then the resulting polarization depends on the radial parts of the wave functions associated with the ↑ and ↓ contributions which complicates its quantitative evaluation.

These considerations give, in principle, an explanation of the photoelectron polarization, but they are far from yielding the results of the aforementioned experiment quantitatively. Owing to the conditions of the experiment, in which a polycrystalline Cs target has been used and the photoelectrons were detected regardless of their energy, the observed polarization was small since averaging over such parameters as the electron energy and the transitions between the energy bands associated with the various directions of k results in a lot of cancellation.

We will now apply our considerations to GaAs, a material of great practical importance. The energy bands, spin-orbit splitting included, are quite well known [7.56]. Relevant $E(k)$ curves along two principal symmetry axes of the Brillouin zone are shown in Fig. 7.15. Let us first consider transitions at the symmetry point Γ which is defined by $k = 0$ (center of the Brillouin zone). At this point GaAs has its minimum band separation $E_g = 1.52$ eV. The wave functions at Γ are known to be of the p type for the valence bands and of the s type for the conduction band. Spin-orbit interaction splits the valence band into $P_{3/2}$ and $P_{1/2}$ levels with the $P_{1/2}$ level lying 0.34 eV lower than the $P_{3/2}$ level. For photon energies $\hbar\omega \approx E_g$, transitions are possible only near Γ and one can use the scheme on the right of Fig. 7.15 which is similar to the scheme used in Fig. 5.2 for the Fano effect turned, however, upside-down.

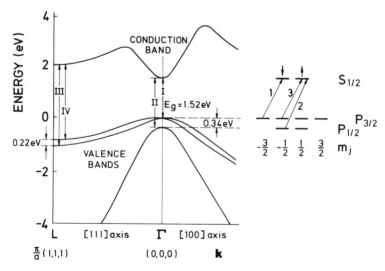

Fig. 7.15. Energy bands in GaAs along the [111] and [100] axes of the Brillouin zone (a = lattice vector) showing the energy gap E_g and the spin-orbit splitting of the valence bands (0.34 eV at Γ). The level diagram on the right holds for the Γ point. 1, 2, 3 label the transitions in accordance with Fig. 5.2 (ratio of transition probabilities 3 : 2 : 1)

If the sample is irradiated with σ^+ light of energy $\hbar\omega \approx E_g$ one has, due to the selection rule $\Delta m_j = +1$, the two transitions from $P_{3/2}$ ($m_j = -3/2, -1/2$) to $S_{1/2}$ ($m_j = -1/2, +1/2$) shown in Fig. 7.15. From the transition matrix elements b_1 and b_3 given by (5.4 and 6) one finds the ratio of the transition probabilities to $S_{1/2}$ ($m_j = -1/2$) and $S_{1/2}$ ($m_j = 1/2$), respectively, to be $b_1^2/b_3^2 = 3$. [In applying (5.4, 6), which have been derived for the $m_j = +3/2, +1/2$ sublevels of $P_{3/2}$, to the $m_j = -3/2, -1/2$ sublevels we have made use of the fact that, for symmetry reasons, the transition probabilities are the same.] Hence the expected electron polarization[2] for transition I at threshold is $P = (1-3)/(1+3) = -0.5$. As long as the photon energy stays below $E_g + 0.34$ eV the polarization does not change very much; the transitions still occur close to the Γ point so that the wave functions are similar. As the photon energy is increased beyond $E_g + 0.34$ eV, transition II from $P_{1/2}$ becomes possible which yields a polarization $+1$. From (5.5) and (5.6) one has $b_2^2/b_3^2 = 2R_1^2/R_3^2 \approx 2$. Neglecting slight differences of the radial matrix elements from $P_{3/2}$ and $P_{1/2}$ we have the relative intensities $1:2:3$ for the three transitions on the right-hand side of Fig. 7.15, so that the polarization of transition I is completely cancelled by transition II.

It is obvious from Fig. 7.15 that, as the photon energy is further increased, more and more transitions away from the Γ point become possible. For these transitions, the foregoing considerations no longer hold. We just mention that for transitions III and IV at the symmetry point L (lying on a face of the Brillouin zone) a similar calculation yields polarizations of -1 and $+1$, respectively, if the light direction coincides with the direction of the wave vector [111].

The theoretical results we have discussed are illustrated by an experiment [7.57] using an ultrahigh-vacuum apparatus similar to that described in Sect. 7.1.1. Needless to say, the magnetizing field was not necessary in the present experiment. Normally, electrons with excitation energies in the range shown in Fig. 7.15 cannot escape from the GaAs sample. Instead, they stay in the conduction band until they recombine. As a matter of fact, the polarization of conduction electrons excited by circularly polarized light has been analyzed in earlier experiments by measuring the polarization of the luminescence light and by resonance techniques. In the present experiment, all the electrons that were excited to the conduction band could escape into the vacuum since the vacuum level E_∞ of the GaAs crystal had been lowered by applying alternating layers of cesium and oxygen to the surface of the sample. Measurement of the polarization with a Mott detector resulted in the curve of Fig. 7.16.

The measured value $P = -40\%$ near threshold is not far from the predicted value of -50%. As the photon energy is increased above 1.86 eV, a strong admixture of $e\uparrow$ due to transition II occurs so that the polarization decreases strongly. With a further enhancement of energy one finds increased cancellations of the positive and negative polarizations produced by the various possible transitions,

---

[2] As in Sects. 5.2–4, the spin polarization of photoelectrons is positive if their spins are preferentially oriented in the direction of light propagation ($=$ direction of photon spins for σ^+ light).

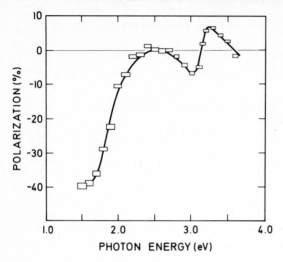

Fig. 7.16. Polarization of photoelectrons obtained from GaAs + CsOCs at $T \lesssim 10$ K with σ^+ light [7.57]

until transition III gives rise to a sudden increase in the number of e↓. As the photon energy goes up by another 0.2 eV, a positive polarization peak occurs due to transition IV. The satisfactory agreement of the experimental results with the theoretical predictions shows that no severe changes of the electron polarization due to spin flips taking place in the transport to and through the surface occur. The polarization spectrum $P(\hbar\omega)$ of Fig. 7.16 has been observed at temperatures $T \leq 10$ K. Since the spin-flip relaxation time decreases with increasing T the polarization is somewhat reduced at room temperature [7.58]. The polarization has been found to depend on the face of the GaAs crystal from which the electrons are emitted. Different results have been obtained by different groups with (110) and (100) faces [7.59, 60].

Photoemission from GaAs is being employed in numerous laboratories as an attractive source of polarized electrons. Postponing the description of the source to Sect. 8.2, we emphasize here that studies of polarized photoemission from nonmagnetic materials less well known than GaAs may help considerably in the understanding of problems like the spin-orbit splitting of energy bands and in improving our knowledge of wave functions in solids. For such investigations experiments with well-defined conditions are necessary: The polarization of photoelectrons emitted within a small solid angle from oriented single crystals should be measured as a function of the photoelectron energy and the photon wavelength λ. As a step in this direction, a cesiated W(100) single crystal has been used as the target and the polarization as a function of photon energy has been measured without energy analysis of the photoelectrons [7.61]. The maximum polarization of 8 % is far below a predicted value of 100 % for tungsten [7.62] for which relativistic band-structure calculations are available. The large discrepancy is explained by hybridization: linear combination of p- and d-type wavefunctions for k values at which the associated bands have the same energy [7.63]. On the other hand, hybridization can also enhance the polarization as has been demonstrated for germanium where the polarization at photothreshold was found to be 50%. From the measured photoelectron polarization the hybridization parameters can be directly determined [7.64]. Other studies of electronic band structure utilizing polarized photoemission have been made for gold [7.65] and platinum [7.66].

In the latter experiment the goal of making an angle- and energy-resolved measurement of the photoelectron polarization obtained with a well-defined crystal orientation has been reached by using the high intensity of circularly polarized vuv radiation from a storage ring. The polarization obtained from Pt(111) has been measured as a function of the photoelectron energy at photon energies between 6.5 and 24 eV for normal incidence and normal emission. The polarization curves, one of which is shown in Fig. 7.17, have a pronounced spectral variation reaching values up to 55%. Comparison of the results with what is anticipated from band-structure calculations leads to an unequivocal identification of the symmetry character of the bands and is a sensitive test of such calculations.

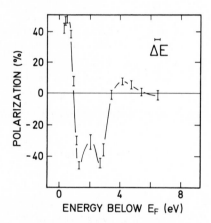

Fig. 7.17. Energy dependence of photoelectron polarization obtained from Pt(111) with circularly polarized radiation of energy 14 eV at normal incidence and normal photoemission. (E_F) Fermi energy [7.66]

It is not necessary to use circularly polarized light in order to eject polarized photoelectrons from unpolarized targets. Linearly polarized or unpolarized radiation will also do if one does not integrate over the angle of emission. This does not only hold for atomic targets as has been discussed before in some detail, but it has also been found with a solid target. The theoretical description is, however, different in the two cases because the atomic and the solid target have different symmetries and because the photoelectrons in the crystal are diffracted, which is essential for the buildup of the polarization (see Sect. 7.3). Polarized photoelectrons have been ejected from a W(001) single-crystal surface by irradiation with unpolarized vuv light of energy $hv = 10.2$ eV [7.67]. The polarization of the photoelectrons was observed at the polar angle of emission $\theta = 70°$. Owing to the diffraction of the photoelectrons in the target, their polarization varied with the azimuthal angle ϕ, reaching values up to 8%. The experimental curve could be quantitatively interpreted by a numerical calculation.

7.3 Low-Energy Electron Diffraction (LEED)

By diffracting slow electrons on crystals, one obtains polarized Bragg reflections if spin-orbit coupling or – in the case of magnetic materials – exchange interaction plays a role. The polarization or – for polarized primary beams – the asymmetry carries information on the surface properties which cannot be obtained from LEED intensities alone. In particular, PLEED (=polarized LEED) is one of the most promising probes of surface magnetism.

In Chaps. 3 and 4 we dealt with electron scattering from single atoms. If there is a coherent superposition of the scattered electron waves originating from the individual atoms of the target, the distribution of the scattering intensity becomes quite different: It is no longer determined by the individual atoms alone but is considerably influenced by a factor which depends on the structure of the target. This is the basis of the method of electron diffraction for investigating the structure of crystal lattices or molecules.

Let us denote the scattering amplitudes f and g given in (3.51, 52) as f_j and g_j when they originate from the jth atom. For the wave scattered from the total target into the direction θ, we then obtain by in-phase superposition (i.e., by taking the path differences into account) the amplitudes

$$F(\theta) = \sum_j f_j e^{i\mathbf{s}\cdot\mathbf{r}_j} \quad \text{and} \quad G(\theta) = \sum_j g_j e^{i\mathbf{s}\cdot\mathbf{r}_j}, \tag{7.1}$$

where $\mathbf{s} = \mathbf{k} - \mathbf{k}'$ (if $\mathbf{k}$ and $\mathbf{k}'$ are the wave vectors of the incident and the scattered beam) and $\mathbf{r}_j$ is the position vector of the jth atom.

From these relations, which are derived from simple geometrical considerations in elementary diffraction theory, it follows that the total scattering intensity from the target is proportional to

$$I(\theta) = |F|^2 + |G|^2 = \sum_{j,k} (f_j f_k^* + g_j g_k^*) e^{i\mathbf{s}\cdot\mathbf{r}_{jk}}, \tag{7.2}$$

where $\mathbf{r}_{jk} = \mathbf{r}_j - \mathbf{r}_k$. Here the incident beam was taken to be unpolarized, i.e., the cross section corresponding to (3.65) was used. According to (3.56 and 73), the polarization of the scattered beam is determined by the expression

$$i\frac{FG^* - F^*G}{|F|^2 + |G|^2} = i\frac{\displaystyle\sum_{j,k}(f_j g_k^* - f_k^* g_j)e^{i\mathbf{s}\cdot\mathbf{r}_{jk}}}{\displaystyle\sum_{j,k}(f_j f_k^* + g_j g_k^*)e^{i\mathbf{s}\cdot\mathbf{r}_{jk}}}, \tag{7.3}$$

which, according to (3.70), also determines the left-right scattering asymmetry, if one has a polarized incident beam.

If the target is composed of identical atoms, (7.2) can be simplified to

$$I(\theta) = (|f|^2 + |g|^2) \sum_{j,k} e^{i\mathbf{s}\cdot\mathbf{r}_{jk}}. \tag{7.4}$$

In a crystal lattice one has appreciable scattering intensities only in those directions for which the Bragg condition

$$s = k - k' = 2\pi h$$

is fulfilled (h is the reciprocal lattice vector). In other directions one has destructive interference. This is a result of the lattice-dependent factor $\sum_{j,k} \exp(is \cdot r_{jk})$ which appears in (7.4) together with the intensity distribution $|f|^2 + |g|^2$ caused by the individual atoms.

For the polarization, we obtain from (7.3)

$$P = i \frac{fg^* - f^*g}{|f|^2 + |g|^2} \hat{n} \tag{7.5}$$

if we have identical atoms. P depends, in this case, only on the scattering atoms and not on their geometrical arrangement – at least in the approximation made here.

According to Sect. 3.6.2, high values of the polarization (7.5) are found at certain angles. If one selects a diffraction maximum that occurs at such an angle, one has combined the high intensity of the diffraction peak with the high polarization of the scattered beam. In doing this the magic rule that high values of polarization are always associated with low scattering intensities (see Sect. 3.4.2) will be broken. It is true that in the scattering of fast electrons, due to the rapid decrease of the cross section with increasing angle, most of the intensity goes into the diffraction maxima at small angles, where $P(\theta)$ is very low; with low-energy electron diffraction (LEED), however, high intensity maxima can also be obtained at large angles where there are high polarization values.

So far, our discussion has given only a rough idea of the polarization effects arising in LEED. A rigorous theoretical treatment is complicated since one has to apply the dynamic theory of electron diffraction instead of the simple kinematic model used here [7.4, 68]. Such a treatment, which has to solve the electron wave equation for the crystal field, shows that there are other effects besides the Bragg condition that determine both the intensity and the polarization: Multiple scattering, inelastic processes, and the surface potential barrier (that is, the detailed shape of the transition from the inner potential level to zero potential in the vacuum) play a crucial role [7.68–71]. Polarizations and intensities in LEED may therefore serve as sensitive probes of these effects.

The first experimental result on PLEED (=polarized LEED) was obtained with tungsten single crystals [7.72]. Early measurements of *Davisson* and *Germer* [7.73] with nickel crystals can be reinterpreted as evidence for electron polarization in LEED [7.74], but it has been pointed out [7.75] that the measured asymmetry could also be caused by very small apparatus misalignment. The interest in the field is manifest by the fact that recently many results have been published by serveral active groups, not only for heavy elements like tungsten, gold, and platinum, where polarization effects are produced by spin-orbit interaction, but also for magnetic materials like nickel where exchange interaction is important [7.76–81]. For a more complete reference list see the reviews [7.4, 68–70].

We will present an experiment where LEED has been used both as polarizer and as analyzer. It is shown in Fig. 7.18. Unpolarized electrons impinge on the (001) surface of the first tungsten crystal. The LEED screen used for observing the electron diffraction pattern contains a hole through which the specularly reflected (0,0) beam can pass. After adjusting the beam to the desired scattering energy of 105 eV for polarization analysis, and after energy analysis in a cylindrical mirror analyzer (CMA), the polarized electrons are focused onto the second tungsten crystal. The left-right asymmetry of the diffraction through equal polar angles is utilized for measuring the polarization. Measurements have been made with fixed primary energy and fixed angle of incidence θ by rotating the polarizer crystal azimuthally about its surface normal so that several diffracted spots move over the hole of the LEED screen ("rotation diagram"). The analyzing power S of the crystal, corresponding to the Sherman function of free atoms, has been found by the method discussed in Sect. 3.7.1: A double scattering experiment with identical angles, energies, and crystals has been done yielding $|S| = 0.28 \pm 0.05$.

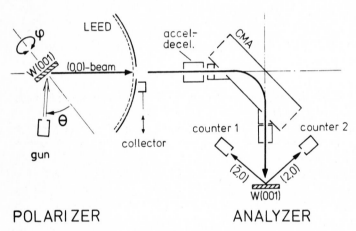

Fig. 7.18. Double scattering LEED experiment [7.82]

Because of high counting rates up to 10^6 counts/s at primary currents of 10^{-7} A, the recording with a x-y plotter of a polarization curve like that shown in Fig. 7.19 took only 5 to 10 min. The theoretical curves given in the figure have been obtained using a relativistic LEED theory. The crystal has been assumed to consist of a finite number of monoatomic layers for which the Dirac equation has been solved to yield the four-spinor amplitudes and thence the intensities and the polarization vectors of the reflected beams. Making appropriate assumptions on the value of the inner potential of the crystal and the shape of the surface potential barrier, and assuming the spacing between the top layer of atoms and

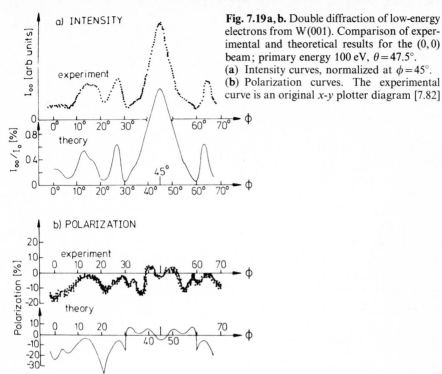

Fig. 7.19a, b. Double diffraction of low-energy electrons from W(001). Comparison of experimental and theoretical results for the (0,0) beam; primary energy 100 eV, $\theta = 47.5°$.
(a) Intensity curves, normalized at $\phi = 45°$.
(b) Polarization curves. The experimental curve is an original x-y plotter diagram [7.82]

the adjacent layer to be 5% smaller than the bulk value, an impressive agreement between theory and experiment has been obtained.

Another approach to PLEED experiments starts with a polarized-electron beam produced by photoemission from GaAs (cf. Sects. 7.2 and 8.2) [7.83]. A current of 10^{-6} A of 43% polarization impinges on the (100) surface of tungsten. A Faraday cup which can scan both the polar and azimuthal scattering angle is positioned on a LEED spot and measures the intensity difference for opposite polarizations P and $-P$ of the incident beam. The spin dependence of the scattering is determined by the intensity asymmetry (cf. Chap. 3)

$$\frac{1}{P}\frac{I(P)-I(-P)}{I(P)+I(-P)} = S(E, \theta) \tag{7.6}$$

for polarization parallel and antiparallel to the normal of the scattering plane. The direction of the polarization is modulated at 37 Hz so that a modulated diffracted signal is obtained whose size measures directly the intensity asymmetry. The dc signal at the Faraday cup gives the spin-averaged LEED intensity. In this experiment a single data run accumulated the asymmetry and the average intensity profile over the incident energy range of 50 to 150 eV in 1-eV steps with an integration time of 1 s per point.

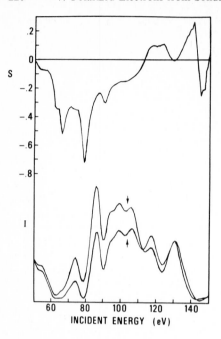

Fig. 7.20. Scattering asymmetry for specular diffraction from W(100) at an angle of incidence of 15° (*upper curve*). The lower curves show the scattered intensities resulting from an incident beam consisting of only e↑ or e↓ [7.83]

Figure 7.20 shows the measured asymmetry for specular scattering from W(100) at an angle of incidence of 15°. The intensity curves that would result if the incident beam were entirely polarized are shown in the bottom part of Fig. 7.20.

It is useful to compare the similarities and differences between the polarization phenomena in electron-atom scattering and in the PLEED experiments discussed so far. In both cases the underlying physical mechanism is the same. The polarization is caused by spin-dependent scattering. The cross sections for the scattering of the e↑ and e↓ in an unpolarized primary beam differ slightly from each other, which results in a polarization of the scattered beam. Furthermore, one obtains in both cases a scattering asymmetry if one has a polarized incident beam. Accordingly, PLEED can be used as a detector of electron polarization in the same way the conventional Mott detector is used. It is an appropriate spin analyzer for low-energy electron diffraction experiments of the type depicted in Fig. 7.18. For many experiments that have to be made in normal vacuum (experiments with gaseous beams, for example), the conventional Mott detector is, however, preferred since LEED works only with an extremely clean surface. Finally, we have the similarity that both in electron-atom scattering and in PLEED there is a rotation of the polarization vector of an initially polarized beam [cf. (3.77)]. Neither theoretical nor experimental studies of this effect in PLEED have yet been made.

In spite of the similarities of the two areas compared, there are significant quantitative differences due to the additional influence of the aforementioned solid-state effects that determine the details of the LEED pattern. A significant example is that, in general, the relation $P=S=A$ discussed in Chap. 3 for single scattering from atoms is not valid in PLEED nor is the polarization vector directed perpendicular to the plane defined by the incident and outgoing beams. The situation in solids is more complicated due to the lower symmetry of the crystal and multiple scattering.

A rough idea of how multiple scattering invalidates the aforementioned simple relations may be obtained from Fig. 7.21. From the positions of the electron gun G and the detector D, the beam appears to have been scattered by the solid target through 90° from the x into the z direction. An initially unpolarized beam (Fig. 7.21a) would therefore be expected to become polarized in the y direction. It is assumed in the figure that the beam has actually been scattered twice. The polarization vector points in the z direction after the first scattering in the x-y plane and is rotated by the second scattering, giving x and y components as well. This means that the polarization is not perpendicular to the directions of incident and outgoing beams. If, on the other hand, the incident beam is (totally) polarized in the z direction (Fig. 7.21b) the first scattering in the x-y plane yields intensities proportional to $1+S$ and $1-S$ for ↑ and ↓ polarization, respectively, i.e., the asymmetry is S according to (7.6). It was shown in Sect. 3.3.3 that the polarization P in this scattering process remains unchanged. P lies thus in the plane of the second scattering process, so that the second scattering does not produce an asymmetry. Accordingly, the asymmetry of the overall process depicted in Fig. 7.21b is S. Judging from the directions of the incident and the outgoing beam alone one would expect no asymmetry at all because the polarization vector is not perpendicular to both these directions (cf. Sect. 3.3.1). It is also obvious that in our example the relation $P=S$ is not fulfilled for the overall process since the magnitude of P changes in the second scattering of Fig. 7.21a from $P=S$ to $P \neq S$ [cf. (3.77)].

This simplified example is to warn the reader against transferring to LEED basic facts discussed in single scattering from atoms, even though in certain cases this may be allowed. If, for instance, the scattering plane is a mirror-symmetry plane of the crystal, the equality between P and S does arise [7.83] as can easily be seen from the results of Sect. 3.5 for the special case where the experiment does not include any elements which violate mirror symmetry. That P equals S in this case can also be seen from our rough picture: If in Fig. 7.21 the apparent scattering plane (x-z plane) is a mirror plane of the target, then the spin-dependent effects leading to $P \neq S$ are compensated by those of the scattering processes which result from mirror reflection and which also reach detector D.

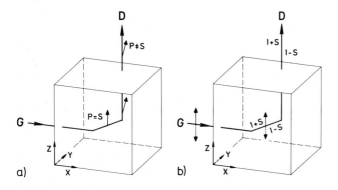

Fig. 7.21a, b. Example of apparent and actual scattering processes in a solid. (**a**) Unpolarized incident beam, (**b**) polarized incident beam

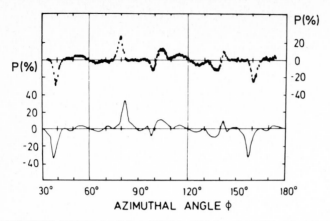

Fig. 7.22. Longitudinal polarization of the specularly reflected beam vs. azimuthal angle (rotation diagram). Angle of incidence on Pt(111) surface 43.5°, electron energy 60 eV. (*Upper curve*) measured results; (*lower curve*) calculated results (for angle of incidence of 44°) [7.80]

The difference between the polarization effects in experimental arrangements which are mirror symmetric and in those which are not has been demonstrated in a PLEED study, the results of which are reproduced in Fig. 7.22. It shows the longitudinal component of the polarization vector versus azimuthal angle ϕ in specular scattering from the Pt(111) surface which has a threefold rotational symmetry. The longitudinal polarization vanishes at the azimuthal angles 60°, 120°, 180°, etc. where the scattering planes are mirror planes of the crystal (cf. Fig. 7.23). This is to be expected from the results of Sect. 3.4.4 where longitudinal components have been shown to be incompatible with mirror symmetry. At other angles of observation ϕ the experimental arrangement is not mirror symmetric so that significant longitudinal polarization components can exist. The numerical results for the longitudinal polarization predicted by relativistic PLEED theory as a consequence of spin-orbit interaction and multiple scattering are seen in Fig. 7.22 to be in remarkable agreement with the experimental data.

The appeal of polarization studies in LEED derives from the fact that they provide detailed information on the parameters which determine the LEED pattern. The polarization is sensitive to the shape of the surface potential and the spacing between top layer and adjacent layer of the crystal; it also depends on the multiple and inelastic processes occurring at and near the crystal surface. The information obtained from polarization studies is not obtainable from studies of

Fig. 7.23. Threefold rotational symmetry resulting in an angular distance of 60° between the mirror planes

the intensity distribution alone; thus these investigations add a new dimension to the study of surfaces. We will illustrate this by a few examples.

Since the polarization is the difference between two quantities (spin-up and spin-down intensities), it tends to exhibit sharp, structure-sensitive peaks. This facilitates comparison of the predictions of theoretical models with experimental results. Figure 7.24 shows, for instance, theoretical predictions [7.84] for both the intensity and polarization of the $(1,0)$ beam diffracted from W(001). The three curves depict the results for an assumed contraction of the spacing between the top layer and the adjacent layer of 0, 5, or 10 percent of the bulk value. One sees clearly that the intensity curves are rather similar, while the polarization curves change dramatically. Consequently, measurement of the polarization is a sensitive method for determining the contraction of the top-layer spacing. In the present example the contraction was found by comparison with experimental date to be $(7 \pm 1.5)\%$.

Another problem where polarization measurements are helpful is the following: PLEED intensities and polarizations depend on the geometrical structure of the target as well as on the potential distribution by which the electrons are scattered. There are certain energy regions where the calculated PLEED data depend strongly upon either the assumed potential or geometry, but not both. Such a decoupling of the influence of the model parameters on the intensity I and polarization P helps to clearly separate the individual model characteristics. Whether P or I allows more accurate conclusions depends on

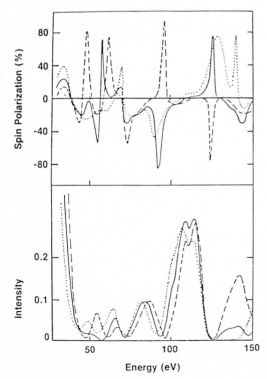

Fig. 7.24. Results of calculations of the polarization and intensity for $(1,0)$ beam from W(001) at normal incidence; assumed contractions of top-layer spacing 0% (——), 5% (······), and 10% (————) [7.84]

energy, angle, and the particular diffracted beam considered. Intensity and polarization data complement each other, one reason being that many pronounced polarization features occur at energies and angles where the intensity is small and insensitive to the relevant parameters.

Polarization experiments are particularly useful for determining the shape of the surface potential. Analogous to the situation we have discussed in electron-atom scattering (cf. Sect. 3.8), an electron incident on a crystal surface can be temporarily trapped by the surface potential [7.71]. Such nonstationary surface states (surface barrier resonances) are observed as narrow fluctuations of the scattering intensity with respect to variation of electron energy or incidence direction. Due to spin-orbit interaction, these fluctuations differ slightly for e↑ and e↓ (spins parallel and antiparallel to the normal of the scattering plane) which means that the intensity curves are split. The spin splitting depends very sensitively on the surface potential so that observation of the fluctuations of the polarization (or of the scattering asymmetry when working with polarized primary electrons) is a new method for testing theoretical surface-potential models [7.85–87].

In addition to the examples considered we mention that the polarization has been found to be sensitive to adsorption of chemically foreign atoms on the surface and to temperature effects like changes of the top-layer spacing or structural phase transitions. For a more detailed presentation of the complementary information obtained in these cases from polarization and intensity measurements we refer to the review papers [7.4, 68–70].

In this section we have so far discussed only processes in which the polarization is caused by spin-orbit coupling. If the scattering takes place on magnetized materials, the polarization can also arise from exchange scattering, analogously to the processes discussed in Chap. 4. We shall explain this with an example calculated by *Feder* [7.68, 88].

Slow electrons are diffracted from a ferromagnetic iron surface. The exchange interaction between an incident electron and the oriented target electrons depends on the spin direction of the incident electron relative to the magnetization axis. Accordingly, one obtains different cross sections for scattering of e↑ and e↓ (magnetic moments parallel and antiparallel to magnetization; definition of polarization resulting from magnetic materials as in Sect. 7.1) which causes the diffracted beam to be polarized.

Although the polarizing interaction is, in principle, the same as that discussed in Chap. 4, one is now confronted with a many-particle problem, and it is certainly not practical to use antisymmetrized wave functions for the infinite system of the incident and surface electrons. Instead, one may treat the problem by using an approximate spin-dependent potential. The results of Fig. 7.25 are based on a statistical description of the target electrons which yields, besides the electrostatic potential, an exchange potential proportional to $\varrho^{1/3}$, where ϱ is the electron density [7.89]. It has been assumed here that not only elastic, but also inelastic scattering is spin dependent, and that spin-orbit interaction can be neglected. Spin-flip processes by exchange between electrons with opposing spin

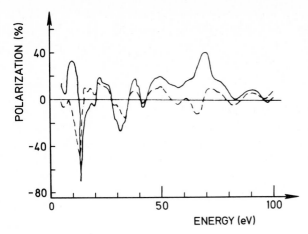

Fig. 7.25. Polarization of specularly reflected beam from ferromagnetic Fe(110) for incidence in a ($\bar{1}$10) plane at 45° against the surface normal. Topmost layer ferromagnetic (——), paramagnetic (–––––) [7.88]

directions have also been excluded because every change of the spin state of the target caused by such processes would mean an excitation of magnons; above 10 eV magnon excitation is, however, an order of magnitude smaller than the other pertinent inelastic processes. Accordingly, polarization of the scattered beam due to spin flip by exchange with target electrons of different spin direction (which was discussed in Chap. 4) does not play a role in the model calculation made here. The polarization of Fig. 7.25 arises because the exchange potential depends on the orientation of the incident spin relative to the excess spin density $\varrho_\uparrow - \varrho_\downarrow$ so that the e$\uparrow$ and e$\downarrow$ of the unpolarized primary beam are scattered from the polarized target with different probabilities.

Since the calculation is based on a model with necessarily simplifying assumptions, one should not take the depicted results too quantitatively. Their value derives from the fact that they clearly show the sensitivity of PLEED to magnetic properties of the target. The polarization curves for two magnetic models differ strongly from each other in height and shape. In contrast, the scattered intensity (not shown here) does not depend on the target magnetization if the primary beam is unpolarized. This is as in electron-atom scattering, where it is seen from (4.36) that the target polarization does not affect the scattered intensity if the incident beam has zero polarization. The information on the magnetic properties is thus coded only in the polarization P if the primary beam is unpolarized, or in the scattering asymmetry A if the primary beam is polarized. Comparison of the calculated polarization and intensity curves [7.88] shows another point of interest: as in many other examples of LEED it is not at all exceptional that maxima of polarization and intensity occur at the same angle and energy.

The example shows how the spin dependence of electron-exchange scattering can be used for studying the structure of magnetic materials. This is analogous to spin-dependent neutron scattering, which led to important discoveries on the structure of magnetic materials. There is, however, an essential difference between the two methods: In neutron scattering the cross sections are very small; they are determined by the short-range nuclear forces and the relatively small interaction of the neutron dipoles with the electron dipoles of the magnetic material. The neutrons can therefore cover great distances in dense material. In electron scattering the cross sections are determined by Coulomb and exchange interactions. Since exchange interaction, which makes the magnetic investigations possible, is significant only at low energies, one must work with slow electrons. The cross sections are then so large that the electrons do not penetrate deep into the material. Electrons are therefore suitable for investigations of the magnetic surface structure, for which neutrons are not appropriate due to their large penetration depths. Accordingly, electron and neutron experiments complement each other in this respect.

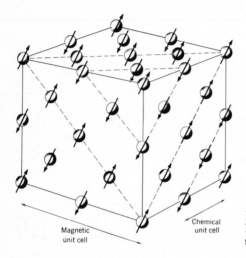

Magnetic
unit cell

Chemical
unit cell

Fig. 7.26. Arrangement of the spins of Ni^{++} ions in NiO. The O^{--} ions are not shown

This application of spin-dependent electron scattering was first recognized and utilized in the late 1960s [7.90, 91]. Positive results were reported for LEED experiments on NiO. The structure of this antiferromagnet is shown in Fig. 7.26. It can be seen that the lattice constant of the magnetic unit cell is twice as large as that of the chemical unit cell. The Coulomb contribution to the scattering is not influenced by the magnetic properties and thus reflects the structure of the chemical unit cell. The spin-dependent exchange scattering is determined by the magnetic unit cell and should therefore yield additional diffraction maxima halfway between those stemming from the chemical unit cell. Observation of these half-order maxima provides a direct means of studying the role of exchange in electron scattering.

Such half-order maxima were indeed observed [7.90, 91] if the temperature of the crystal was kept below the Néel temperature, below which the antiferromagnetism occurs. These investigations on antiferromagnets lie outside the scope of our general topic: They do not have to be carried out with polarized

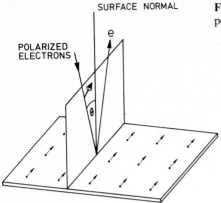

SURFACE NORMAL

e

POLARIZED
ELECTRONS

θ

Fig. 7.27. Scattering geometry for PLEED experiment on surface magnetism

primary beams and the diffraction maxima are not polarized – contrary to what has occasionally been claimed in the literature (see Problem 7.1).

In a later PLEED experiment [7.92] exchange scattering of electrons was directly applied to observing the magnetic field dependence and the temperature dependence of the magnetization at a Ni(110) surface. Figure 7.27 indicates the scattering geometry. The polarization P of the transversely polarized electrons was placed in the scattering plane. Scattering asymmetries caused by spin-orbit interaction can therefore be neglected. As in the experiment described in connection with (7.6), the polarization of the electrons coming from a GaAs source was modulated between P and $-P$. The scattering asymmetry thus produced by exchange interaction is determined by the component of P along the direction of the target polarization, cf. (4.36). This component was made large by operating near normal incidence ($\theta = 12°$). The intensity asymmetry of the specularly reflected beam for the two directions of P is shown in Fig. 7.28 as a function of the magnetic field. The hysteresis curve measured proves clearly that the small asymmetries observed are in fact caused by magnetic scattering. In such

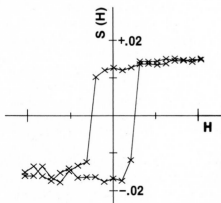

S (H)

+.02

H

−.02

Fig. 7.28. Hysteresis curve of intensity asymmetry in PLEED from magnetic Ni(110) surface; $E = 125$ eV, angle of incidence 12°, specular reflection [7.92]

an experiment it is crucial to minimize stray magnetic fields which can deflect the electrons and cause precession of the electron spins and depolarization. This has been achieved by magnetizing the sample parallel to the surface and providing a closed path for the magnetic flux. The apparatus has also been used for studying elastic and inelastic scattering of electrons from amorphous ferromagnetic glasses where effects of crystal diffraction are negligible [7.93].

The import of such experiments derives from the fact that there are few reliable methods for measuring surface magnetization even though it can be expected to be different from that of the bulk due to the reduced number of neighboring atoms or the possibility of surface reconstruction (i.e., the surface structure differs from that of the truncated bulk lattice). PLEED is therefore applied in several laboratories as an appealing new tool for that purpose. In recent PLEED studies with ferromagnetic Fe(110) films and Ni crystals [7.94–97], the scattering asymmetries of incident polarized electrons caused by spin-orbit interaction and exchange scattering have been determined separately. In contrast to the geometry of Fig. 7.27, both the incident electron spins and the magnetization of the sample were oriented perpendicular to the scattering plane. Reversal of the magnetization at fixed incident polarization causes asymmetries owing to exchange interaction only. Simultaneous reversal of incident polarization and magnetization does not affect exchange scattering but causes an asymmetry owing to spin-orbit interaction which is sensitive to the relative orientation between the polarization and the normal to the scattering plane. In this way the two spin-dependent mechanisms can be experimentally decoupled.

In heavier elements, where both spin-orbit and exchange interaction give rise to significant effects, interference terms between these two interactions may play a role, a situation which was described for electron-atom scattering in Sect. 4.5.2. It has recently been shown that even for the case of unpolarized electrons the interplay between spin-orbit and exchange interaction can cause a sizeable scattering asymmetry if a ferromagnetic gadolinium surface is used as the target [7.97, 98]. This implies that one does not even need polarized electrons in order to study magnetic surface structure by LEED.

Though exchange scattering of polarized electrons from magnetic materials is still in its infancy, its power is obvious. It carries information on both exchange interaction and magnetic surface properties. It has, for instance, been found that for Fe(110) the surface magnetic moment is enhanced by about 30 % with respect to the bulk [7.99], and that the effect of chemisorption on the exchange-induced scattering asymmetry is a sensitive probe for adsorbate layers [7.100]. One may even consider studies of magnon excitation by observing the change of the polarization of the incident beam on scattering.

We must, however, point out that it is not an easy task to extract the surface magnetization from the measured asymmetry since, in general, the asymmetry is not proportional to the magnetization [7.69]. The analysis of surface properties like the temperature dependence of the magnetization is complicated by nonlinear and thermal effects [7.101] and therefore requires comparison of the experimental results with theoretical calculations in which these effects have been taken into account.

Let us mention in passing that the spin orientation of the electrons in magnetic surfaces can also be determined by studying the polarization of heavy particles (e.g., deuterons) which have captured electrons by interaction with the surface [7.102] or by the asymmetry in triplet positronium formation when the electrons are captured by polarized positrons [7.103]. Since the subject of our discussion is polarized free electrons we shall not go into the details of these possibilities.

Problem 7.1. An unpolarized beam of slow electrons impinges on an antiferromagnetic crystal (Fig. 7.26). Is the resulting diffraction pattern determined by the magnetic unit cell (i.e., can one expect to find half-order maxima) or by the chemical unit cell?

Solution. It can immediately be seen that with a polarized beam one would get half-order maxima. This is because the scattering amplitude coming from a particular atom is – as indicated schematically in Fig. 7.29 – generally different according to whether the spins of the electron and atom are parallel or antiparallel to each other (if we had free alkali atoms, then according to Sect. 4.1 the intensities would be $|f|^2 + |g|^2$ in the cases characterized by the solid lines and $|f-g|^2$ for the dashed lines). Accordingly, there can never be complete destructive interference of the bundles coming from two neighboring atoms since their amplitudes are not equal. The lattice constant determining the interference pattern is therefore twice the interatomic distance. If the incident beam consists of e↓ the scattered waves indicated by the solid and dashed lines in Fig. 7.29 must be interchanged. Everything else remains the same; one obtains the same diffraction maxima as before.

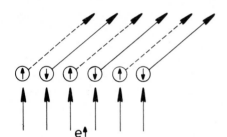

Fig. 7.29. Schematic diagram of the scattering of a polarized electron beam by an antiferromagnet

For an unpolarized incident beam, which consists of e↑ and e↓, one obtains the same scattering intensity from every atom (dashed plus solid lines). Nevertheless, there will be no complete destructive interference between the bundles coming from two neighboring atoms. This would be possible only if the two bundles were coherently superimposed. This does not, however, occur because the part of the scattered wave originating from the e↑ half of the incident wave is not coherent with that originating from the e↓ half, since the incident beam already is an incoherent superposition of e↑ and e↓. The diffraction pattern is therefore the same whether one works with polarized or with unpolarized electrons. Half-order maxima occur in both cases.

8. Further Applications and Prospects

8.1 Polarization Analysis

The motion of polarized electrons in electric and magnetic fields that are used as polarization transformers is described. Methods employing spin-dependent scattering for polarization analysis are discussed with a view toward elimination of systematic errors in asymmetry measurements.

In our treatment of the physics of polarized free electrons we have met many processes which, in principle, are suitable for measurement of electron polarization, but only a few of them have turned out to be really practical. Møller scattering and Compton scattering have been found to be advantageous in the high-energy regime, where other methods fail (cf. Sects. 4.8, 6.1, and 8.4). Production of circularly polarized bremsstrahlung by longitudinally polarized electrons was used in early polarization measurements on β particles (cf. Sect. 6.1). Methods based on spin-dependent electron-atom scattering are widely used in many areas of physics.

In the majority of polarization measurements, techniques are employed which are sensitive to transverse components only, so that transformation of longitudinal to transverse polarization is required. This can be done with the aid of electric and magnetic fields. Let us therefore start with a survey of the motion of polarized electrons in static fields.

8.1.1 Polarization Transformers

Figure 8.1 shows the motion of a polarized electron beam in an electrostatic field that rotates the velocity vectors of the electrons through 90° but does not affect their magnetic moments, so that in the end the longitudinal polarization becomes transverse.

It is, however, only in the nonrelativistic approximation that the electrostatic field does not affect the magnetic moments. To illustrate this we recall the picture used when discussing spin-orbit coupling. An electron experiences in its rest frame a magnetic field equal to $\boldsymbol{E} \times \boldsymbol{v}/c$. In the example of Fig. 8.1 this field is perpendicular to the plane of the diagram. Hence the spins precess in this field so that the polarization becomes slightly rotated when passing through the electric field. Exact relativistic treatments can be found in [8.1, 2]. Neglecting the anomalous magnetic moment of the electron we find from these calculations that

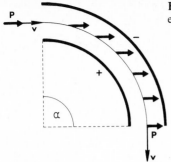

Fig. 8.1. Motion of a polarized electron beam in an electrostatic field

the sector angle α of the electric field in Fig. 8.1 must be $\gamma\pi/2$ if the emerging polarization is to be transverse $(\gamma = 1/\sqrt{1-(v/c)^2})$.

Needless to say, magnetic fields also can be used for spin rotation. Since a rigorous relativistic calculation of the spin rotation in arbitrary fields is complicated, we only discuss the most important special cases (a survey was given by *Farago* [8.3]). If the electrons move along the direction of the magnetic field and the polarization is perpendicular to it (see Fig. 8.2a), their direction of motion remains unchanged and $\boldsymbol{P}$ precesses about the field direction with a frequency which in the first approximation is $\omega = eB/m\gamma c$. The exact expression for the precession frequency in this particular case is

$$\omega = \frac{g}{2} \cdot \frac{eB}{m\gamma c}, \qquad (8.1)$$

which is identical to the previous expression if $g = 2$. A more exact value of g is, however, $g = 2(1 + a)$ where $a = 1.16 \cdot 10^{-3}$ (see Sect. 8.3).

We now consider the case of Fig. 8.2b where $\boldsymbol{B}$, $\boldsymbol{v}$, and $\boldsymbol{P}$ are mutually perpendicular to each other. Then the electrons move in a circular orbit with the cyclotron frequency

$$\omega_c = \frac{eB}{m\gamma c}. \qquad (8.2)$$

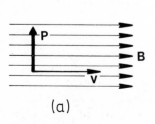

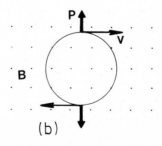

(a)   (b)

Fig. 8.2a,b. Motion of a transversely polarized electron beam (a) parallel and (b) perpendicular to a homogeneous magnetic field

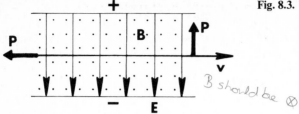

Fig. 8.3. Wien filter as a spin rotator

B should be ⊗

If the g factor were exactly 2, then precession frequency and cyclotron frequency would be equal so that the angle between P and v would remain unchanged. Since g is slightly larger than 2, the magnetic moment of the electrons is somewhat larger than the Bohr magneton so that P precesses slightly faster. The precession frequency differs by $a \cdot eB/mc$ from the value for $g=2$:

$$\omega_p = \frac{eB}{m\gamma c} + a \cdot \frac{eB}{mc} = \frac{eB}{mc}\left(\frac{1}{\gamma}+a\right). \tag{8.3}$$

We note only briefly that the difference from the frequency in (8.1) for motion along the magnetic field is due to the fact that with a circular motion, a Lorentz transformation does not lead directly to the rest frame but to a precessing frame (Thomas precession).

Due to the gradual advance of the polarization caused by $g>2$, there is a gradual change in the angle between P and v, so that after nearly 10^3 revolutions of the electrons the polarization vector has made one extra revolution (when $\gamma \approx 1$). This means that after approximately 200 revolutions, the polarization has been transformed from longitudinal to transverse (or vice versa).

Transformation of longitudinal into transverse polarization, or vice versa, is also possible without affecting the orbit. This happens in the Wien filter shown in Fig. 8.3. In the crossed electro- and magnetostatic fields, the electrons experience no resultant force if $eE - eBv/c = 0$, i.e., $v = Ec/B$. Their spins, however, precess through the angle $eLB^2/m\gamma^2c^2E$ ($L=$ length of the Wien filter), as can immediately be seen for the nonrelativistic limiting case.

Instead of macroscopic fields one may also use atomic fields to transform the polarization. After scattering through 90° a longitudinally polarized beam has transverse polarization, provided that the spins are not affected by the scattering process. It is true that scattering does change the polarization, but the quantitative results of Sect. 3.6 show that, at many energies and scattering angles, the change of the polarization vector P is small even for target atoms of high atomic number. By scattering from atoms of low atomic number, P remains practically unaffected. Scattering from carbon has therefore been used as a simple though reliable polarization transformer [8.4]. Even scattering from a gold foil is a satisfactory method for transformation of polarization [8.5].

8.1.2 Polarization Analysis Based on Scattering Asymmetry

The overwhelming majority of electron polarization studies have exploited the left-right asymmetry of scattering. Most of them use Mott scattering in the 100-keV region, but low-energy scattering [8.6–9] and the PLEED analyzer presented in the preceding chapter [8.10, 11] have turned out to be very suitable as well. Because of its practical importance, we will now describe in more detail experimental aspects of polarization analysis by measurement of the left-right scattering asymmetry.

Though the method is quite simple, there are quite a few problems to overcome if the results are to be reliable. An experimental setup as shown in Figs. 3.7 or 3.28 usually possesses a purely instrumental asymmetry which may be caused by different efficiencies of the detectors on the right and left, by not strictly axial alignment of the incident polarized beam, by inhomogeneity of the target or of the beam, etc. Either the instrumental asymmetry must be eliminated, or the measurements have to be corrected for this spurious asymmetry. In what follows, we will discuss three basic cases in order of increasing complication.

a) Perfect Alignment of Incident Beam. Let us first assume a case of limited instrumental asymmetry, where there is no misalignment of the incident beam and the scattering angle θ in the analyzer is well adjusted, but where the solid angles $\Omega_{1,2}$ subtended by the two detectors and the detector efficiencies $E_{1,2}$ are different. For an incident $e\uparrow$ beam the numbers of counts in the detectors on the left and right are, according to (3.70),

$$L_\uparrow = nNE_1\Omega_1 I(\theta)\,[1 + PS(\theta)] \tag{8.4}$$

$$R_\uparrow = nNE_2\Omega_2 I(\theta)\,[1 - PS(\theta)],$$

where n is the number of incident particles, N is the number of target atoms per unit area, and P is the polarization of the incident beam.

If $E_1\Omega_1 \neq E_2\Omega_2$ one has a spurious asymmetry which must be eliminated. This can be done by reversing the polarization direction of the incident beam. In Fig. 3.7 this means, for example, rotating the primary beam through an azimuthal angle ϕ of 180°. This leads to reversal of the vector $\hat{\boldsymbol{n}}_1 = \boldsymbol{k} \times \boldsymbol{k}'/|\boldsymbol{k} \times \boldsymbol{k}'|$ normal to the scattering plane and thus to reversal of the polarization $S(\theta_1)\hat{\boldsymbol{n}}_1$. Such a flip of the polarization results in an inversion of the scattering asymmetry and one obtains

$$L_\downarrow = n'N'E_1\Omega_1 I(\theta)\,[1 - PS(\theta)]$$

$$R_\downarrow = n'N'E_2\Omega_2 I(\theta)\,[1 + PS(\theta)], \tag{8.5}$$

where the primes indicate that, in this run, the number of incident particles and the effective target thickness can be different (the effective target thickness may change due to inhomogeneities of the target and the beam without a change of the overall alignment of the beam).

From

$$\sqrt{L_\uparrow R_\downarrow} \equiv N^+ = \sqrt{nn'NN'E_1 E_2 \Omega_1 \Omega_2} I(1+PS), \tag{8.6a}$$

$$\sqrt{R_\uparrow L_\downarrow} \equiv N^- = \sqrt{nn'NN'E_1 E_2 \Omega_1 \Omega_2} I(1-PS), \tag{8.6b}$$

one obtains the asymmetry

$$A = \frac{N^+ - N^-}{N^+ + N^-} = PS \tag{8.7}$$

independent of detector efficiencies and solid angles and of the variation of target thickness and number of incident particles. The measurements just discussed enable us to determine the instrumental asymmetry: The ratio of the quantities

$$\sqrt{L_\uparrow L_\downarrow} = \sqrt{nn'NN'[1-(PS)^2]} IE_1 \Omega_1$$
$$\sqrt{R_\uparrow R_\downarrow} = \sqrt{nn'NN'[1-(PS)^2]} IE_2 \Omega_2 \quad \text{is} \tag{8.8}$$

$$\frac{\sqrt{L_\uparrow L_\downarrow}}{\sqrt{R_\uparrow R_\downarrow}} = \frac{E_1 \Omega_1}{E_2 \Omega_2} \equiv \varrho.$$

If this ratio differs from 1 we have, according to (8.4) or (8.5), an instrumental asymmetry.

In discussing the procedure for determining PS independent of such an instrumental asymmetry, we have tacitly assumed that ϱ does not vary in time. This is, however, not necessarily warranted. Different ϱ for the two runs can, for instance, occur at high counting rates when the difference in the dead time corrections between the high- and low-intensitity directions is not accounted for. Monitoring $\sqrt{L_\uparrow L_\downarrow}/\sqrt{R_\uparrow R_\downarrow}$ provides a check on the performance of the apparatus. If this quantity is not constant in time the measured asymmetry cannot be claimed to be accurate.

b) Misalignment of Incident Beam; Cancellation of Instrumental Asymmetry.
Let us now assume that we have a misalignment of the incident beam. Let the beam be displaced an amount Δx and rotated by an angle δ with respect to the symmetry axis of the detector system (see Fig. 8.4). Instead of (8.4) we then obtain

$$L_\uparrow = nNE_1 \Omega_1(\Delta r_1, \Delta\theta_1) I(\theta+\Delta\theta_1) [1+PS(\theta+\Delta\theta_1)]$$
$$R_\uparrow = nNE_2 \Omega_2(\Delta r_2, \Delta\theta_2) I(\theta+\Delta\theta_2) [1-PS(\theta+\Delta\theta_2)], \tag{8.9}$$

where Δr and $\Delta\theta$ are illustrated in Fig. 8.4 and, to first order, are given by

Fig. 8.4. Misalignment of incident beam

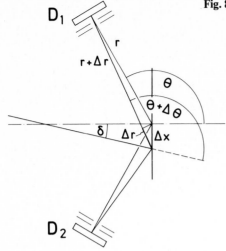

$$\Delta r_1 = \Delta x \sin\theta = -\Delta r_2 \equiv \Delta r$$

$$\Delta\theta_1 = \frac{\Delta x}{r}\cos\theta + \delta = -\Delta\theta_2 \equiv \Delta\theta. \tag{8.10}$$

Using the first-order expansion

$$S(\theta \pm \Delta\theta) \approx S(\theta)\left(1 \pm \frac{1}{S}\frac{\partial S}{\partial\theta}\,\Delta\theta\right), \tag{8.11}$$

one obtains

$$L_\uparrow = nNE_1\Omega_1(\Delta r, \Delta\theta)I(\theta + \Delta\theta)\left(1 + PS + P\frac{\partial S}{\partial\theta}\,\Delta\theta\right)$$

$$R_\uparrow = nNE_2\Omega_2(-\Delta r, -\Delta\theta)I(\theta - \Delta\theta)\left(1 - PS + P\frac{\partial S}{\partial\theta}\,\Delta\theta\right). \tag{8.12}$$

A flip of the polarization yields instead of (8.5)

$$L_\downarrow = n'N'E_1\Omega_1(\Delta r, \Delta\theta)I(\theta + \Delta\theta)\left(1 - PS - P\frac{\partial S}{\partial\theta}\,\Delta\theta\right)$$

$$R_\downarrow = n'N'E_2\Omega_2(-\Delta r, -\Delta\theta)I(\theta - \Delta\theta)\left(1 + PS - P\frac{\partial S}{\partial\theta}\,\Delta\theta\right). \tag{8.13}$$

In forming $\sqrt{L_\uparrow R_\downarrow}$ and $\sqrt{R_\uparrow L_\downarrow}$ the first-order terms of the Sherman function cancel. Neglecting second-order errors one obtains the asymmetry (8.7),

$$A = \frac{N^+ - N^-}{N^+ + N^-} = PS.$$

The essential assumption underlying this result is that the beam position is not changed when the polarization is reversed. The misalignment must stay the same during the whole measurement. This can be checked by monitoring the beam position when the polarization is flipped.

c) Misalignment of Incident Beam; Correction for Instrumental Asymmetry. It is important to point out that the conclusions made above do not hold if, in order to eliminate the instrumental asymmetry, the detectors are interchanged while the beam remains fixed in space (cf. Fig. 8.5). This can be done by rotating the part of the scattering chamber which contains the detectors through 180°. We use the first-order expansions

$$I(\theta \pm \Delta\theta) \approx I(\theta)\left(1 \pm \frac{1}{I}\frac{\partial I}{\partial \theta}\Delta\theta\right) \quad \text{and}$$

$$\Omega(\pm \Delta r, \pm \Delta\theta) \approx \Omega \cdot \left(1 \pm \frac{\Delta\Omega}{\Omega}\right),$$

where Ω without an argument denotes again the solid angle for perfect alignment of the beam. Hence one has for the first run

$$L_\uparrow = nNE_1\Omega_1 I(\theta)\left(1 + PS + P\frac{\partial S}{\partial \theta}\Delta\theta\right)\left(1 + \frac{\Delta\Omega}{\Omega}\right)\left(1 + \frac{1}{I}\frac{\partial I}{\partial \theta}\Delta\theta\right)$$

$$R_\uparrow = nNE_2\Omega_2 I(\theta)\left(1 - PS + P\frac{\partial S}{\partial \theta}\Delta\theta\right)\left(1 - \frac{\Delta\Omega}{\Omega}\right)\left(1 - \frac{1}{I}\frac{\partial I}{\partial \theta}\Delta\theta\right)$$

$$(8.14a)$$

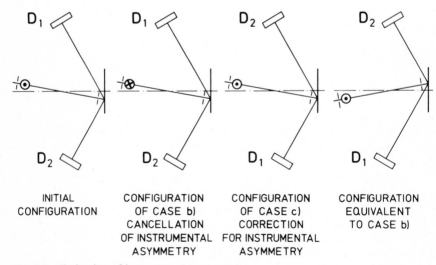

| INITIAL CONFIGURATION | CONFIGURATION OF CASE b) CANCELLATION OF INSTRUMENTAL ASYMMETRY | CONFIGURATION OF CASE c) CORRECTION FOR INSTRUMENTAL ASYMMETRY | CONFIGURATION EQUIVALENT TO CASE b) |

Fig. 8.5. Elimination of instrumental asymmetries

and for the second run

$$L'_\uparrow = n'N'E_2\Omega_2 I(\theta) \left(1 + PS + P\frac{\partial S}{\partial \theta}\Delta\theta\right)\left(1 + \frac{\Delta\Omega}{\Omega}\right)\left(1 + \frac{1}{I}\frac{\partial I}{\partial \theta}\Delta\theta\right)$$

$$R'_\uparrow = n'N'E_1\Omega_1 I(\theta) \left(1 - PS + P\frac{\partial S}{\partial \theta}\Delta\theta\right)\left(1 - \frac{\Delta\Omega}{\Omega}\right)\left(1 - \frac{1}{I}\frac{\partial I}{\partial \theta}\Delta\theta\right),$$

(8.14b)

where L and R still denote the counting rates on the left and the right, whereas E and Ω are correlated with the physical detectors and not with their position in space (detectors 1 and 2 have been interchanged now).

From

$$\sqrt{L_\uparrow L'_\uparrow} \equiv N^+, \qquad \sqrt{R_\uparrow R'_\uparrow} \equiv N^-,$$

and (8.14) we now obtain the measured asymmetry

$$A_m = \frac{N^+ - N^-}{N^+ + N^-} = \frac{(1 + PS + P\Delta S)(1 + A_f) - (1 - PS + P\Delta S)(1 - A_f)}{(1 + PS + P\Delta S)(1 + A_f) + (1 - PS + P\Delta S)(1 - A_f)}$$

$$= \frac{PS + A_f}{1 + P\Delta S + A_f PS},$$

where

$$\Delta S = \frac{\partial S}{\partial \theta}\Delta\theta \quad \text{and} \quad A_f = \frac{\Delta\Omega}{\Omega} + \frac{1}{I}\frac{\partial I}{\partial \theta}\Delta\theta,$$

while second-order terms have been omitted. Since analyzers are usually operated at angles where $\Delta S \ll 1$, one has

$$A_m = \frac{PS + A_f}{1 + A_f PS}. \tag{8.15}$$

For an unpolarized beam, $P = 0$, one has $A_m = A_f$, i.e., the measured asymmetry equals the false asymmetry. This enables one to determine the false asymmetry A_f if it is certain that beam positions and directions are the same for the polarized and the unpolarized beam. The polarization P can then be determined from (8.15).

An arrangement such as described in Case b) is certainly preferable, since PS can be evaluated without determination of spurious asymmetries. It is also possible, however, to achieve this in the case where the detectors are interchanged, if the apparatus is rotated in such a way that beam position and direction remain invariant with respect to each of the detectors. This is illustrated in the last configuration of Fig. 8.5. It is obtained by turning the configuration of Case b) upside down and is thus equivalent to that arrangement. In order to accomplish such a rotation it is necessary that the slits which define the beam in the analyzer rotate together with the detectors. In addition, the polarization and the beam current readings on these slits must remain the same in the initial and the second run.

The above considerations show that, since the ideal Case a) of a perfectly aligned beam cannot be accomplished in practice, one should attempt to realize Case b) in a polarization measurement, either by flipping the polarization or, as shown in the last configuration of Fig. 8.5, by rotating the complete analyzing system. If the arrangement is not close to the Case b) of exact cancellation of the instrumental asymmetries, then part of these asymmetries has to be corrected for. In a situation like this, one should adjust the apparatus so that the instrumental asymmetries are small; errors in their evaluation are then of little influence on the final result for the polarization. Even if one works fully in the less favorable Case c), the effect of the false asymmetry A_f on the measured asymmetry is shown by (8.15) to be small if A_f is small compared to PS.

Sometimes instrumental asymmetries are checked simply by reducing the electron polarization to zero. Let us assume that this can be done in good approximation without affecting the beam characteristics, for instance by switching from circular to linear light polarization in studies with photoemission sources of polarized electrons (cf. Sect. 8.2). If the numbers of counts which follow from (8.12) for $P=0$ are denoted by L_0 and R_0, the instrumental asymmetry A_i is given by

$$A_i = \frac{L_0 - R_0}{L_0 + R_0}.$$

Since, as mentioned above, $\Delta S = (\partial S / \partial \theta) \Delta \theta \ll 1$ is usually fulfilled, one obtains from (8.12) for the asymmetry measured in one run with the polarized beam

$$A'_m = \frac{L_\uparrow - R_\uparrow}{L_\uparrow + R_\uparrow} = \frac{L_0(1 + PS) - R_0(1 - PS)}{L_0(1 + PS) + R_0(1 - PS)} = \frac{PS + A_i}{1 + A_i PS}. \tag{8.16}$$

A'_m has the form of (8.15) though, due to the different measurement procedure, A_i is now the full instrumental asymmetry which is generally larger than A_f. Accordingly, this method does not reach the accuracy of the others unless A_i can be determined with high precision. However, evaluation of PS from (8.16) may serve as an additional check of the measurement by the spin-flip method discussed above.

In order to check whether the changes of the beam polarization made to eliminate instrumental asymmetries affect position and angle of the beam, it is useful to monitor spurious asymmetries by two additional detectors set up symmetrically at small angles (see Fig. 5.10). If the angles are chosen such that $S(\theta) \approx 0$, any asymmetries observed are purely instrumental. In cases where the reversal of the polarization must be carried out by reversal of a magnetic field, such as in experiments with magnetic materials, the influence of the field on the beam trajectory is otherwise hard to control.

The above discussion of a few basic examples of asymmetry elimination is meant to illustrate how the practical situation in a realistic experiment may be reduced to a simple idealized case. Idealizations such as disregarding the fact that the beam has a finite diameter induce additional errors. The experimentalist has to make sure that in each specific situation these errors are only of second order. In the example of finite beam diameter, the beam can be considered to be made up of individual trajectories which are displaced with respect to the axis and can be treated according to Case b). If the beam does not deviate too much from axial symmetry the corrections are of second order. For polarization measurements of high precision it does not suffice to rely on a general discussion such as the foregoing; instead the specific experimental arrangement has to be reanalyzed.

One error source we have not yet discussed is random variation of the beam, which is a common problem. The best way of coping with it is the rapid reversal of the polarization in regular short intervals so that, on average, the random fluctuations cancel out.

One can certainly imagine further sources of spurious asymmetry which have not been discussed here, for instance, uncertainties of the angular position of the two detectors or errors of second order. Since a mechanical apparatus can be constructed nowadays with great accuracy and since the detectors are usually positioned in directions where one has only a slight angular dependence of the analyzing power S, such errors remain, in general, below the limits reached in polarization experiments.

In order to determine P from the asymmetry $PS(\theta)$ observed in a Mott analyzer, one needs an accurate value of S. One cannot simply use the theoretical value which was calculated for single scattering by one atom. Every real target contains so many atoms that plural or multiple scattering processes also occur; this becomes more likely, the thicker the target foil is (see end of Sect. 3.7.1). Electrons which arrive at the counters after several consecutive scattering processes usually reduce the intensity asymmetry.

Consequently, instead of using the ideal Sherman function one must use an effective Sherman function which depends on the target thickness. It also depends on other conditions of the experiment, for example, on the range of the scattering angle θ which is recorded by the electron detectors. Since there are rather large uncertainties in the theoretical treatment of plural scattering, it is advisable to experimentally ascertain the effective Sherman function for the Mott detector chosen. This can be done by calibrating the Mott detector with an electron beam of known polarization. It can also be done with a polarized beam of unknown polarization by measuring the increase of the scattering asymmetry with decreasing foil thickness d. If the amount of plural scattering is very small with the thinnest of the foils used, one can extrapolate from such a measured curve to the asymmetry for the foil thickness $d=0$. This can be related to the Sherman function for $d=0$ which is well established theoretically as well as experimentally for a suitable choice of scattering angle, electron energy, and target material (e.g., at $\theta=120°$, $E=120$ keV, $Z=79$ the value $S=-0.37$ has been established to within $\pm3\%$ [8.12]). The measured curve then yields the

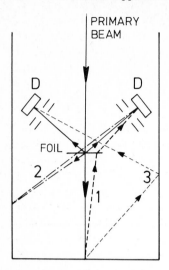

PRIMARY
BEAM

Fig. 8.6. Distortion of the measurement by backscattered electrons

effective Sherman function also for the other foils used in the series of measurements.

Spurious electrons can arrive at the counters not only due to plural scattering in the foil but also due to reflection at the walls of the scattering chamber. Figure 8.6 shows a few typical cases:

1) Unscattered electrons from the incident beam are reflected on the chamber wall and after hitting the foil are scattered into a detector.
2) Scattered electrons hit the wall and are reflected into the direction of a detector.
3) Electrons from the incident beam reach a detector after double reflection at the wall.

In order to suppress these background electrons several of the following preventative measures are usually required: coating the inside of the chamber with a material having a small backscattering coefficient (small atomic number, e.g., carbon), suitable arrangement of diaphragms which capture the backscattered electrons, and good energy resolution of the counters since the electrons lose part of their energy due to the reflection.

To obtain significant left-right asymmetries, large values of the effective Sherman function S_{eff} are desirable. We have seen, however, in Sect. 3.4.2 that at angles where the Sherman function is large, the cross sections, and thus the scattering intensities, are small. Therefore one must find a compromise between high scattering intensity I and high asymmetry. It can easily be seen (see Problem 8.1) that one should choose $S_{\text{eff}}^2 I$ to be as large as possible, in order to make the statistical error as small as possible. As I also depends on the incident intensity I_0, it is reasonable to use the quantity $S_{\text{eff}}^2 I / I_0$ as a figure of merit when comparing different Mott detectors.

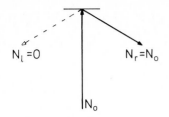

Fig. 8.7. Ideal Mott detector

$N_l = 0$ $N_r = N_0$

N_0

The decrease of the effective Sherman function with increasing foil thickness is rather slow. Even for an infinitely thick scatterer, $S_{eff} = -0.027$ has been found at 261 keV and 105° [8.12]. This is still about 9 % of the value for an infinitely thin foil. More detailed measurements at bulk targets have been made [8.13]. Due to the slow decrease of S_{eff} and a stronger increase of the scattering intensity, $S_{eff}^2 I/I_0$ increases with increasing foil thickness within wide limits. Nevertheless one should not use a really thick scattering foil for polarization analysis as one could be inclined to do when paying attention to the statistical error only: owing to the small left-right asymmetry of the thick foil with its small S_{eff}, the instrumental asymmetry predominates and may give rise to comparatively large systematic errors of the polarization measurement. An example of a favorable choice of foil thickness is the following. For electron energies of 120 keV and scattering angles of about 120°, gold foils with area densities between 0.1 and 0.5 mg/cm² (i.e., thicknesses of approximately 50–250 nm) are suitable.

Even if one chooses a foil thickness in the upper range of the example just given, the values of I/I_0 still do not exceed 10^{-3} to 10^{-2}. This means that from a thousand polarized electrons, less than ten can be detected. In comparison with this an analyzer for polarized light functions practically without loss. This is one of the reasons why experiments with polarized electrons are so much more difficult. For a Mott detector the efficiency $S_{eff}^2 I/I_0$ is of order 10^{-4} [8.14]. The corresponding value for an analyzer of polarized light, which, according to its position, almost totally absorbs or transmits polarized light is near 1. A Mott detector of this efficiency would have to scatter all the electrons of a totally polarized incident beam off to one side (see Fig. 8.7)!

Mott detectors can also be operated at low energies, for instance in the range 5–1000 eV. Then scattering foils are no longer employed because due to the much larger scattering cross sections at these energies multiple scattering would affect the Sherman function considerably. One therefore uses Hg-vapor beams of moderate density. The advantage of low-energy operation is that, when experimenting with slow polarized electrons, it is not necessary to accelerate them afterwards to higher energies in order to measure their polarization so that the apparatus becomes more compact. The efficiency is comparable to that of Mott detectors working at higher energies, a values of $4 \cdot 10^{-5}$ having been reported at a detection energy of 15 eV [8.7].

In another version which is occasionally employed, scattering from the gold foil occurs at high energy whereas the scattered electrons are detected at low energy [8.15, 16]. The compactness of the device and the improved rejection of inelastically scattered electrons is offset by a somewhat lower efficiency: According to Liouville's theorem, the cone of scattered electrons is inflated in the course of deceleration from scattering energy to detection energy so that the number of electrons reaching the counters decreases with increasing discrimination against inelastically scattered electrons.

Since the PLEED detector (cf. page 220) exploits the same concept of asymmetry measurement, our considerations on elimination of instrumental asymmetries may also be applied to this analyzer, although specific features require attention. Because of the angular sensitivity of the analyzing power S, the diffraction conditions must be carefully chosen so as to stay within the first-order corrections discussed above; the energy dependence of S requires that the polarized electrons have a small energy width. A calibration of S is mandatory. Efficiencies of 10^{-4} can be reached without difficulty [8.10]. The necessity of a vacuum in the low 10^{-10} mbar range makes the PLEED detector an appropriate device for experiments which already require ultrahigh vacuum for other reasons.

Problem 8.1. Calculate the statistical error of the polarization measurement with a Mott detector and establish the assertion that $S_{\text{eff}}^2 I$ should be as large as possible (if systematic errors are ignored).

Solution. Since $A = (N_l - N_r)/(N_l + N_r) = P S_{\text{eff}}$, the error of the polarization measurement for a given S_{eff} is

$$\Delta P = \frac{1}{S_{\text{eff}}} \cdot \Delta A.$$

From the law of propagation of errors, the error ΔA of the measured asymmetry is expressed in terms of the errors of the individual measurements ΔN_l and ΔN_r by

$$\Delta A = \sqrt{\left(\frac{\partial A}{\partial N_l}\right)^2 (\Delta N_l)^2 + \left(\frac{\partial A}{\partial N_r}\right)^2 (\Delta N_r)^2}$$

$$= \sqrt{\left(\frac{2 N_r}{(N_l + N_r)^2}\right)^2 N_l + \left(\frac{-2 N_l}{(N_l + N_r)^2}\right)^2 N_r}$$

(for the errors ΔN_i of the individual measurements, the statistical errors $\sqrt{N_i}$ have been substituted). Setting $N_l + N_r = N$ one obtains

$$\Delta A = \sqrt{\frac{4 N_r N_l}{N^3}}.$$

Since $1 - P^2 S_{\text{eff}}^2 = 4 N_l N_r / N^2$, it follows that

$$\Delta A = \sqrt{\frac{1}{N} (1 - P^2 S_{\text{eff}}^2)} \quad \text{and}$$

$$\Delta P = \sqrt{\frac{1}{N} \left(\frac{1}{S_{\text{eff}}^2} - P^2\right)}.$$

With the Mott detectors used in practice, the effective Sherman functions are not very large so that $1/S_{eff}^2 > 10\,P^2$; thus

$$\Delta P = \sqrt{\frac{1}{N S_{eff}^2}}.$$ (8.17)

Since N, the number of particles observed, is (under otherwise identical conditions) proportional to the scattering intensity I, ΔP becomes smaller as $S_{eff}^2 I$ becomes larger.

8.1.3 Spin-Dependent Absorption

Another idea that utilizes spin-dependent scattering for polarization analysis has been put forward more recently [8.17–19]. It exploits the fact that, due to spin-dependent interaction, the absorption of an electron beam in matter depends on its polarization. Figure 8.8 illustrates the basic idea. The current registered by the ammeter (which will be called absorbed current I_a for simplicity) differs from the current I_0 hitting the sample; this is because part of the incident electron beam is scattered backward by the sample and because secondary electrons leave the sample. At certain energies the number of electrons leaving the sample happens to equal the number of incident electrons, so that the absorbed current is zero. Let us assume that, at some energy $E_{0\uparrow}$, this situation occurs for $e\uparrow$ (polarization parallel to normal of plane of incidence) and that the sample has a high atomic number Z causing a significant left-right asymmetry of the scattered electrons. If the same experiment is then made with $e\downarrow$ one cannot expect the absorbed current to be zero again. This is because the numbers of electrons scattered to the right and to the left which, owing to the oblique incidence of the primary beam, have different chances of escaping from the sample are then interchanged. The left-right asymmetry of scattering thus results in different numbers of escaping electrons for incidence of $e\uparrow$ or $e\downarrow$ so that the absorbed current cannot be zero for both spin directions at the same electron energy.

The spin-dependent absorption is illustrated in Fig. 8.9 for the situation where the electrons hit a polycrystalline gold foil at an angle of 35°. At an energy E_0 of 130.3 eV, where one finds $I_a = 0$ for unpolarized electrons, totally polarized electrons produce an absorbed current of about 0.2 % of the incident current. Reversal of the polarization reverses the direction of the absorbed current. The difference between the absorbed currents at E_0 is proportional to the degree of polarization and may, after calibration, be used for polarization measurement. The difference $E_{0\uparrow} - E_{0\downarrow}$ limits the tolerable energy spread of the beam to be spin analyzed.

Spin-dependent absorption has been found not only in materials like gold and tungsten where spin-orbit interaction is responsible for the effect, but also in ferromagnetic material like the amorphous glass $Ni_{40}Fe_{40}B_{20}$ where it is caused by spin-dependent exchange interaction [8.17]. Absorption in magnetic solids can therefore be used as a polarization analyzer as well.

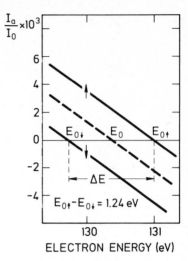

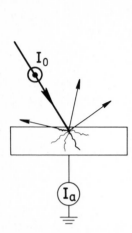

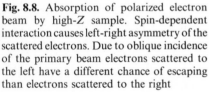

Fig. 8.8. Absorption of polarized electron beam by high-Z sample. Spin-dependent interaction causes left-right asymmetry of the scattered electrons. Due to oblique incidence of the primary beam electrons scattered to the left have a different chance of escaping than electrons scattered to the right

Fig. 8.9. Fraction I_a/I_0 of absorbed current vs. electron energy for totally polarized and unpolarized (– – –) electron beams impinging on a clean polycrystalline gold foil. Angle of incidence 35°. Polarization parallel ↑ and antiparallel ↓ to normal of plane of incidence [8.19]

The values of I_a/I_0 and $\overline{\Delta E}$ given in Fig. 8.9 are typical ones. By suitable choice of the parameters of the experiment (such as angle of incidence, structure, and contamination of the surface) more favorable values can be found as can be seen from the references given, so that efficiencies comparable to the best Mott and PLEED analyzers can be attained. In order to obtain stable and reproducible analyzer parameters annealed gold films evaporated in ultrahigh vacuum had to be used [8.20]. The absorption detector can be employed when the current of polarized electrons is in the measurement range of sensitive ammeters ($> 10^{-17}$ A), since the method does not permit the counting of single electrons. For a detailed discussion of polarization analysis by spin-dependent absorption we refer to [8.20].

Though Mott scattering is presently the standard technique for polarization analysis, various polarization studies have been made with each of the other methods explained. Since different experiments put different demands on the analyzer, it is worth considering which is the most suitable device for a specific project. Generally speaking, though, electron polarimeters, with their low efficiency, are less advanced than polarized-electron sources, which will be discussed in the next section. Consequently, in those cases where an asymmetry measurement with a polarized beam and a polarization measurement with an unpolarized (incident) beam give equivalent results, the asymmetry measurement is certainly preferable.

8.2 Sources of Polarized Electrons

Some of the processes discussed in this monograph have been used to build polarized-electron sources. Characteristics of the sources are given and a comparison of their utility is made.

The discussions throughout this book have shown that an efficient source of polarized electrons is an attractive device to have for novel investigations in various fields of physics. Indeed, there have been many attempts to utilize as polarized-electron sources many of the processes we have described. Sometimes workers in the field gave an overoptimistic prognosis for the efficiency, as a source, of the particular process they were studying or proposing. From all the processes we have met, only a few have turned out to be an adequate basis for sources of high efficiency.

We shall now give a brief account of the state of the art of polarized-electron sources. In order to be able to compare the various sources with each other, one needs common criteria of performance. To establish such criteria may at first sight appear trivial since a source is better, the higher its polarization and current are and the better its beam is collimated. One does not, however, get much further with such general ideas when one has, for example, to make the following simple decision: Would one rather have a source that yields a totally polarized beam with a moderate current or one with ten times as much current and a polarization of, say, 20 %.

To answer this question, we start with the fact that in experiments with polarized-electron beams the information is mostly drawn from the relative difference or "asymmetry" A of the counting rates obtained with opposing polarizations[1] P and $-P$. The asymmetry $A = PS$ is determined by the polarization P and the quantity S which describes the spin dependence of the process to be studied. If the polarization of the electrons emitted by the source is known, the error of the desired quantity S is determined by the error of the asymmetry measurement:

$$\Delta S = \Delta A / P.$$

By referring to Problem 8.1, one can immediately see that the statistical error is

$$\Delta S \approx \frac{1}{\sqrt{P^2 N}}. \tag{8.18}$$

(In Problem 8.1, S was the known quantity and P had to be measured. The situation is reversed here so that P and S are also interchanged in the final result.) Since the observed number of particles N, under otherwise identical conditions,

---

[1] This includes measurements of left-right asymmetry, if the direction of P is referred to the direction of $\hat{n}$.

is proportional to the incident current I, it follows from (8.18) that the error will decrease as $P^2 I$ increases. This quantity is therefore often taken to be a measure of the quality of a source of polarized electrons. In the example above, if the polarization is five times smaller, one needs twenty-five times as much current in order to obtain the same error limits in the same time.

There is, however, no point in using $P^2 I$ in every case as a figure of merit, expecially when counting statistics is not the main source of error. The polarization must be large enough so that the spin-dependent asymmetries one wants to study do not become completely masked by asymmetries of instrumental origin which would lead to a situation where the systematic errors outweigh the statistical errors. Polarizations below a few percent are of no interest for most purposes. Another case in which it would be inappropriate to use $P^2 I$ as a figure of merit is high-energy electron scattering on polarized targets (see Sect. 8.4) where the electron bombardment reduces the polarization of the target. In such an experiment it would be pointless to compensate for small polarizations by using high intensities. When using $P^2 I$ as a somewhat rough figure of merit in the following, we should bear such restrictions in mind. If two sources have the same value of $P^2 I$, the source with the higher polarization is usually to be preferred.

One also needs to know whether a source yields a well-collimated beam which can easily be handled by electron optical devices, that is, can be sent through lenses, filters, or spectrometers without much loss of intensity. This can be suitably described by the brightness b which is conventionally used to describe normal electron sources. It is defined as the current density per unit solid angle. A source that concentrates a high current density into a small solid angle has high brightness. If r_0 is the radius of a beam-cross-section minimum (for example, at the exit of the source) and α_0 the corresponding semi-aperture of the beam, one has the brightness (see Problem 8.2)

$$b_0 = \frac{I}{\pi^2 r_0^2 \alpha_0^2} \tag{8.19}$$

if α_0 is not too large (I = beam current).

One should also take into account that the angular divergence of an electron beam is reduced if the beam is accelerated. This is obvious because of the increase of the longitudinal momentum components during acceleration and is quantitatively described by Lagrange's law

$$r_0 \alpha_0 \sqrt{E_0} = r_f \alpha_f \sqrt{E_f} \tag{8.20}$$

(the indices refer to the states before and after acceleration). Consequently, if two sources have the same brightness b_0 but different energies, the source with the lower energy is superior: After acceleration of the electrons to the energy E_f at which the experiment is to be carried out, this source will yield the larger brightness, since

$$b_f = \frac{I}{\pi^2 r_f^2 \alpha_f^2} = \frac{I E_f}{\pi^2 r_0^2 \alpha_0^2 E_0} = b_0 \frac{E_f}{E_0}.$$

Thus for a given brightness b_0 of the source and a given energy E_f at which the experiment is to be carried out, the brightness b_f is inversely proportional to the energy of the electrons leaving the source.

When one considers that for polarized electrons it is not I but rather $P^2 I$ that is an adequate figure of merit for the source, and that E_f is not a property of the source but rather a parameter of the experiment, it seems, from the previous discussion, sensible to use the quantity

$$q = \frac{P^2 I}{r_0^2 \alpha_0^2 E_0} \tag{8.21}$$

for comparison of the beam quality of various polarized-electron sources. The quantity q takes into account the polarization, intensity, and collimation of the beam.[2]

In sources that utilize strong magnetic fields, the off-axis trajectories become skewed as the electrons pass through the inhomogeneous field region on their way to the field-free region where the polarized beam is to be used. This deteriorates the emittance $r\alpha$ of the beam [8.21]. Comparison of different sources should therefore be made under comparable conditions, for instance, in regions free of magnetic fields.

Trying to define a universal figure of merit which comprises all characteristics of a source would not make much sense, since there are quite a few properties of polarized-electron sources that are relevant only in certain experiments. Investigations with slow electrons, for example, often require the energy spread of the incident beam to be small, whereas in high-energy experiments even an energy spread of more than 1 keV is irrelevant. Whether a source with a continuous or pulsed, transversely or longitudinally polarized beam is more suitable also depends on the particular experiment. On the other hand it is desirable for most experiments to have no change of the beam properties when the electron polarization is reversed, and to have good stability of the current and polarization.

We shall now consider the different methods for producing polarized electrons with respect to their performance as sources of polarized electrons. We shall proceed in the order in which the methods were discussed in this book and only select those cases where development of the method as a usable source of polarized electrons has actually been achieved.

According to Chap. 3, electron scattering from unpolarized targets with high atomic numbers yields considerable polarization at energies from a few electron volts up to the MeV region. Numerous combinations of energy and scattering angle yield nearly total polarization. However, the intensities that can

---

[2] The subscript in (8.21) will be dropped in the following.

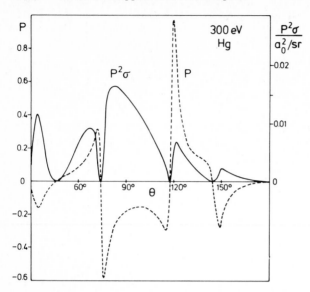

Fig. 8.10. $P(\theta)$ and $P^2(\theta)\sigma(\theta)$ for 300-eV electrons elastically scattered by Hg (a_0 = Bohr radius)

be obtained are moderate because the polarization maxima lie near cross-section minima and are moreover very narrow when they approach 1, as has been explained in Sects. 3.4 and 3.6.

Polarization (or Sherman function) diagrams such as those given in Sect. 3.6 do not provide the best survey of the regions in which scattering yields the most favorable values. Since the scattered current I is proportional to the differential cross section $\sigma(\theta)$, diagrams of the quantity $P^2\sigma(\theta)$ or of the quantity $P^2(\theta, E)\sigma(\theta, E)\big/E$ are more suitable. The latter expression takes into account that, as the electron energy decreases, it becomes increasingly difficult to produce a high-intensity primary beam because the current of a space-charge-limited electron gun is proportional to $E^{3/2}$. Since the scattered current I is directly proportional to the primary current and the quality of the source is, according to (8.21), inversely proportional to the energy E of the electrons leaving the source, it follows that the quality of the source based on scattering is governed by the quantity

$$\frac{P^2\sigma E^{3/2}}{E} = P^2\sigma\sqrt{E}.$$

Figure 8.10 or the comparison of Fig. 8.11 with Fig. 3.21 shows that the criteria used here give a picture that is quite different from that obtained when only the polarization values are considered.

Table 8.1 gives a comparison of various sources of polarized electrons. The value of P^2I given therein for scattering from unpolarized targets has been

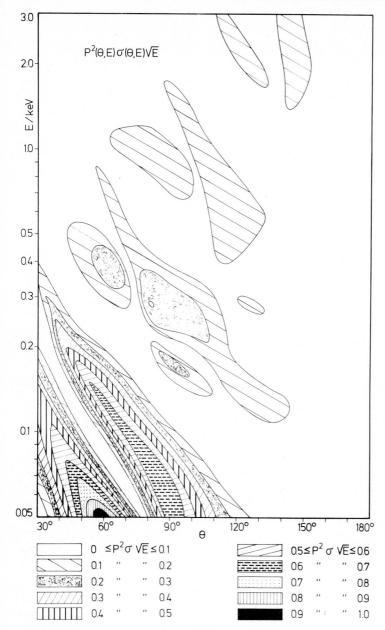

Fig. 8.11. Contours of $P^2(\theta, E)\sigma(\theta, E)\sqrt{E} = $ const. in units of $a_0^2\sqrt{eV}/sr$ for elastic scattering by Hg

Table 8.1. Comparison of various sources of polarized electrons

Method	P	I d.c. [μA]	I el/pulse rep.rate	$P^2 I$ [μA]	Polarization reversal	ΔE [eV]	$r\alpha$ [mrad cm]	at energy E [eV]	Beam quality q (8.21)
Scattering from unpolarized targets	0.2	$3.5 \cdot 10^{-2}$		10^{-3}	angle or energy	≈ 1	20	300	medium
Photoionization of polarized atoms	0.85		$2 \cdot 10^9/1.5\,\mu s$ 180 Hz	$5 \cdot 10^{-2}$	magnetic field	1500	<10	$70 \cdot 10^3$	medium
Fano effect	0.63 0.84 0.65	$2.5 \cdot 10^{-2}$ $5 \cdot 10^{-3}$	$2.2 \cdot 10^9/12\,ns$ 50 Hz	10^{-2} $3.5 \cdot 10^{-3}$ $7 \cdot 10^{-3}$	optical $\left\{ \begin{matrix} < 3 \\ < 1.5 \\ < 500 \end{matrix} \right.$		<20 10 1	10^3 10^3 $120 \cdot 10^3$	medium
Optically pumped He discharge	0.8 0.4	$5 \cdot 10^{-2}$ 50		$3 \cdot 10^{-2}$ 8	optical	0.15	2	400	high
Field emission from W-EuS	0.85	10^{-2}		$7 \cdot 10^{-3}$	magnetic field	0.1	10^{-3}	3	very high
Photoemission from GaAs	0.4 0.37	10	$4 \cdot 10^{11}/1.5\,\mu s$ 120 Hz[a]	3.2 1	optical	0.15 0.2	<20 < 7	0.25 $60 \cdot 10^3$	high medium

[a] This source is being constantly improved.

obtained with mercury-vapor beams to within the same order of magnitude in three different laboratories (Mainz, Stanford, Münster). This indicates that a considerable improvement in this value is hardly possible, which makes this method inferior to the other sources listed. Another disadvantage is that for a reversal of the polarization, which is necessary in almost all applications of polarized electrons, one has to change either the scattering angle or the electron energy. We shall see that polarization reversal in other sources is easier to perform. Instead of the beam scattered from a mercury target one may use a LEED spot diffracted from a single crystal as polarized-electron source [8.22]. This does not, however, increase P^2I dramatically and has no advantage with respect to polarization reversal.

Exchange scattering of slow electrons by polarized atoms, as discussed in Chap. 4, may also be used to build a source of polarized electrons. Equation (4.24) shows that the polarization P_e' of the scattered electrons approaches the value P_A of the atoms if $|f(\theta)|^2/\sigma(\theta)$ approaches zero, that is, if virtually only exchange scattering takes place. Since the exchange processes have appreciable cross sections only at low electron energies, this technique – contrary to scattering by spinless targets – is restricted to slow primary electrons. The values of P_e'/P_A which can (in theory) be obtained have been calculated as a function of scattering angle and electron energy for rubidium between 0 and 7 eV [8.23]. For certain ranges of these parameters (e.g., $E \approx 0.03$ eV, $\theta > 90°$, and $E \approx 2$ eV, $\theta \approx 100°$) one obtains $P_e'/P_A > 0.8$, that is, the attainable electron polarization amounts to more than 80 % of the polarization of the rubidium beam. Similar results have been obtained for the other alkali atoms [8.24].

While nobody has attempted to build a continuous source of polarized electrons by scattering a beam of slow electrons from a polarized atomic beam, it seems that the best results with exchange scattering can be obtained if the electrons are trapped for a while in the region containing the polarized atoms so that they have plenty of opportunity for exchange collisions. In this way their polarization gradually builds up. A typical experimental setup [8.25] works as follows.

A hot cathode emits a pulse of slow electrons of a few eV. The electrons are stored in a trap which is a combination of an electric potential well and a magnetic field. The trapped electrons collide with polarized potassium atoms which are produced using a six-pole magnet and which flow continuously through the electron trap. At the end of the trapping period of approximately 20 ms the electrons are extracted along the direction of the magnetic field. Thus 1-μs pulses of 10^4 longitudinally polarized electrons with polarizations of up to 50 % were obtained (up to 10^5 electrons per pulse for smaller P).

In a later attempt [8.26] to optimize the method, where polarized hydrogen atoms were used as the target, 10^7 electrons per pulse with 20 % polarization were reported. Comparing the results of the exchange method, which is described in detail in [8.27], with those of the other techniques presented in Table 8.1 one finds that even the optimized version has failed to live up to intensity and polarization expectations as a useful laboratory instrument.

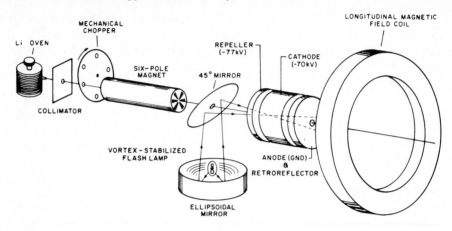

Fig. 8.12. Polarized electrons from photoionization of polarized lithium atomic beam [8.5]

Much work has gone into the development of polarized-electron sources from the photoionization of polarized atoms (see [8.5, 21, 28]). This was the first method used to tackle the problem of building a source. Figure 8.12 shows the scheme of an apparatus with which good results have been obtained. Uv light is reflected into a longitudinally polarized atomic beam that is produced by a six-pole magnet. The photoionization takes place in a magnetic field of 20 mT which is strong enough to decouple electron and nuclear spins from each other (cf. Sect. 5.1). The magnitude of the photoionization cross section and that of the hyperfine interaction between electrons and nuclei both play an important role in the choice of the atoms to be used. In the latest version of the experiment lithium was used because of all the alkali atoms it has the largest photoionization cross section. The experiment was carried out with ^{6}Li because this isotope has a smaller hyperfine interaction than ^{7}Li. Lithium also has the advantage that the spin-orbit coupling which, according to Sect. 5.2, can drastically change the initial polarization during the photoionization process is negligibly small. In addition, one need not be afraid of an appreciable reduction of the polarization due to Li_2 molecules because their photoionization cross section is – contrary to what was said in Sect. 5.2.2 for Cs – merely of the same order of magnitude as that of the Li atoms.

The highest intensities were obtained by using a pulsed light source, since pulsed light sources such as sparks have a peak radiance in the uv which is many orders of magnitude larger than that of continuous lamps. The source is thus suitable for operation with a pulsed high-energy accelerator for which it was in fact constructed (see Sect. 8.4). The injector of the accelerator is designed to accept ≈ 70-keV electrons with an energy spread of ≤ 1.5 keV and an emittance $r\alpha \approx 7$ mrad cm. The ionization region is therefore maintained at a potential of -70 keV, and the photoelectrons are extracted to ground potential. The chopper wheel shown in the figure is to limit the lithium accumulation in the six-pole magnet. The purpose of the repeller electrode is to prevent photoelectrons from leaving the ionization region in the backward direction. The direction of the electron polarization could be reversed in ≤ 1 s by reversing the current through the cylindrical magnetic field coil. In high-energy scattering experiments the polarization was reversed every three minutes.

In addition to the characteristics of the source given in Table 8.1 we mention that the number of atoms in the photoionization region was 10^{11} (atom density on axis $3 \cdot 10^{11}$ atoms/cm^3) and that 2 % of the atoms were photoionized. The oven temperature was 875 °C because of the high boiling point of lithium. At this temperature the oven load of 750 g lithium lasted ≈ 175 hours.

A glance at the original papers cited above shows that the method is considerably more complicated in practice than would appear from the schematic diagram in Fig. 8.12. The discovery of the Fano effect (see Sect. 5.2) about 40 years after the proposal of this well-known method opened up the possibility of obtaining polarized electrons with somewhat less effort by photoionizing *un*polarized atoms. Such sources have been developed in various laboratories. Their setup is, in principle, not complicated. All that is required is a strong source of (near-) uv radiation and a strong alkali atomic beam. One needs neither a six-pole magnet like that shown in Fig. 8.12 for the selection of atomic spin states nor a magnetic field for the decoupling of electron and nuclear spins.

Let us compare the Fano-effect source built by the Yale group [8.29] with the source of Fig. 8.12 built by the same group. It is a dc source which uses a cesium atomic beam. With 60 g of Cs in the oven, operation at a density of 10^{12} atoms/cm^3 can be maintained for 75 h. The light of a 1000-W high-pressure Hg-Xe arc lamp is circularly polarized in the usual way, shown in Fig. 5.5. The helicity of the light is reversed upon rotation of either the linear polarizer or quarter-wave plate by 90°, which results in a reversal of the electron polarization. The longitudinally polarized beam has the characteristics given in Table 8.1. The second data set applies to a similar source at Münster [8.4, 30].

The third set of data refers to a pulsed source [8.31] employing twenty rubidium atomic beams in a recirculating oven arranged along the path of the ionizing radiation. The uv light of 266 nm is provided by a quadrupled Nd-YAG laser which gives typically 5–8 mJ per 12-ns pulse. The source has been built for use at the Bonn synchrotron.

An advantage of the Fano-effect source is that, unlike in the case of photoionization of polarized atoms, a magnetic field which determines the direction of the electron polarization can be dispensed with. The presence of a strong magnetic field in the source region cannot only increase the emittance of the beam, as mentioned before, but also produce significant changes in beam intensity and position if the field is reversed for reversal of the electron polarization. From the discussion of the preceding section we know that such changes of the beam characteristics will limit the precision of an asymmetry measurement. For crossed-beam electron-atom scattering experiments, for example, small changes in the position of the electron beam can result in significant changes in beam overlap and hence cause appreciable systematic errors in the measured asymmetry. The optical method of polarization reversal, which is used in the Fano effect as well as in the other sources utilizing circularly polarized light, is less likely to result in variation of the beam properties. Moreover, the polarization may be easily modulated: if the quarter-wave plate rotates at a frequency ω, the polarization is modulated sinusoidally at 2ω. This is a great advantage for all experiments where small asymmetries are to be measured and where drifts may easily cause systematic errors.

Optical polarization reversal also is possible in a source based on collisional ionization in an optically pumped helium discharge. We have seen in Sect. 5.5.2 that with this method the largest polarization is obtained in the afterglow of the discharge. This has been utilized to develop an efficient polarized-electron source [8.32–34].

The discharge was maintained in flowing helium. The metastable helium atoms, polarized by optical pumping with circularly polarized radiation (1.08 μm, 40 mW) from a cw color-center laser, were chemi-ionized[3] by collisions with a reactant gas injected downstream, beyond the region of active discharge (cf. Fig. 8.13). The resultant electrons, which were polarized as a result of spin conservation during the chemi-ionization reactions [8.35], were extracted

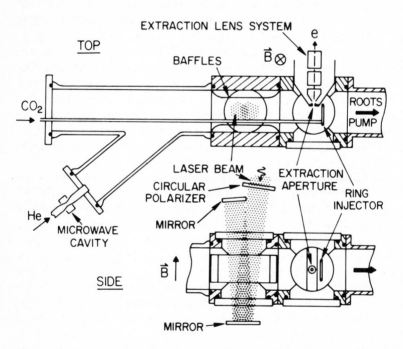

Fig. 8.13. Source of polarized electrons from chemi-ionization of optically pumped metastables in a flowing helium afterglow [8.34]

---

[3] Ionization processes, like $He(2\,{}^3S_1) + AB \rightarrow He(1\,{}^1S_0) + A + B^+ + e^-$, in which the chemical composition of the reactants is changed.

from the afterglow through a differentially pumped aperture and formed into a collimated beam by a series of electron lenses. In order to obtain high metastable densities, the operating pressure in the helium flow tube was typically 0.065–0.2 mbar, necessitating a 500 l/s Roots pump for exhaustion.

All changes in comparison to the arrangement of Fig. 5.26 were made in order to optimize the electron polarization. A microwave discharge was used because this form of excitation provides a low singlet (2^1S_0) to triplet (2^3S_1) metastable ratio in the afterglow. The number of unpolarized electrons produced in chemi-ionization of singlets was thus minimized. The distance between the microwave cavity and the extraction aperture was made sufficiently large so that electrons produced in the microwave discharge did not contribute significantly to the extracted current. Molecular gases like N_2 and CO_2 were particularly suitable as reactant gases because they have large cross sections both for chemi-ionization and for rotational and vibrational excitation by low-energy electrons. The chemi-ionization reactions raised the electron-production rate by about a factor of 100, compared to He-He metastable collisions, and the rapid thermalization of the electrons due to rotational and vibrational excitation of the molecules reduced the energy spread of the extracted beam to values ≤ 0.15 eV.

Table 8.1 shows that this source can provide a sizeable current with high polarization. Since, in the apparatus shown, the polarization decreases strongly for currents above 1 µA, another arrangement with a modified flow tube was developed in order to optimize the polarization at the highest currents. The second data set refers to the latter version. All data hold for CO_2 as reactant gas. In the arrangement of Fig. 8.13, transversely polarized electrons are produced. An earlier version [8.32] in which the direction of the pumping light had been rotated through 90° about the flow tube axis yielded longitudinal polarization. Comparison with the other techniques listed in Table 8.1 shows that the performance of the source is very good indeed.

A source with high polarization, small energy width, and very small emittance has been obtained by field emission from ferromagnetic EuS [8.36]. The apparatus is depicted in Fig. 7.8. The performance characteristics given in Table 8.1, in particular the energy width of less than 0.1 eV, depend crucially on the annealing conditions of the field-emitting tip. The narrow energy distribution was found only for an annealing temperature of 840 K and an annealing time of less than one second. Accordingly, this source, which requires low temperatures and ultrahigh vacuum, needs quite a lot of know-how in producing EuS-coated tungsten tips of optimum structure. In order to impose a well-defined direction on the spontaneous magnetization in the EuS tip and thus on the polarization of the emitted electrons, a magnetic field of about 5 mT is needed. The polarization can be reversed by raising the temperature above the Curie point and cooling in a field of appropriate direction. The source does not, however, suffer from the aforementioned (see page 247) deterioration of the emittance caused by magnetic fields: since the electrons emerging from the sharp field-emission tip are produced very close to the axis, skewing of the off-axis trajectories by the magnetic field is negligible.

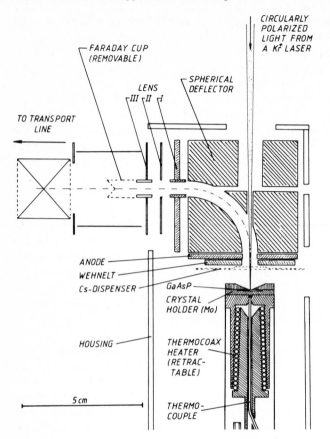

CIRCULARLY
POLARIZED
LIGHT FROM
A Kr⁺ LASER

FARADAY CUP
(REMOVABLE)

LENS
III II I

SPHERICAL
DEFLECTOR

TO TRANSPORT
LINE

ANODE
WEHNELT
Cs-DISPENSER

GaAsP
CRYSTAL
HOLDER (Mo)

HOUSING

THERMOCOAX
HEATER
(RETRAC-
TABLE)

5 cm

THERMO-
COUPLE

Fig. 8.14. GaAs polarized-electron source [8.41]

This influence does, however, affect the beam quality in photoemission from ferromagnetic materials like EuO in a magnetic field (cf. Sect. 7.1.1). In that method one has either to cut down the current of the source by using only a small area near the center of the photocathode or be content with a rather large emittance $r\alpha$ [8.37].

This problem does not occur in photoemission from nonmagnetic materials induced by circularly polarized light (see Sect. 7.2), which does not require a magnetic field for producing the polarization. Of this type is the most popular source of polarized electrons, which is based on photoemission from GaAs and which is utilized in a great number of laboratories [8.38]. As an example, Fig. 8.14 shows a source which is not of the elaborate type used in high-energy experiments [8.39, 40] but is simple enough to be constructed with moderate effort in any laboratory.

The GaAs crystal is mounted on a molybdenum holder in ultrahigh vacuum. It can be heated to 650 °C, which is close to the melting point of GaAs, in order to clean the crystal. By coating with Cs and O_2, the p-doped ($\approx 10^{19}$ Zn atoms per cm^3) GaAs cathode acquires a negative electron affinity so that the conduction band lies above the vacuum level and the photoexcited polarized electrons can escape into the vacuum. This activation process, which is monitored by observing the photocurrent from the cathode, is crucial for the current and the polarization finally available from the source.

Because polarized photoemission from a GaAs crystal is exploited by many groups, there are several versions of such sources. Both GaAs and GaAsP (as in the example of Fig. 8.14) are used, since by variation of the cathode material the polarization curve may be shifted on the energy scale (cf. Fig. 7.16), allowing the maximum polarization to be chosen near a wavelength delivered by a familiar laser type. GaAs$_{0.62}$P$_{0.38}$, for instance, has an energy gap of 1.82 eV, so that the 633-nm line of a He-Ne laser is close to the maximum of the polarization curve [8.42, 43]. There are also several versions of cleaning (e.g., by cleaving in vacuo [8.44]), activating, and operating the source. Instead of holding the cathode at room temperature it may be cooled to liquid-nitrogen temperature which results in a slight increase of the polarization [8.38, 45]. A vacuum in the low 10^{-10}-mbar range is, however, indispensable for all sources. In order to reverse the electron polarization, one has to reverse the helicity of the circularly polarized light by means of the quarter-wave retarder ($\lambda/4$ plate or Pockels cell), whereas rotation from longitudinal to transverse polarization is achieved with a 90° electrostatic deflector.

The sources built by different groups vary somewhat in maximum polarization, quantum yield, photothreshold, and wavelength dependence of the polarization. The results have also been found to depend on the face of the GaAs crystal. The data given in the table for the pulsed source are taken from [8.46], though this type of source is being constantly improved [8.39, 40], while those for the dc source are conservative values as obtained by several groups. More favorable values of one quantity or another have been reported. Typically, the emitted current decays from its initial value at activation with time constants between 10 and 100 hours depending on the kind of cathode and the residual gas by which it is surrounded. The yield can, however, be returned to its original value by renewed cesiation. The lifetime of the cathode decreases as the current is increased. A notable point is the narrow energy distribution of the GaAs source which is definitely better than that of the conventional thermoionic source of unpolarized electrons. By choosing a small negative electron affinity and a photon energy only slightly greater than the band gap, an energy width of 31 meV at a current of 1 μA has been obtained with the GaAs cathode at liquid-nitrogen temperature [8.47].

Despite the limited polarization of about 40 %, the GaAs gun seems at present to be the most acceptable source for most purposes since it has good characteristics and may be set up with moderate effort in a reasonable time. Attempts to increase the polarization and/or the current are certainly worthwhile, but there is also hope of finding materials yielding higher polarization than GaAs [8.39, 48–50].

Summarizing, one can say that several successful approaches to the problem of building an intense source of polarized electrons have been made. After a long history of cumbersome polarized-electron sources, a breakthrough has been made in the past few years. Still, research will continue toward improving the source characteristics.

Problem 8.2. Show that the brightness (=current density/solid angle) is given by the expression (8.19), if the aperture of the electron beam is not too large.

Solution.

$$b = \frac{i}{\Omega} = \frac{I}{\pi r_0^2 \cdot 2\pi \int\limits_0^{\alpha_0} \sin \alpha d\alpha} = \frac{I}{\pi^2 r_0^2 \alpha_0^2}, \quad \text{if} \quad \sin \alpha \approx \alpha.$$

Problem 8.3. Verify the entries on beam quality in Table 8.1 by finding the numerical values of q defined by (8.21).

Solution. In the order given in the table, one obtains the following rounded-off values:

$$q = 10^{-4}; \; 10^{-4}; \; 3 \cdot 10^{-4}, \; 4 \cdot 10^{-4}, \; 6 \cdot 10^{-4}; \; 0.2, \; 50; \; 2 \cdot 10^7;$$
$$3 \cdot 10^2, \; 3 \cdot 10^{-3} \; \text{Am}^{-2} \; \text{sterad}^{-1} \; \text{eV}^{-1}.$$

These numbers must not be considered as an absolute measure for the comparison of the sources because in certain experiments properties which have not been taken into account in q may be important (for example, stability, reliability, ease and speed of polarization reversal, and beam variation with such reversal). If one took the numbers too seriously one would come to the conclusion that even with a polarization of 0.01 the field-emission source would be superior ($q \approx 10^{-1} \text{Am}^{-2} \text{sterad}^{-1} \text{eV}^{-1}$) to all the other sources, which shows that a figure of merit can only serve as a general guide.

8.3 Experiments for Measuring the Anomalous Magnetic Moment of the Electron. Electron Maser

Spin-precession experiments with polarized electrons have yielded very precise results for the anomalous magnetic moment of free electrons. The g-factor anomaly has been measured most accurately using resonance transitions between Landau levels with different spin orientations. The selection of Landau levels with a specific spin orientation by means of inhomogeneous magnetic fields yields polarized electrons and can be utilized to build a tunable maser in the millimeter wavelength range.

One of the most impressive experiments that has been made so far with polarized electrons is the precision measurement of the magnetic moment of free electrons [8.51].

According to the Dirac theory (Sect. 3.1), the magnetic moment μ of the electron is $-\mu_B$, where $\mu_B = e\hbar/2\,mc$. Hence the g factor, which is defined as

$$g = \frac{-\mu/\mu_B}{s} \tag{8.22}$$

(s = spin quantum number = $\frac{1}{2}$), has the value 2. By observation of the shift of the fine-structure levels $2\,^2S_{1/2}$ and $2\,^2P_{1/2}$ of hydrogen and deuterium (Lamb shift), evidence was found in 1947 that the g factor differed slightly from 2 and that the Dirac theory was not completely satisfactory. This was one of the reasons for the development of quantum electrodynamics which predicts for the g factor the value

$$g = 2(1+a) \tag{8.23}$$

with the anomaly [8.52]

$$a = 0.5(\alpha/\pi) - 0.328479(\alpha/\pi)^2 + (1.1765 \pm 13)(\alpha/\pi)^3 - \ldots$$
$$= 1\,159\,652.4 \cdot 10^{-9} \tag{8.24}$$

where only the final digit contains an uncertainty ($\alpha =$ fine-structure constant). A precision measurement of the g factor enables quantum electrodynamics to be tested. Of particular interest are the (very small) higher terms in the expansion of a in terms of α/π, since slight deficiencies of the theory should first become noticeable there. These terms would be the first to indicate when further refinement of the theory were necessary. It is therefore desirable to make as precise a measurement as possible of the g factor.

A method yielding extremely accurate results is based on a fact that was discussed in Sect. 8.1.1: If a polarized electron beam circulates in a homogeneous magnetic field as shown in Fig. 8.2b, the polarization alternates periodically between being longitudinal and transverse. As we have seen, the reason for this is that, since $|\mu| > \mu_B$, the precession frequency of the spins

$$\omega_p = \frac{eB}{mc}\left(\frac{1}{\gamma} + a\right) \tag{8.3}$$

is slightly greater than the cyclotron frequency

$$\omega_c = \frac{eB}{m\gamma c}. \tag{8.2}$$

Accordingly, the change between longitudinal and transverse polarization is determined by the difference frequency

$$\omega_D = \omega_p - \omega_c = a\frac{eB}{mc}. \tag{8.25}$$

By measuring the frequency ω_D, one can therefore determine the g-factor anomaly a if the magnetic field B is known. A great advantage for the accuracy of the measurement is that it is not the total g factor that is directly measured but rather the small anomaly a.

The principle of the experiment is shown in Fig. 8.15. A short pulse ($\approx 10^{-7}$ s) of polarized electrons is injected into the magnetic field. Simultaneously a timer is switched on. The electrons circulate for a certain time T in the magnetic field,

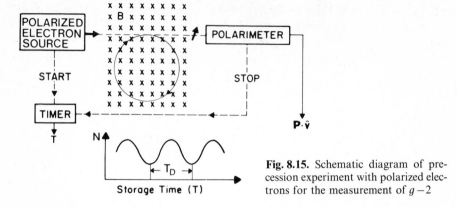

Fig. 8.15. Schematic diagram of precession experiment with polarized electrons for the measurement of $g-2$

their polarization alternating periodically between longitudinal and transverse. After the storage time T which is measured by the timer, the electrons are injected into a Mott detector to analyze their polarization. The pulse height in the Mott detector depends on the state of the polarization. For transverse polarization, that is $\boldsymbol{P} \cdot \boldsymbol{v} = 0$, there is a maximum or minimum of the scattering intensity according to whether $\boldsymbol{P}$ is parallel or antiparallel to the normal of the scattering plane. This scattering asymmetry was discussed in Sect. 3.3.1 [see (3.70)]. For longitudinal polarization one finds the mean value of the scattering intensity. If the procedure just described is repeated with many different storage times T one obtains a sinusoidal curve (Fig. 8.15) for N, the number of scattered electrons observed in the Mott polarimeter:

$$N \propto I(\theta)\,[1 + S(\theta)\,P \sin \varphi_{\mathrm{D}}],$$

where φ_{D} is the angle between $\boldsymbol{P}$ and the velocity $\boldsymbol{v}$. Since $\varphi_{\mathrm{D}} = \omega_{\mathrm{D}} T + \varphi_0$ [the angle $\varphi_{\mathrm{D}}(T=0) = \varphi_0$ depends on the kind of polarized-electron source used], the observation of a complete period of this sine curve yields the desired difference frequency

$$\omega_{\mathrm{D}} = 2\pi/T_{\mathrm{D}}, \tag{8.26}$$

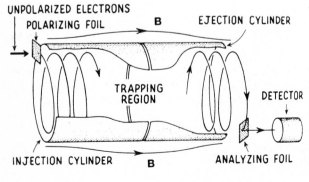

Fig. 8.16. Precession experiment with polarized electrons for the measurement of $g-2$ [8.53]

where T_D (see Fig. 8.15) is the time in which the polarization vector makes one complete rotation more than the velocity vector. By determining the time T_D, one obtains, with (8.25), the anomaly a.

Figure 8.16 gives a general idea of the actual experimental setup. The polarized electrons were produced by scattering from a gold foil. The electron energy was in the 100-keV region and the polarization was approximately 20 %. Instead of selecting the electrons scattered exactly at 90°, those which still had a small forward component of the momentum were sent through an aperture; they spiralled towards the detector with a pitch of approximately 1°. As the number of rotations of the polarization vector increases, the error in the measurement of T_D decreases because one has an increasing number of periods to use in determining T_D. It is therefore important to keep the electrons in the magnetic field as long as possible. This was achieved with the help of the cylindrical electrodes shown in Fig. 8.16. As the electrons drift across the gap between the cylinders, a momentary retarding voltage applied to the cylinders causes the electrons to lose sufficient axial velocity so that they can no longer spiral out of the magnetic field. The electrons are trapped because the magnetic field is slightly inhomogeneous and thus has field components that prevent the electrons from leaving the field ("magnetic bottle"). In this manner, stable helical paths are formed in the area denoted as the trapping region.

After a few milliseconds (which corresponds to several million rotations and several thousand periods T_D), a momentary acceleration voltage between the cylinders gives the electrons sufficient axial velocity to leave the magnetic field and to reach the analyzing foil of the Mott detector.

For a precision measurement of the anomaly a it does not suffice to determine T_D as accurately as possible. Equation (8.25) shows that the magnitude and local variation of the magnetic field must also be known accurately. Furthermore, the observed frequency is affected by stray electric fields which may arise from contact potentials, static charges, or the space charge of the beam. Although these stray fields are small, they determine the uncertainty in the final digits of the experimental value

$$a = (1\,159\,656.7 \pm 3.5) \cdot 10^{-9}.$$

The accuracy in the anomaly a of 3 ppm corresponds to an accuracy in the g factor, which is almost a thousand times larger, of approximately $4 \cdot 10^{-9}$. This accuracy is only possible because one does not measure the total g factor as in conventional spin-resonance methods but instead measures the small deviation $g - 2$ directly. Comparison with (8.24) shows very good agreement with the prediction of quantum electrodynamics.

The successful technique discussed here has also been applied to other elementary particles. At CERN, for example, polarized muons of a few GeV have been injected into a storage ring in order to measure the g factor of free muons with the same method [8.54].

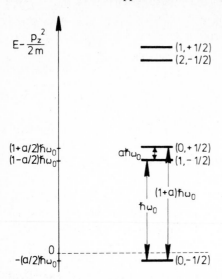

Fig. 8.17. The lowest Landau levels for an electron (not drawn to scale)

Precision measurements of the anomalous electron moment are also made with microwave techniques. We shall outline the basic idea which was put forward by *Bloch* [8.55].

The energy of a free electron moving in a homogeneous magnetic field $\boldsymbol{B} = B\hat{\boldsymbol{e}}_z$ is given in the nonrelativistic limit by

$$E = \frac{p_z^2}{2m} + \left(n + \frac{1}{2}\right)\hbar\omega_0 \pm \mu B, \tag{8.27}$$

where ω_0 is the cyclotron frequency (8.2) for $\gamma = 1$. The first term in this equation corresponds to the free electron motion in the z direction, the second term corresponds to the cyclotron motion in the magnetic field (this can be considered to be composed of harmonic oscillations of frequency ω_0 in two mutually perpendicular directions); the last term is the energy for the two possible orientations of the magnetic moment $\boldsymbol{\mu}$ in the magnetic field. A rigorous derivation of relation (8.27) can be found in *Landau* and *Lifshitz* [8.56].

With the use of (8.2, 22, 23), one obtains from (8.27) the energy terms

$$E - \frac{p_z^2}{2m} = \left(n + \frac{1}{2}\right)\hbar\omega_0 \mp \frac{1}{2}g\mu_B B = \left(n + \frac{1}{2} \mp \frac{1+a}{2}\right)\hbar\omega_0$$

which are shown in Fig. 8.17 for the lowest values of n. With suitable microwave frequencies, transitions of the electrons can be induced between the energy levels shown. With the frequency ω_0 one obtains transitions $\Delta n = \pm 1$ between different cyclotron levels; the frequency $(1 + a)\omega_0$ causes a spin flip in the same cyclotron orbit ($\Delta n = 0$), and the frequency $a\omega_0$ causes both a spin flip and a change of the cyclotron orbit. The anomaly a can be determined by measuring these frequencies, e.g., by evaluating the ratio $(1 + a)\omega_0/a\omega_0$.

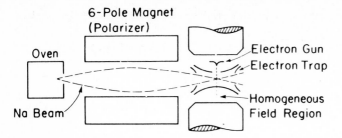

Fig. 8.18. Polarized-electron resonance experiment for measurement of $g-2$ [8.57]

Transitions between the various levels can be observed only if the levels have different populations. Even in high magnetic fields the distances between the levels are very small in comparison with the spread of the energy distribution that free electrons usually have ($\mu_B B \approx 6 \cdot 10^{-5}$ eV at 1 T). Accordingly, the population differences of the various levels are generally very small. Large population differences can, however, be obtained if polarized electrons are used, since then the spin-up and spin-down states ($+\frac{1}{2}$ and $-\frac{1}{2}$ in Fig. 8.17) are differently populated.

We shall now explain with an example how the idea outlined here can be realized in practice. In the experiment shown in Fig. 8.18 an electric quadrupole field is superimposed on the magnetic field so that the electrons can be confined both radially and axially [8.57]. A hot cathode outside this trap generates a pulsed electron beam which passes through the trap and ionizes the residual gas molecules therein, so that slow electrons are obtained in the storage region.

By exchange collisions with polarized sodium atoms, which also pass through the trap, the trapped free electrons become polarized in the direction of the magnetic field. This means that the spin-up and spin-down Landau levels attain different populations so that one can induce transitions between them by irradiating with the spin-flip frequencies. These transitions diminish the population differences, thus causing a drop in the electron polarization. This decrease of the polarization is utilized to indicate when spin-flip processes occur, i.e., when a suitable frequency is being used.

To monitor the decrease in the polarization, the authors exploited the fact that the cross section for inelastic scattering of polarized electrons by polarized Na atoms depends on the mutual spin orientation of the collision partners. This was discussed in detail in Chap. 4. From the relations derived there, one obtains the cross section Q' for excitation of a ^2P state:

$$Q' = \tfrac{1}{4} Q^s + \tfrac{3}{4} Q^t - \tfrac{1}{4}(Q^s - Q^t) P_e P_A \tag{8.28}$$

(see Problem 8.4). Q^s and Q^t are the excitation cross sections, integrated over the solid angle, for the antisymmetric (singlet) and symmetric (triplet) spin states of the two colliding electrons; P_e and P_A are the respective polarizations of the electrons and the atoms.

In the experiment discussed, excitation of the $3\,^2P$ state is the main cause of the energy loss of the electrons. The magnitude of the corresponding cross section thus determines how fast the electrons lose their energy due to collisions with the Na atoms. Therefore, according to (8.28), the energy-loss rate of the electrons depends on their polarization. Consequently, monitoring the electron energy distribution provides a means of detecting the decrease in P_e induced by spin-flip transitions. The specific method used is to lower the trap voltage by a certain amount. Then the electrons that can overcome the potential barrier will escape. Measurement of the number of electrons remaining in the trap yields the desired information on the polarization. Finally the trap is cleared by a negative voltage pulse before another pulse from the electron gun starts the next measuring cycle.

Determination of the anomaly a from the ratio of the frequencies $(1+a)\omega_0$ and $a\omega_0$ would be possible only if there were no electric field superimposed on the magnetic field. In actual fact, one must take into account the influence of the electric quadrupole field used to trap the electrons, which is ignored in our formulae. This correction can easily be made, since the frequency shift caused by this field can be calculated theoretically and tested experimentally.

The accuracy of the value obtained for a is determined in this experiment by the precision with which the resonance frequencies can be measured, that is, by the width of the resonance lines. Furthermore, uncertainties from corrections for electric fields caused by the space charge in the trap play a role. The uncertainty of the result was 260 ppm, not reaching the accuracy of the precession experiment.

A dramatic improvement in precision has been obtained by *Dehmelt* and coworkers who pioneered work with the resonance technique [8.58] and who have carried out several versions of such experiments. They succeeded in storing a single electron in a trap and measuring the cyclotron frequency and the difference between spin-flip frequency and cyclotron frequency [8.59]. Working with a single electron has the advantage that there is no space charge of other electrons causing an electric field which would affect the electron motion and shift its frequency. Needless to say, when working with one electron one need not generate spin polarization, i.e., a population difference between spin levels, since a single electron is necessarily "totally polarized" (although it is not very appropriate to transfer the concept of spin polarization, defined for an ensemble of electrons, to one electron). Our treatise on polarized electrons will therefore not discuss the single-electron experiments in detail. We will, however, mention the outstanding accuracy of the result: $a=(1\,159\,652\,200\pm40)\cdot10^{-12}$.

Comparison of this numerical result which has been obtained with very slow electrons with the value obtained by the precession method in the GeV regime [8.60] makes possible a precision experimental test of special relativity because both measurements should give the same result.

Suitable conditions for resonance experiments can also be obtained by selecting electrons in some of the Landau states and thereby producing population differences. This can be done with the aid of inhomogeneous magnetic fields. The lowest state $(0, -\frac{1}{2})$ in Fig. 8.17 is seen to be paramagnetic: its energy decreases in the magnetic field. Accordingly, electrons in this state are drawn toward regions of high field strength in an inhomogeneous magnetic field. The electrons in all the other states are drawn into regions of low field strength since they behave diamagnetically (increase of energy in the field).

Consequently, a magnetic bottle such as that indicated (in a different connection) in Fig. 8.16 can be used for selecting diamagnetic Landau states. Electrons in these states are reflected from both ends into the middle and thus remain in the bottle. Electrons in the paramagnetic ground state are drawn toward the ends and are lost at the chamber walls. We must, however, keep in mind that a change in the magnetic field of 2 kG corresponds to a change in the energy of the lowest diamagnetic state of about 10^{-5} eV (for the paramagnetic ground state the energy change is only $\approx 10^{-8}$ eV). Consequently, a magnetic field with a gradient of 0.2 T can only prevent an electron in this state from escaping if the electron's maximum kinetic energy in the axial direction lies below 10^{-5} eV. For higher excited states this limit is correspondingly higher.

Apart from making possible the measurement of the electron magnetic moment by resonance transitions to the depopulated paramagnetic level, the technique discussed also opens up other possibilities: By depopulating the ground state in the magnetic bottle as just described, one might generate a population inversion which would make maser operation possible. Stimulated rf transitions could then produce a gain in rf energy. One thus would have a maser in the 50 GHz range which could easily be tuned by changing the level distances through variation of the magnetic field.

A group at Stanford [8.61] has succeeded in selecting the paramagnetic ground state as follows: An electron source is located in an inhomogeneous magnetic field that decreases in the direction in which the electrons emerge. From a maximum value of 0.6 T near the electron source, the field changes to a homogeneous 0.4-T region, approximately 1 m long, which is used for time-of-flight measurements. Electrons in the paramagnetic ground state are decelerated as they pass through the inhomogeneous region of the magnetic field, while electrons in the diamagnetic higher states are accelerated. The energy loss of the ground-state electrons due to the deceleration is only 10^{-8} eV since the magnetic field decreases by approximately 0.2 T. The electrons in the higher states gain at least 10^{-5} eV due to the acceleration. Electrons of 10^{-5} eV need approximately 0.5 ms to pass through the 1-m drift region. All those electrons which take considerably longer time must therefore be in the ground state. Consequently, ground-state electrons can be identified by their time of flight. To select polarized electrons for further experiments one must use only those electrons which are still in the drift region more than 1 ms after the electron pulse has started from the cathode.

In this experiment one can, of course, separate only those ground-state electrons whose thermal energy is not large enough to make them cross the drift region in less than 1 ms. Since the mean thermal energy of the electrons used was approximately 0.5 eV, only very few electrons (≈ 1 electron/pulse) at the extreme lower end of the thermal energy distribution fulfilled this condition.

Without going further into the numerous difficulties of such an experiment, it should be mentioned that the suppression of stray electric fields also represents a considerable problem. Contact-potential differences were suppressed so much that they had no appreciable effect on the axial motion of the 10^{-8}-eV electrons. The authors actually succeeded in detecting electrons in the ground state.

These examples show that one can make very interesting, though difficult, experiments with polarized electrons in Landau levels.

Problem 8.4. In [8.57], the total cross section for excitation of the 3^2P state by collisions of polarized electrons with polarized Na atoms is given in the form (8.28). Deduce this expression from the results in Chap. 4.

Solution. We start from the differential cross section (4.36) which has been derived for elastic scattering. With the abbreviations

$$\sigma^s = |f+g|^2, \qquad \sigma^t = |f-g|^2, \tag{8.29}$$

the cross section (4.36) can be written

$$\sigma(\theta) = \tfrac{1}{4}\sigma^s + \tfrac{3}{4}\sigma^t - \tfrac{1}{4}(\sigma^s - \sigma^t)\, P_e P_A \tag{8.30}$$

where use has been made of (4.14) and of

$$|f|^2 + |g|^2 - |f-g|^2 = fg^* + f^*g = \tfrac{1}{2}(|f+g|^2 - |f-g|^2);$$

P_e is the polarization component of the electrons in the direction of the atomic polarization. For an inelastic process leading to the sublevel $m_l = i$ $(i=0, \pm 1)$ one has

$$\sigma_i^s = \frac{k'}{k}\,|f_i+g_i|^2, \qquad \sigma_i^t = \frac{k'}{k}\,|f_i-g_i|^2$$

(see Sect. 4.5.2).

The differential cross section for excitation of a P state is therefore given by (8.30) with

$$\sigma^s = \frac{k'}{k}\sum_{i=-1}^{1}|f_i+g_i|^2, \qquad \sigma^t = \frac{k'}{k}\sum_{i=-1}^{1}|f_i-g_i|^2$$

so that (8.28) follows from integration of (8.30) when the notation of (4.59b) is used.

8.4 High-Energy Physics

High-energy experiments with polarized electrons put theoretical parton models to a strong quantitative test and gave evidence of parity violating interactions in electron scattering.

The use of polarized electrons also opens up new possibilities in high-energy physics. When describing electron scattering in the GeV region in which the large electron accelerators operate, the atomic nucleus can no longer be conceived of as a point charge. Its finite size becomes significant, and the distribution of the electric charge density within the nucleus can be determined from electron-scattering experiments. At these energies, the interaction between the magnetic electron moment and the nuclear moment also plays an important role in the scattering process. One of the consequences is that the cross section depends on the mutual orientation of the electron and nuclear spins. If one scatters polarized electrons by polarized nuclei, one obtains different scattering intensities according to whether the spins of the collision partners are parallel or antiparallel to each other. In analogy to what was said for atoms in Chaps. 3 and 4, this difference reveals details of the structure of the nucleus which cannot be found from cross-section measurements alone. Another reason for the interest in polarization is the relative growth of weak interaction effects at high energy. The fundamental couplings of the Z^0 and $W^\pm$ bosons can be investigated by measurement of parity-violating amplitudes in polarization studies with electron-positron colliders.

Quite a few suggestions for high-energy experiments with polarized electrons have already been made. We will not, however, give a survey here of the numerous possibilities and suggestions that have not yet been realized in practice (for a recent review article see [8.62]), but rather we will pick out two experiments that have been carried out at the Stanford linear accelerator (SLAC).

One experiment is designed to study electron scattering by protons at short distances, i.e., with large momentum transfer. This deep inelastic scattering provides particularly interesting information about the structure of the proton and its electromagnetic interaction at high energies. From the various proton models that exist, the results obtained from scattering of unpolarized electrons favor the parton models: the electrons appear to be scattered by point-like constituents within the proton.

The purpose of the experiment under consideration is to investigate the spin dependence of deep inelastic scattering of longitudinally polarized electrons by polarized protons. The main reason for using longitudinally polarized electrons is that, at high energies, transverse polarization components no longer contribute very much to the spin-dependent effects. The goal of the experiment is to measure the relative difference between the differential inelastic cross sections for parallel and antiparallel mutual orientations of the electron and proton spins. Such an asymmetry in deep inelastic scattering was first predicted by *Bjorken* [8.63].

Theoretical results for the asymmetry which were obtained on the basis of various models differ significantly. Parton models based on different assumptions about the point charges within the proton (their number, spin, mass, etc.; one of the most widely known is the quark model) predict that the cross section for antiparallel spins of electron and proton should be larger than for parallel spins. The exact values of the asymmetry, however, depend on the particular model used. The experiment yields information on the spin distribution of quark constituents inside the proton and is thus important for tests of models of nuclear structure [8.64].

The measurements were made at primary energies up to 22.7 GeV for several discrete values of the four-momentum transfer [8.65, 66]. The polarized-electron source used is based upon photoionization of a state-selected ^{6}Li atomic beam as described in detail in Sect. 8.2. The raw asymmetries observed were of the order of a few tenths of a percent, necessitating careful studies of false asymmetries. The experimental results are consistent with the predictions of certain quark-parton models, but disagree strongly with other theories. The experiments were remarkably successful even though, because of their difficulty, only a modest amount of data on polarized e–p scattering was obtained. Major advances are to be anticipated.

From the numerous experimental problems that arise with high energies we want to mention only the polarization analysis, since it is of interest in connection with our general theme. One cannot simply assume that the polarization of the electrons is the same after acceleration as it was at the time of injection into the accelerator, since it is generally affected by the electromagnetic acceleration fields. In order to make sure that there is negligible depolarization during acceleration, a measurement of the polarization of the high-energy electrons is indispensable. Mott scattering is unsuitable for this purpose because, at these energies, it no longer produces enough asymmetry. One might use the relative differences between the cross sections for elastic scattering of polarized electrons by polarized protons with parallel and antiparallel spins [8.67]. If, by suitable choice of energy and scattering angle, one works in the region where the asymmetry is largely independent of the still existing uncertainties of the proton form factors, then exact enough measurements of the electron polarization can be carried out. In the experiments made so far, Møller scattering from a magnetized iron foil as discussed in Sect. 4.8 has been used for polarization analysis [8.68].

High-energy experiments with polarized electrons do not necessarily have to be made with polarized targets. Possible contributions to high-energy electron scattering from weak, parity-violating interactions have, for example, been studied by using polarized electrons and unpolarized protons or deuterons [8.46]. This experiment has been made to find evidence for or against the Weinberg-Salam model which unifies two of the fundamental interactions in physics. One of the consequences of the unified theory of electromagnetic and weak interactions is that the scattering cross section for electrons is determined by both these interactions. The cross section for inelastic electron scattering from an unpolarized target contains an interference term between the weak and electromagnetic amplitudes which depends on the helicity of the incident electron beam, resulting in different scattering intensities for beams with polarization parallel and antiparallel to the direction of propagation. From previous chapters of this book we are quite familiar with such situations where asymmetry or polarization effects are caused by interference of amplitudes describing different processes.

The fact that the cross section for scattering from an unpolarized target depends on the electron helicity means violation of parity conservation as is

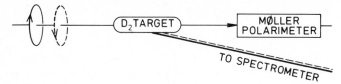

Fig. 8.19. Parity nonconservation in inelastic electron scattering. Mirror inversion interchanges the electron helicities which yield the two different intensities

typical of weak interactions. This is illustrated by Fig. 8.19 which gives the basic idea of the experiment. If a reversal of the incident polarization results in different scattering intensities (indicated by a full and a broken line) then the mirror image of the experiment differs from what is found in the laboratory, because reflection at the plane of drawing changes only the helicity of the incident beam, but not the scattered intensity. Hence, one has different results in the laboratory and in the mirror image, implying that parity conservation does not hold.

In the experiment under discussion a polarized-electron beam of energy between 16.2 and 22.2 GeV was scattered from an unpolarized deuterium (and later a hydrogen) target. Of crucial importance to this experiment was the development of an intense source of longitudinally polarized electrons. It was based on photoemission from GaAs by circularly polarized light as discussed in Sect. 8.2. The light source was a dye laser operated at 710 nm and pulsed to match the linear accelerator (1.5 µs pulses at 120 pulses per second). The average beam polarization as determined by Møller scattering was 37 % and the beam intensity at the target varied between 1 and $4 \cdot 10^{11}$ electrons per pulse. The helicity of the electrons was reversed by reversing the circular light polarization.

The target was a 30-cm cell of liquid deuterium. The electrons scattered through 4° passed through a spectrometer whose setting was about 20 % below the primary energy. Approximately 1000 scattered electrons per pulse entered the counters. The high rates were handled by integration rather than by counting individual particles.

The main difficulty of such an experiment which is looking for an interference between weak and electromagnetic forces is the small size of the asymmetry as given in Fig. 8.20. The energy dependence shown there is caused by the fact that the direction of the electron polarization at the target depends on energy because of the spin precession in the magnetic field that deflects the beam before reaching the target. Owing to the anomalous magnetic moment of the electron, the electron spin direction precesses relative to the momentum direction with a frequency

$$\omega_D = \omega_p - \omega_c = a \, \frac{eB}{mc} \qquad (8.25)$$

as has been discussed in the preceding section. Accordingly, the electron spin precesses relative to the momentum by the angle

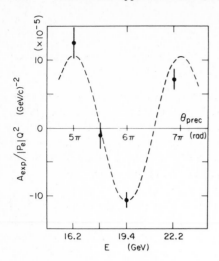

Fig. 8.20. Observed scattering asymmetry in inelastic electron scattering from unpolarized target. The curve shows the expected cosine variation as a function of beam polarization changes as a function of beam energy owing to the $g-2$ precession in the beam-transport system. The experimental asymmetries are expected to be proportional to Q^2 ($=$ four-momentum transfer) and have therefore been divided by Q^2 which varied between 1.05 and 1.91 $(\text{GeV}/c)^2$. P_e is the electron polarization [8.46]

$$\theta_D = \frac{\omega_D}{\omega_c}\,\theta_c = a\gamma\theta_c = a\,\frac{E}{mc^2}\,\theta_c$$

where (8.2) has been used and θ_c denotes the change of the momentum direction by the deflecting magnet. Since $\theta_c = 24.5°$ in the present experiment, one has

$$\theta_D = \frac{1.16\cdot 10^{-3}E}{0.511\text{ MeV}}\,\frac{24.5}{180}\,\pi\text{ rad} = \frac{E(\text{GeV})}{3.237}\,\pi\text{ rad}.$$

At 19.4 GeV $= 6\cdot 3.237$ GeV, for instance, the precession is equivalent to three full revolutions of the electron spins, so that the polarization is not changed by the deflection magnet. If the energy is decreased (or increased) by 3.237 GeV, the spins will make 2.5 (3.5) revolutions so that the beam polarization is reversed. Taking into account that only the longitudinal polarization component gives rise to the scattering asymmetry, one therefore expects from the last formula a cosine variation of the observed asymmetry versus E as shown in Fig. 8.20. That the experimental results follow the expected behavior is taken as very strong evidence that the observed effects are due to electron spin. In an experiment like this, which requires an accuracy of 1 in 10^5 in order to determine quantitatively the asymmetry of $\approx 10^{-4}$, the danger of spurious asymmetries is extremely great. Several consistency checks and null measurements as well as stabilization of beam position, angle, and energy by a microcomputer-driven feedback system were necessary to reduce the experimental uncertainties to the point that the minute asymmetries given in the figure could be reliably measured.

The observed asymmetry not only shows that the electron-scattering cross section contains a contribution caused by weak interaction, but it also gives its strength and allows determination of the only free parameter of the Weinberg-Salam theory.

Now that high-energy accelerators of polarized electrons are operational, there will be an increasing interest in this area of polarized-electron physics.

8.5 Electron Microscopy

While in conventional transmission electron microscopy spin polarization has no significance at present, it may be utilized in scanning electron microscopy.

The previous discussions of the applications of spin polarization in electron diffraction and electron scattering suggest the idea that polarization effects might also be utilized in electron microscopy. In fact, electron microscopists were among the physicists who made the first quantitative studies of electron polarization. While it is hard to see any practical applications of the polarization effects in conjunction with the conventional methods of electron microscopy, they may be utilized in scanning electron microscopy.

It is true that the image contrast in conventional transmission electron microscopy comes from electron scattering in the object. This scattering is in principle spin dependent, but the spin dependence is of no significance here: in order to keep the effects of lens aberrations small, one must work with aperture angles of $0.1°-1°$. Electrons that are scattered in the object at larger angles do not pass through the apertures and thus do not contribute to the image intensity. Because of the small scattering angles of the imaging electrons, the Sherman function $S(\theta)$ is virtually zero in the angular range relevant to conventional transmission electron microscopy (see Sect. 3.6). Thus the electrons which contribute to the image in the conventional electron microscope are almost entirely unpolarized. For the same reason, the left-right asymmetry would be negligible if polarized electrons were used in such an electron microscope. Any polarization effects which could be expected are substantially below 10^{-6}.

The spin dependence of the exchange scattering in magnetized materials which was discussed in Chap. 7 likewise cannot be used in conventional transmission electron microscopy. Since one works with fast electrons and small scattering angles, no observable effects can be expected.

On the other hand, a practical application of spin-dependent interactions is possible in scanning electron microscopy by exploiting the polarization of backscattered and secondary electrons. In a scanning microscope with an unpolarized primary beam, such polarization is characteristic of high-Z elements in a nonmagnetic sample if caused by the spin-orbit interaction [8.69–71]. Even more interesting seems to be the application to imaging magnetic materials by making use of the fact that the local magnetization of the target results in an polarization of the secondary electrons [8.72]. By scanning an iron (001) surface with a 10-keV electron beam and analyzing the polarization of the secondary electrons, a picture of the magnetic domain structure has been obtained [8.73]. Figure 8.21 shows this impressive result which represents a first step in the direction of polarized-electron microscopy. Unlike the normal image obtained in electron microscopy, such an image is independent of the topography of the surface since it is based on asymmetry (i.e., difference) measurements. A scanning electron polarization microscope with high spatial resolution is currently being developed [8.74].

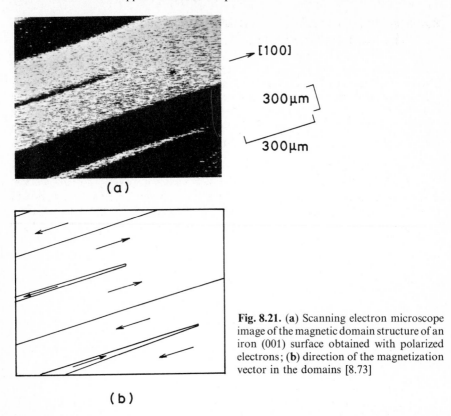

(a)

[100]

300μm

300μm

(b)

Fig. 8.21. (a) Scanning electron microscope image of the magnetic domain structure of an iron (001) surface obtained with polarized electrons; **(b)** direction of the magnetization vector in the domains [8.73]

If the scanning electron microscope is operated with a polarized primary beam, one can dispense with a polarization analysis of the emitted electrons. Instead, spin-dependent contrast may be obtained by monitoring the change of the backscattered intensity when the polarization of the primary beam is reversed. In this case, however, the incident energies must be clearly below the 1-keV limit, because exchange scattering must play a dominant role if one relies on the asymmetry of the backscattered intensity alone. This restriction does not play a role in the above method which takes advantage of the polarization of the secondary electrons.

The recent development justifies the hope that the electron polarization microscope, up to now only a vision of imaginative physicists, will soon be a practical instrument allowing the study of magnetic structure below the wavelength limit of optical studies.

8.6 Electron-Molecule Scattering.
Why Isn't Nature Ambidextrous?

Electrons scattered from high-Z atoms of a molecule are labeled by their polarization which can facilitate the analysis of electron-molecule scattering. Irradiation of certain organic molecules with polarized electrons hopefully may give a clue as to the origin of the one-handedness of nature. The lack of reflection symmetry of optically active molecules should give rise to minute polarization effects which are below the present detection limit.

From what has been said about the polarization of electrons that have been scattered from atoms and solids one will expect that spin polarization occurs also in electron-molecule scattering. This has in fact been observed as can be seen from the results presented in Fig. 8.22. They illustrate the following general situation: Appreciable polarization occurs if there are atoms like iodine of fairly

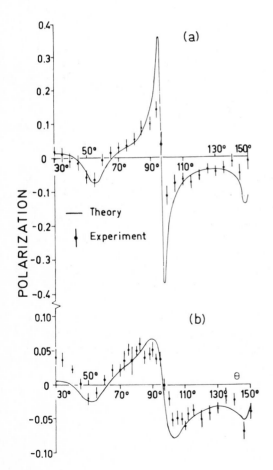

Fig. 8.22a, b. Experimental and calculated values of the polarization of 300-eV electrons after elastic scattering by (a) I_2 and (b) C_2H_5I molecules [8.75–77]

high atomic number (causing considerable spin-orbit coupling) within the molecule. If the molecule also contains light atoms, as in organometallic compounds or halogenated hydrocarbons like C_2H_5I, the contribution to the scattering intensity from these atoms is practically unpolarized. Then the total observed polarization P does not reach the value P_1 which is caused by the heavy atoms of the molecule alone. It is instead determined by the ratio of the scattering intensities I_1 and I_2 which come from the heavy and light parts of the molecule, respectively. According to (2.16), one has

$$P = \frac{I_1 P_1 + I_2 P_2}{I_1 + I_2}. \tag{8.31}$$

The interference terms which were taken into account in (7.3) have been omitted here since they are negligible in molecular scattering at large angles where the polarization effects arise [8.78]. If $P_2 \approx 0$ we obtain from (8.31)

$$P = \frac{P_1}{1 + I_2/I_1} \tag{8.32}$$

giving the reduction of the polarization in Fig. 8.22b in comparison to Fig. 8.22a.

The ratio I_2/I_1 can be determined by comparing the measured value P with the polarization P_1 caused by the heavy atoms alone. To do this, one can refer to the theoretical values which are available for most atoms. These calculations are reliable when the energies are not too low, as we saw in Sect. 3.6.

Let us apply our consideration to scattering from $Bi(C_6H_5)_3$, where only the electrons scattered from the heavy bismuth atoms ($Z = 83$) are labeled by spin polarization, and can thus be distinguished from the virtually unpolarized electrons scattered from the phenyl groups. It can be seen from Fig. 8.23 that the polarization measured at low electron energies is small. It is much smaller than for scattering by free Bi atoms, as can be easily seen by comparison with tabulated values for Bi. With increasing energy, the polarization tends toward the values that one obtains in scattering by free Bi atoms. Thus Fig. 8.23 shows that at low energies the fraction of the electrons scattered from the Bi atom is very small; with increasing energy this fraction continually increases, as is apparent from the increasing polarization (for a more accurate analysis, see [8.78]).

This is another example of the method of "labeling" electrons by their polarization. The fact that some of the scattered electrons are labeled by their spin orientation simplifies the analysis of electron-molecule scattering by giving information about the proportion of the scattering intensities which come from the various parts of the molecule.

In the examples given, calculation of the polarization curves is not difficult because, at the relatively large angles of more than 30° where the polarization is appreciable and at the energies considered, the influence of chemical bonding (modification of the atomic scattering potentials!), interference terms, and

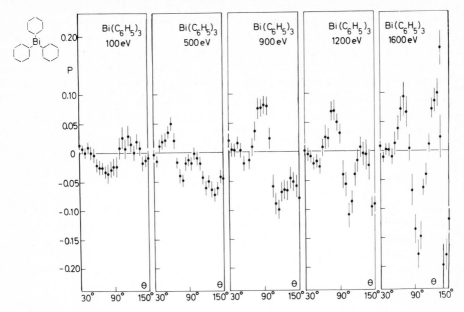

Fig. 8.23. Polarization $P(\theta)$ for scattering of slow electrons by Bi$(C_6H_5)_3$ at different energies [8.78]

intramolecular plural scattering on the polarization is small. Discrepancies between the theoretical values and experimental results occur merely at pronounced peaks of some polarization curves, e.g., with I_2 and Sb_4 [8.79].

Less well understood are the electron polarization effects related to the optical activity of molecules. Compounds containing carbon atoms bonded to four atoms or groups no two of which are alike are capable of existing in two optically active forms. They are distinguished by being respectively left- and right-handed (L- and D-enantiomers). Such molecular species are found in the terrestrial biosphere in only one of the two possible forms. One finds, for example, only L-amino acids in natural proteins and only D-sugars in carbohydrates and nucleic acids. The question of the origin of this dissymmetry has intrigued scientists since the time of Pasteur.

Among the possible causes which have been suggested are: Active seeds of organic compounds that reached the Earth from the universe, enantiomorphous minerals (e.g., left- or right-handed quartz crystals) that catalyzed or adsorbed organic compounds in a stereoselective way, or the Earth's magnetic and electric fields that influenced chemical reactions. Another explanation would be possible if an excess of either left- or right-circularly polarized vuv radiation could be detected on Earth or had existed for some time during chemical evolution, since it has been shown that optical activity can be induced through stereoselective decomposition of racemic substances by circularly polarized radiation. (Racemic substances are mixtures of equal amounts of L- and D-enantiomers.)

There was not much progress in the development of these and other hypotheses until the discovery of parity violation in weak interactions provided a stimulus for further discussion, and molecular dissymmetry was hypothesized to be a result of the dissymmetry of interactions within the nucleus. The transfer of the dissymmetry from the nuclei to the molecules might be achieved by the longitudinally polarized electrons emitted in β decay or by circularly polarized bremsstrahlung produced by these electrons. This idea can be examined by irradiating racemic mixtures or optically inactive substances with β particles and analyzing any stereoselective degradation or synthesis that might occur. The experimental results hitherto obtained are controversial. Positive results that had been claimed by a few authors could not be confirmed by other groups [8.80, 81].

More recently, there was some hope that the situation could be clarified by utilizing in such studies modern sources of polarized electrons having higher intensities and well-defined beam characteristics. In the first of these experiments a racemic mixture of D- and L-leucine has been partially degraded by irradiation with a beam of longitudinally polarized 130-keV electrons [8.82]. The electron polarization has been produced by scattering of slow electrons on a mercury target, as described in Chap. 3. The polarized electrons with energies in the 100-eV region were then accelerated to 130 keV and either sent into a Mott detector for polarization analysis or fired on the leucine target, after conversion of their transverse polarization into longitudinal polarization.

The polarized-electron source which provided a varying polarization between 10 and 23 % did not come up to the present state of the art. Nevertheless, the results were reported to be clearly beyond the statistical error limits. After irradiation times that produced 50–75 % degradation of the leucine sample, the enantiomorphous composition of the undecomposed leucine was analyzed. It turned out that electrons of negative helicity (spin antiparallel to momentum), as they are emitted in β decay, bring about more extensive degradation of D-leucine than of L-leucine. Reversal of the direction of polarization showed that electrons with positive helicity engender the asymmetric decomposition of D,L-leucine in strictly the opposite sense. The extent of the asymmetry in degradation was claimed to be in the 1 % range.

In an attempt to reproduce the results, the experiment has been repeated by a different group [8.83] using an electron beam of considerably higher polarization. The increased polarization of typically 43 % was expected to result in a more pronounced asymmetric degradation. In spite of a wide variation of experimental parameters and attempts to duplicate as closely as possible the conditions of the first experiment, no single case of preferential decomposition of one of the enantiomers was observed within the experimental detection limit of $\approx 10^{-3}$. This agrees with first attempts to estimate the expected asymmetry which yielded values of the order of 10^{-6} or even less [8.84, 85].

Other polarization effects derive directly from the lack of reflection symmetry of the enantiomers. As shown in Sect. 3.5, scattering of unpolarized electrons from such targets may yield polarization components parallel to the scattering plane [8.86]. An experiment has been performed to search for such in-plane components after scattering of 25-eV electrons at angles of 40°–70° from L- and D-camphor $C_{10}H_{16}O$. No polarization could be detected. The upper limit of the polarization, as given by the error limits of the measurement, was found to be $5 \cdot 10^{-3}$ [8.87]. A rough estimate suggests the polarization to be $\lesssim 10^{-5}$ [8.85].

The same order of magnitude is to be expected for the following polarization phenomena which also are related to the lack of reflection symmetry of optically active molecules. The attenuation of a longitudinally polarized electron beam traversing an optically active substance depends on the beam polarization. As a consequence of differing attenuation of electrons with positive and negative helicity, an initially unpolarized beam emerges from such a substance with longitudinal polarization. If the incident beam is transversely polarized, its polarization undergoes a rotation about the beam axis, the angle of rotation having opposite signs for L- and D-enantiomers [8.88]. Furthermore, electrons resulting from ionization of optically active molecules by photons or charged particles may have some small polarization components of the kind excluded in our earlier discussion of targets possessing reflection symmetry (cf. Table 5.1) [8.89, 90].

The majority of papers that treat the minute polarization phenomena ensuing from the lack of reflection symmetry of enantiomorphous targets or from the even smaller influence of parity-violating interactions in electron-atom scattering [8.91] are theoretical investigations. This is not surprising since measurements in this field make sense only if electron polarization of order $\lesssim 10^{-5}$ can be detected. From the discussion of polarization analysis throughout this book it is clear that here we encounter the limits of present experimental possibilities. This is a good reason for finishing our considerations.

8.7 Prospects

A survey of our present knowledge about the physics of polarized electrons is given along with an assessment of the next steps that should be taken.

We conclude our treatment of the physics of polarized electrons with a few remarks on the state of the art and an evaluation of the present knowledge in this relatively new field. It has been rewarding to see how many of the unsolved problems mentioned in the first edition of this book have since been successfully studied. We hope that the present edition will also stimulate research leading to improved understanding of the interactions of the electron spin.

The spin polarization arising from elastic electron scattering by unpolarized targets is quite well understood in the whole energy range between about 100 eV and a few MeV. The theoretical knowledge is well developed, and despite severe discrepancies between theory and experiment in the early stages of these studies, the increasing knowledge of how to avoid experimental pitfalls has finally led to complete confirmation of the theoretical results. "Triple" scattering experiments which are necessary if all the parameters governing the scattering process are to be found have been successfully started so that complete information on the conceptually simple process of elastic scattering can now be obtained. At small energies the agreement between theory and experiment breaks down as discussed in Sect. 3.6.2. That is why the elastic scattering entry in the survey given in Table 8.2 extends from "good" through "moderate". Further work is necessary for a better understanding of the scattering process at these energies.

Inelastic scattering is more involved than elastic scattering and therefore needs more transition amplitudes for its description. On the other hand, there are more observables: In addition to the polarization behavior of the projectile, as in elastic scattering, one may study the various types of light polarization emitted

Table 8.2. Present knowledge of electron polarization in various fields of physics

Field	Knowledge		
	Good	Moderate	Poor
Electron Scattering			
Elastic	■■■■■■■		
Inelastic		■■■■	
Exchange		■■■■	
Impact excitation, light emission	■■■■■■	■■	
From optically active molecules			■■■
Bremsstrahlung		■■■■■	■
Ionization			
Polarized atoms			
Photoionization	■■		
Collisional ionization		■■■■■■	
Unpolarized atoms			
Fano effect	■■■■■		
Unpolarized radiation	■■■■■	■■	
Excited atomic states	■■■■	■■■■	
Multiphoton ionization		■■■	
Solids and Surfaces			
Emission from magnetic materials	■■■■■	■■	
Photoemission from nonmagnetic materials	■■■■■	■■	
LEED	■■■■■	■■	
Nuclei, elementary particles			
β decay	■■■■■		
$g-2$ experiments	■■■		
High-energy scattering		■■■■	

by the excited atoms. The correlations between the properties of the emitted radiation and the polarization of the projectile can be studied in great detail by coincidence experiments. If one has polarized targets and projectiles, measurement of the difference between the cross sections for spins parallel or antiparallel yields further information. Certain observables allow one to disentangle effects of exchange and spin-orbit interaction, both of which usually contribute to the inelastic process. The larger number of independent observables requires more experimental effort so that it is quite natural that knowledge of inelastic scattering lags behind that of elastic scattering. Owing to the advent of efficient polarized-electron sources there has been considerable progress in the decade between the two editions of this book. The numerous open problems in this area may now be successfully attacked, promising better insight into atomic interactions. It is, however, questionable whether measuring all the independent observables that completely describe inelastic processes is worth the great effort required. Instead, it seems advisable to not strive for the perfect inelastic experiment but to pick out those observables that are essential for determining the dynamics of the process and that can be measured with the greatest accuracy.

The numerous resonance features found not only in cross sections but also in polarization and asymmetry curves certainly imply an additional complication of electron-impact work. On the other hand, their systematic investigation opens the possibility of classifying and understanding the compound states causing the resonances. This has so far been done only in a few cases.

The generation of polarized electrons by photoionization of polarized atoms is one of the areas in which present knowledge can be rated as good. It is, however, worth noting that the discovery of the Fano effect has cast new light also on this process. The polarization of the photoelectrons may be quite different from that of the atoms owing to a significant number of spin-flip processes during photoionization at certain wavelengths. The commonly used argument [8.92] that photoionization causes no spin flip since it is brought about by the electric vector of the light wave is too simple. With this argument one would not find polarization effects in electron-atom scattering either, since it is caused by the "electric" field of the atom!

The polarized electrons originating from polarized atoms can also be used as a diagnostic tool for the analysis of collision processes. Researchers are only beginning to make use of this appealing technique. Numerous such processes are conceivable; their exploitation will yield detailed knowledge on particle collisions – knowledge which so far has been obscured in the averaged results obtained with unpolarized collision partners.

A detailed analysis of photoionization processes has been obtained by investigating the polarization of photoelectrons from unpolarized targets. Measurements of the Fano effect and its extension to molecules and solid targets are now facilitated by the advent of circularly polarized synchrotron radiation of high intensity. In the first edition of this book we emphasized that not a single experiment had studied the polarization of photoelectrons released from unpolarized atoms by unpolarized light. This situation has completely changed and rapid progress in experimental and theoretical work has much promoted our knowledge of spin dependence of photoionization. Polarization effects in multiphoton ionization seem, however, to have had more appeal to theoreticians than to experimentalists though a number of interesting problems for laser-equipped laboratories can be found here.

The effectiveness of polarization studies in revealing the structure of solids and surfaces has led to a great expansion of the field and to development of elaborate experimental methods. Much of the attention has been focused on magnetic materials, but also with nonmagnetic substances more insight has been obtained from polarization measurements. A further stimulus for such research is certainly the hope of discovering a material yielding totally polarized electrons of high intensity. This goal has been reached only half-way with the GaAs cathode, though the impact of this source on polarized-electron physics cannot be overestimated. The main problem in polarized-electron studies is now no longer the source intensity, but rather the low efficiency of electron polarimeters.

As to LEED experiments with polarized electrons, we stated "the first encouraging results" in the preceding edition. In the meantime we have seen a most inspiring development of this field. Experimental and theoretical investigations with remarkable accuracy and agreement have been published complementing the conventional LEED technique and showing the possibility of studying even subtle effects of surface structure. Spin-dependent effects based on spin-orbit and on exchange interaction have been demonstrated to be easily separable by suitable arrangements. The chances of PLEED will certainly be exploited in the future both for nonmagnetic and magnetic surfaces.

The utility of polarized-electron studies in nuclear and elementary-particle physics is well known to a broad community because of the famous experiments on parity violation in weak interactions. Much experimental and theoretical work on electron polarization in β decay has been done in the past decades and has been treated in comprehensive monographs; thus, we have not felt it necessary to deal with this topic. The more recent parity-violation experiments in the GeV regime clearly confirmed the theory of electroweak interaction. No doubt polarized electrons will play a crucial part in future high-energy experiments. When talking about elementary particles one has certainly to recognize the fantastic accuracy of the measurements determining the g factor of the electron (and of other elementary particles) which is likely to be further improved in the future.

The host of impressive results brought about in the past decade should not make us forget that there are subjects that have been scarcely cultivated, such as experiments on polarization correlations in bremsstrahlung or on electron scattering from optically active molecules. What is generally true for the entire field of polarized-electron physics is especially true for the latter area: Progress depends crucially on major advances in the methods for producing and – above all – for analyzing electron polarization.

References

Complete reference lists are to be found in the review articles and monographs given below. The primary sources given are either directly referred to in the text or have appeared later than the review papers listed.

Chapter 1

1.1 N. F. Mott, H. S. W. Massey: *The Theory of Atomic Collisions* (Clarendon, Oxford 1965) Chap. IX
1.2 P. S. Farago: Adv. Electron. Electron Phys. **21**, 1 (1965)

Chapter 2

2.1 H. A. Tolhoek: Rev. Mod. Phys. **28**, 277 (1956)
2.2 U. Fano: Rev. Mod. Phys. **29**, 74 (1957)
2.3 K. Blum: *Density Matrix Theory and Applications* (Plenum, New York 1981)

Chapter 3

3.1 N. F. Mott, H. S. W. Massey: *The Theory of Atomic Collisions* (Clarendon, Oxford 1965) Chap. IX
3.2 D. M. Fradkin, R. H. Good: Rev. Mod. Phys. **33**, 343 (1961)
3.3 L. I. Schiff: *Quantum Mechanics*, 3rd ed. (McGraw-Hill, New York 1968) Chap. 13
3.4 H. A. Bethe, E. E. Salpeter: In *Handbuch der Physik*, Vol. XXXV (Springer, Berlin, Göttingen, Heidelberg 1957) p. 133ff.
3.5 N. Sherman: Phys. Rev. **103**, 1601 (1956)
3.6 H. F. Schopper: *Weak Interactions and Nuclear Beta Decay* (North-Holland, Amsterdam 1966);
     H. Frauenfelder, A. Rossi: In *Methods of Experimental Physics*, ed. by L. C. L. Yuan, C. S. Wu, Vol. 5, Pt. B, (Academic, New York 1963) p. 214;
     H. Frauenfelder, R. M. Steffen: In *Alpha, Beta, and Gamma Ray Spectroscopy*, ed. by K. Siegbahn, Vol. II (North-Holland, Amsterdam 1968) p. 1431;
     L. A. Page: Rev. Mod. Phys. **31**, 759 (1959)
3.7 P. S. Farago: J. Phys. **B13**, L 567 (1980)
3.8 R. Feder, J. Kirschner: Surf. Sci. **103**, 75 (1981)
3.9 J. W. Motz, H. Olsen, H. W. Koch: Rev. Mod. Phys. **36**, 881 (1964)
3.10 J. Kessler, N. Weichert: Z. Phys. **212**, 48 (1968); W. Bühring: Z. Phys. **212**, 61 (1968) and references therein

282   References

3.11 H. Überall: *Electron Scattering from Complex Nuclei*, Pt. A (Academic, New York 1971)
3.12 K. Jost, J. Kessler: Z. Phys. **195**, 1 (1966)
3.13 D. W. Walker: Adv. Phys. **20**, 257 (1971)
3.14 W. Bühring: Z. Phys. **208**, 286 (1968)
3.15 H. Deichsel, E. Reichert, H. Steidl: Z. Phys. **189**, 212 (1966);
      M. Düwecke, N. Kirchner, E. Reichert, S. Schön: J. Phys. **B9**, 1915 (1976)
3.16 G. F. Hanne, K. J. Kollath, W. Wübker: J. Phys. **B13**, L 395 (1980)
3.17 I. E. McCarthy, C. J. Noble, B. A. Phillips, A. D. Turnbull: Phys. Rev. **A15**, 2173 (1977)
3.18 C. B. Lucas, I. E. McCarthy: J. Phys. **B11**, L 301 (1978)
3.19 L. T. Sin Fai Lam: J. Phys. **B15**, 119 (1982)
3.20 B. Awe, F. Kemper, F. Rosicky, R. Feder: J. Phys. **B16**, 603 (1983)
3.21 L. Fritsche, J. Noffke, H. Gollisch: J. Phys. **B17**, 1637 (1984)
3.22 O. Berger, W. Wübker, R. Möllenkamp, J. Kessler: J. Phys. **B15**, 2473 (1982)
3.23 R. J. v. Duinen, J. W. G. Aalders: Nucl. Phys. **A115**, 353 (1968)
3.24 W. Wübker, R. Möllenkamp, J. Kessler: Phys. Rev. Lett. **49**, 272 (1982)
3.25 R. Möllenkamp, W. Wübker, O. Berger, K. Jost, J. Kessler: J. Phys. **B17**, 1107 (1984)
3.26 K. Schackert: Z. Phys. **213**, 316 (1968)
3.27 M. Klewer, M. J. M. Beerlage, M. J. Van der Wiel: J. Phys. **B12**, 3935 (1979)
3.28 M. J. M. Beerlage, Zhou Qing, M. J. Van der Wiel: J. Phys. **B14**, 4627 (1981)
3.29 M. Klewer, M. J. M. Beerlage, M. J. Van der Wiel: J. Phys. **B12**, L 525 (1979)
3.30 F. Kemper, B. Awe, F. Rosicky, R. Feder: J. Phys. **B16**, 1819 (1983)
3.31 J. Kessler: Rev. Mod. Phys. **41**, 1 (1969)
3.32 W. Eckstein: Internal Rept. IPP 7/1, Institut für Plasmaphysik, Garching bei München (1970)
3.33 G. Holzwarth, H. J. Meister: *Tables of Asymmetry, Cross Sections and Related Functions for Mott Scattering of Electrons by Screened Au and Hg Nuclei* (University of Munich, Munich 1964)
3.34 M. Fink, A. C. Yates: At. Data **1**, 385 (1970);
      M. Fink, J. Ingram: At. Data **4**, 129 (1972);
      D. Gregory, M. Fink: At. Data Nucl. Data Tables **14**, 39 (1974)
3.35 J. van Klinken: Nucl. Phys. **75**, 161 (1966)
3.36 W. Eitel, K. Jost, J. Kessler: Phys. Rev. **159**, 47 (1967); Z. Phys. **209**, 348 (1968)
3.37 T. Heindorff, J. Höfft, E. Reichert: J. Phys. **B6**, 477 (1973)
3.38 T. Suzuki, H. Tanaka, M. Saito, H. Igawa: J. Phys. Soc. Jpn. **39**, 200 (1975)
3.39 W. Franzen, R. Gupta: Phys. Rev. Lett. **15**, 819 (1965)
3.40 A. Albert, C. Christian, T. Heindorff, E. Reichert, S. Schön: J. Phys. **B10**, 3733 (1977)
3.41 W. Eitel, J. Kessler: Z. Phys. **241**, 355 (1971)
3.42 D. H. Madison, W. N. Shelton: Phys. Rev. **A7**, 514 (1973)
3.43 C. B. O. Mohr, F. H. Nicoll: Proc. Roy. Soc. (London), Ser. A **138**, 229 (1932);
      H. S. W. Massey, E. H. S. Burhop: *Electronic and Ionic Impact Phenomena*, Vol. 1 (Oxford University Press, Oxford 1969) p. 559ff.
3.44 R. A. Bonham: J. Electron Spectrosc. **3**, 85 (1974)
3.45 K. Bartschat, K. Blum: J. Phys. **B15**, 2747 (1982)
3.46 J. Kirschner: *Polarized Electrons at Surfaces*, in *Springer Tracts Mod. Phys.*, Vol. 106 (Springer, Berlin, Heidelberg 1985)
3.47 K. Bartschat, G. F. Hanne, A. Wolcke, J. Kessler: Phys. Rev. Lett. **47**, 997 (1981)
3.48 K. Franz, G. F. Hanne, K. H. Hiddemann, J. Kessler: J. Phys. **B15**, L 115 (1982)
3.49 G. F. Hanne: Phys. Rep. **95**, 95 (1983)
3.50 G. F. Hanne, Cz. Szmytkowski, M. Van der Wiel: J. Phys. **B15**, L 109 (1982)

Chapter 4

4.1   P. S. Farago: Rep. Prog. Phys. **34**, 1055 (1971)
4.2   K. Blum: *Density Matrix Theory and Applications* (Plenum, New York 1981)
4.3   P. G. Burke, H. M. Schey: Phys. Rev. **126**, 163 (1962)
4.4   U. Fano: Rev. Mod. Phys. **55**, 855 (1983)
4.5   B. Bederson: Comments At. Mol. Phys. **1**, 41 (1969)
4.6   H. G. Dehmelt: Phys. Rev. **109**, 381 (1958)
4.7   R. E. Collins, B. Bederson, M. Goldstein: Phys. Rev. **A3**, 1976 (1971)
4.8   B. Bederson: In *Atomic Physics 3*, ed. by S. J. Smith, G. K. Walters (Plenum, New York 1973) p. 401 and references therein
4.9   D. Hils, M. V. McCusker, H. Kleinpoppen, S. J. Smith: Phys. Rev. Lett. **29**, 398 (1972)
4.10   E. M. Karule, R. K. Peterkop: *Atomic Collisions III*, ed. by Y. Ia. Veldre (Latvian Academy of Sciences, Riga, USSR 1965). Available through SLA Translation Service, John Crerar Library, 86 East Randolph St., Chicago, IL. Translation No. TT-66-1239
4.11   G. D. Fletcher, M. J. Alguard, T. J. Gay, V. W. Hughes, C. W. Tu, P. F. Wainwright, M. S. Lubell, W. Raith, F. C. Tang: Phys. Rev. Lett. **48**, 1671 (1982)
4.12   M. R. C. McDowell, P. W. Edmunds, R. M. Potvliege, C. J. Joachain, R. Shingal, B. H. Bransden: J. Phys. **B17**, 3951 (1984)
4.13   P. G. Burke, L. F. B. Mitchell: J. Phys. **B7**, 214 (1974)
4.14   P. S. Farago: J. Phys. **B7**, L28 (1974)
4.15   D. W. Walker: J. Phys. **B7**, L489 (1974)
4.16   M. J. Alguard, V. W. Hughes, M. S. Lubell, P. F. Wainwright: Phys. Rev. Lett. **39**, 334 (1977);
       T. J. Gay, G. D. Fletcher, M. J. Alguard, V. W. Hughes, P. F. Wainwright, M. S. Lubell: Phys. Rev. **A26**, 3664 (1982)
4.17   D. Hils, H. Kleinpoppen: J. Phys. **B11**, L283 (1978)
4.18   G. Baum, E. Kisker, W. Raith, W. Schröder, U. Sillmen, D. Zenses: J. Phys. **B14**, 4377 (1981);
       G. Baum, M. Moede, W. Raith, W. Schröder: J. Phys. **B18**, 531 (1985)
4.19   D. Hils, W. Jitschin, H. Kleinpoppen: J. Phys. **B15**, 3347 (1982)
4.20   M. H. Kelley, W. T. Rogers, R. J. Celotta, S. R. Mielczarek: Phys. Rev. Lett. **51**, 2191 (1983)
4.21   C. H. Greene, A. R. P. Rau: Phys. Rev. Lett. **48**, 533 (1982)
4.22   A. Temkin: Phys. Rev. Lett. **49**, 365 (1982)
4.23   K. Rubin, B. Bederson, M. Goldstein, R. E. Collins: Phys. Rev. **182**, 201 (1969)
4.24   B. Bederson: Comments At. Mol. Phys. **1**, 65 (1969); **2**, 160 (1971)
4.25   W. Lichten, S. Schultz: Phys. Rev. **116**, 1132 (1959)
4.26   W. Jitschin, S. Osimitsch, H. Reihl, H. Kleinpoppen, H. O. Lutz: J. Phys. **B17**, 1899 (1984)
4.27   G. Baum, M. Moede, W. Raith, W. Schröder, U. Sillmen: In *Electronic and Atomic Collisions*, Abstracts of Papers, XIV ICPEAC, Stanford, 1985
4.28   E. U. Condon, G. H. Shortley: *The Theory of Atomic Spectra* (Cambridge Univ. Press, Cambridge 1967)
4.29   G. F. Hanne: Comments At. Mol. Phys. **14**, 163 (1984)
4.30   G. F. Hanne: Phys. Rep. **95**, 95 (1983)
4.31   N. S. Scott, K. Bartschat, P. G. Burke, W. B. Eissner, O. Nagy: J. Phys. **B17**, L191 (1984)
4.32   K. Bartschat, K. Blum, P. G. Burke, G. F. Hanne, N. S. Scott: J. Phys. **B17**, 3797 (1984)
4.33   G. F. Hanne, Cz. Szmytkowski, M. Van der Wiel: J. Phys. **B15**, L109 (1982)
4.34   J. Macek, D. H. Jaecks: Phys. Rev. **A4**, 2288 (1971)
4.35   L. I. Schiff: *Quantum Mechanics*, 3rd ed. (McGraw-Hill, New York 1968)
4.36   H. Kleinpoppen: Phys. Rev. **A3**, 2015 (1971) (several formulae in this paper have to be corrected)

4.37 F. S. Crawford, Jr.: *Waves; Berkeley Physics Course*, Vol. 3 (McGraw-Hill, New York 1968)
4.38 E. Hecht, A. Zajac: *Optics* (Addison-Wesley, Reading, MA 1974)
4.39 K. Bartschat, K. Blum: Z. Phys. **A304**, 85 (1982)
4.40 K. Bartschat, K. Blum, G. F. Hanne, J. Kessler: J. Phys. **B14**, 3761 (1981)
4.41 U. Fano, J. Macek: Rev. Mod. Phys. **45**, 553 (1973)
4.42 G. F. Hanne, J. Kessler: J. Phys. **B9**, 791 (1976)
4.43 G. F. Hanne: J. Phys. **B9**, 805 (1976)
4.44 J. C. Steelhammer, S. Lipsky: J. Chem. Phys. **53**, 1445 (1970)
4.45 R. A. Bonham: J. Chem. Phys. **57**, 1604 (1972)
4.46 B. L. Moiseiwitsch: J. Phys. **B9**, L245 (1976)
4.47 R. A. Bonham: J. Phys. **B15**, L361 (1982)
4.48 A. Wolcke, K. Bartschat, K. Blum, H. Borgmann, G. F. Hanne, J. Kessler: J. Phys. **B16**, 639 (1983)
4.49 K. Bartschat, N. S. Scott, K. Blum, P. G. Burke: J. Phys. **B17**, 269 (1984)
4.50 P. S. Farago. J. S. Wykes: J. Phys. **B2**, 747 (1969)
4.51 J. Wykes: J. Phys. **B4**, L91 (1971)
4.52 M. Eminyan, G. Lampel: Phys. Rev. Lett. **45**, 1171 (1980)
4.53 J. Goeke, G. F. Hanne, J. Kessler, A. Wolcke: Phys. Rev. Lett. **51**, 2273 (1983)
4.54 K. Blum, H. Kleinpoppen: Adv. At. Mol. Phys. **19**, 187 (1983)
4.55 J. Slevin: Rep. Prog. Phys. **47**, 461 (1984)
4.56 A. Wolcke, J. Goeke, G. F. Hanne, J. Kessler, W. Vollmer, K. Bartschat, K. Blum: Phys. Rev. Lett. **52**, 1108 (1984)
4.57 V. D. Ob'edkov, I. Kh. Mossallami: Vestn. Leningr. Univ. **22**, 43 (1971)
4.58 E. Karule: J. Phys. **B5**, 2051 (1972)
4.59 D. M. Campbell, H. M. Brash, P. S. Farago: Phys. Lett. **36A**, 449 (1971)
4.60 R. Krisciokaitis, W. Y. Tsai: Nucl. Instrum. Meth. **83**, 45 (1970);
      R. Krisciokaitis, W. K. Peterson: In *Electronic and Atomic Collisions*, Abstracts of Papers, VIII ICPEAC, Belgrade 1973 (Institute of Physics, Belgrade 1973) p. 257
4.61 A. M. Bincer: Phys. Rev. **107**, 1434 (1957)
4.62 C. Møller: Ann. Phys. (Leipzig) **14**, 531 (1932)
4.63 G. Holzwarth: Z. Phys. **191**, 354 (1966)
4.64 K. Ulmer: Z. Phys. **135**, 232 (1953)
4.65 J. J. Kepes, B. Waldmann, W. C. Miller: Ann. Phys. **6**, 90 (1959)
4.66 A. Ashkin, L. A. Page, W. M. Woodward:Phys. Rev. **94**, 357 (1954)

Chapter 5

5.1 E. Fues, H. Hellmann: Phys. Z. **31**, 465 (1930)
5.2 G. Baum, U. Koch: Nucl. Instrum. Meth. **71**, 189 (1969)
5.3 V. W. Hughes, R. L. Long, Jr., M. S. Lubell, M. Posner, W. Raith: Phys. Rev. **A5**, 195 (1972)
5.4 M. J. Alguard, J. E. Clendenin, R. D. Ehrlich, V. W. Hughes, J. S. Ladish, M. S. Lubell, K. P. Schüler, G. Baum, W. Raith, R. H. Miller, W. Lysenko: Nucl. Instrum Meth. **163**, 29 (1979)
5.5 G. Baum, E. Kisker, W. Raith, W. Schröder, U. Sillmen, D. Zenses: J. Phys. **B14**, 4377 (1981);
      G. Baum, M. Moede, W. Raith, W. Schröder: J. Phys. **B18**, 531 (1985)
5.6 D. Hils, W. Jitschin, H. Kleinpoppen: J. Phys. **B15**, 3347 (1982)
5.7 M. H. Kelley, W. T. Rogers, R. J. Celotta, S. R. Mielczarek: Phys. Rev. Lett. **51**, 2191 (1983)
5.8 W. Jitschin, S. Osimitsch, H. Reihl, H. Kleinpoppen, H. O. Lutz: J. Phys. **B17**, 1899 (1984)

5.9   U. Fano: Phys. Rev. **178**, 131 (1969); Addendum: Phys. Rev. **184**, 250 (1969)
5.10 U. Heinzmann, J. Kessler, J. Lorenz: Z. Phys. **240**, 42 (1970)
5.11 G. Baum, M. S. Lubell, W. Raith: Phys. Rev. **A5**, 1073 (1972)
5.12 J. C. Weisheit: Phys. Rev. **A5**, 1621 (1972)
5.13 D. W. Norcross: Phys. Rev. **A7**, 606 (1973)
5.14 K.-N. Huang, A. F. Starace: Phys. Rev. **19**, 2335 (1979)
5.15 W. R. Johnson, G. Soff: Phys. Rev. Lett. **50**, 1361 (1983)
5.16 W. von Drachenfels, U. T. Koch, Th. M. Müller, W. Paul, H. R. Schaefer: Nucl. Instrum. Meth. **140**, 47 (1977)
5.17 V. L. Jacobs: J. Phys. **B5**, 2257 (1972)
5.18 N. A. Cherepkov: Zh. Eksp. Teor. Fiz. **65**, 933 (1973) [English transl.: Sov. Phys. – JETP **38**, 463 (1974)]
5.19 C. M. Lee: Phys. Rev. **A10**, 1598 (1974)
5.20 U. Heinzmann, G. Schönhense, J. Kessler: J. Phys. **B13**, L153 (1980)
5.21 N. A. Cherepkov: J. Phys. **B12**, 1279 (1979)
5.22 K.-N. Huang, W. R. Johnson, K. T. Cheng: Phys. Rev. Lett. **43**, 1658 (1979); F. A. Parpia, W. R. Johnson, R. Radojević: Phys. Rev. **A29**, 3173 (1984)
5.23 G. Schönhense, F. Schäfers, U. Heinzmann, J. Kessler: Z. Phys. **A304**, 31 (1982)
5.24 G. Schönhense, U. Heinzmann, J. Kessler, N. A. Cherepkov: Phys. Rev. Lett. **48**, 603 (1982)
5.25 G. Schönhense: Phys. Rev. Lett. **44**, 640 (1980)
5.26 K. Kollath: J. Phys. **B13**, 2901 (1980)
5.27 K.-N. Huang: Phys. Rev. Lett. **48**, 1811 (1982)
5.28 H. Klar, H. Kleinpoppen: J. Phys. **B15**, 933 (1982)
5.29 Ch. Heckenkamp, F. Schäfers, G. Schönhense, U. Heinzmann: Phys. Rev. Lett. **52**, 421 (1984)
5.30 U. Heinzmann: J. Phys. **B13**, 4353, 4367 (1980)
5.31 G. Schönhense, U. Heinzmann: Phys. Rev. **A29**, 987 (1984)
5.32 F. Schäfers, G. Schönhense, U. Heinzmann: Z. Phys. **A304**, 41 (1982)
5.33 H. Kaminski, J. Kessler, K. Kollath: Phys. Rev. Lett. **45**, 1161 (1980)
5.34 K.-N. Huang, W. R. Johnson, K. T. Cheng: At. Data Nucl. Data Tables **26**, 33 (1981)
5.35 W. R. Johnson, V. Radojević, P. Deshmukh: Phys. Rev. **A25**, 337 (1982)
5.36 N. A. Cherepkov: Adv. At. Mol. Phys. **19**, 395 (1983)
5.37 U. Heinzmann, F. Schäfers, B. A. Hess: Chem. Phys. Lett. **69**, 284 (1980); U. Heinzmann, B. Osterheld, F. Schäfers, G. Schönhense: J. Phys. **B14**, L79 (1981); F. Schäfers, M. A. Baig, U. Heinzmann: J. Phys. **B16**, L1 (1983); G. Schönhense, V. Dzidzonov, S. Kaesdorf, U. Heinzmann: Phys. Rev. Lett. **52**, 811 (1984)
5.38 L. C. Chiu: Phys. Rev. **154**, 56 (1967)
5.39 G. V. Marr, R. Heppinstall: Proc. Phys. Soc. London **87**, 293 (1966)
5.40 J. Berkowitz, W. A. Chupka: J. Chem. Phys. **45**, 1287 (1966)
5.41 U. Heinzmann, H. Heuer, J. Kessler: Phys. Rev. Lett. **34**, 441 (1975)
5.42 J. P. Connerade, M. A. Baig: J. Phys. **B14**, 29 (1981)
5.43 N. A. Cherepkov: J. Phys. **B13**, L181 (1980)
5.44 U. Heinzmann, H. Heuer, J. Kessler: Phys. Rev. Lett. **36**, 1444 (1976)
5.45 U. Heinzmann, F. Schäfers, K. Thimm, A. Wolcke, J. Kessler: J. Phys. **B12**, L679 (1979)
5.46 W. R. Johnson, K. T. Cheng, K.-N. Huang, M. LeDourneuf: Phys. Rev. **A22**, 989 (1980)
5.47 F. Schäfers, G. Schönhense, U. Heinzmann: Phys. Rev. **A28**, 802 (1983)
5.48 R. Clauberg, W. Gudat, E. Kisker, E. Kuhlmann, G. M. Rothberg: Phys. Rev. Lett. **47**, 1314 (1981)
5.49 M. Landolt, D. Mauri: Phys. Rev. Lett. **49**, 1783 (1982)
5.50 K. H. Bennemann: Phys. Rev. **B28**, 5304 (1983)

5.51   A. Kotani, H. Mizuta: Solid State Commun. **51**,727 (1984)
5.52   H. Klar: J. Phys. **B13**, 4741 (1980)
5.53   N. M. Kabachnik: J. Phys. **B14**, L337 (1981)
5.54   K.-N. Huang: Phys. Rev. **26**, 2274 (1982)
5.55   N. M. Kabachnik, I. P. Sazhina: J. Phys. **B17**, 1335 (1984)
5.56   U. Hahn, J. Semke, H. Merz, J. Kessler: J. Phys. **B18**, L 417 (1985)
5.57   P. Lambropoulos: Adv. At. Mol. Phys. **12**, 87 (1976)
5.58   M. R. Teague, P. Lambropoulos, D. Goodmanson, D. W. Norcross: Phys. Rev. **A14**, 1057 (1976)
5.59   M. R. Teague, P. Lambropoulos: J. Phys. **B9**, 1251 (1976)
5.60   A. Declemy, G. Laplanche, M. Jaouen, A. Rachman: J. Phys. **B14**, 3377 (1981)
5.61   P. S. Farago, D. W. Walker: J. Phys. **B6**, L280 (1973)
5.62   H. D. Zeman: In *Electron and Photon Interactions with Atoms*, ed. by H. Kleinpoppen, M. R. C. McDowell (Plenum, New York 1976) p. 581
5.63   P. S. Farago, D. W. Walker, J. S. Wykes: J. Phys. **B7**, 59 (1974)
5.64   E. H. A. Grannemann, M. Klewer, G. Nienhuis, M. J. Van der Wiel: J. Phys. **B10**, 1625 (1977)
5.65   P. Lambropoulos, M. Lambropoulos: In *Electron and Photon Interactions with Atoms*, ed. by H. Kleinpoppen, M. R. C. McDowell (Plenum, New York 1976) p. 525
5.66   S. N. Dixit, P. Lambropoulos, P. Zoller: Phys. Rev. **A24**, 318 (1981)
5.67   M. J. Van der Wiel, E. H. A. Grannemann: Comments At. Mol. Phys. **7**, 59 (1977)
5.68   G. Nienhuis, E. H. A. Grannemann, M. J. Van der Wiel: J. Phys. **B11**, 1203 (1978)
5.69   H. Kaminski, J. Kessler, K. Kollath: J. Phys. **B12**, L383 (1979)
5.70   B. Donnally, W. Raith, R. Becker: Phys. Rev. Lett. **20**, 575 (1968)
5.71   H. A. Bethe, E. E. Salpeter: In *Handbuch der Physik*, Vol. XXXV (Springer, Berlin, Göttingen, Heidelberg 1957) p. 370ff.
5.72   M. V. McCusker, L. L. Hatfield, G. K. Walters: Phys. Rev. **A5**, 177 (1972)
5.73   J. C. Hill, L. L. Hatfield, N. D. Stockwell, G. K. Walters: Phys. Rev. **A5**, 189 (1972)
5.74   P. J. Keliher, F. B. Dunning, M. R. O'Neill, R. D. Rundel, G. K. Walters: Phys. Rev. **A11**, 1271 (1975)
5.75   B. L. Donnally, R. Faber, J. Gates, C. Volk: Bull. Am. Phys. Soc. **18**, 141 (1973)
5.76   G. F. Drukarev, V. D. Ob'edkov, R. K. Janev: Phys. Lett. **42A**, 213 (1972)
5.77   V. D. Ob'edkov: Zh. Eksp. Teor. Fiz. Pisma. Red. **21**, 220 (1975) [English transl.: Sov. Phys. – JETP Lett. **21**, 98 (1975)]
5.78   L. D. Schearer: Phys. Rev. **A10**, 1380 (1974)

Chapter 6

6.1   J. Kessler: Rev. Mod. Phys. **41**, 1 (1969)
6.2   H. F. Schopper: *Weak Interactions and Nuclear Beta Decay* (North-Holland, Amsterdam 1966);
H. Frauenfelder, A. Rossi: In *Methods of Experimental Physics*, ed. by L. C. L. Yuan, C. S. Wu, Vol. 5, Pt. B (Academic, New York 1963) p. 214;
H. Frauenfelder, R. M. Steffen: In *Alpha, Beta, and Gamma Ray Spectroscopy*, ed. by K. Siegbahn, Vol. II (North-Holland, Amsterdam 1968) p. 1431;
L. A. Page: Rev. Mod. Phys. **31**, 759 (1959)
6.3   H. K. Tseng, R. H. Pratt: Phys. Rev. **A7**, 1502 (1973)
6.4   A. Aehlig: Z. Phys. **A294**, 291 (1980)
6.5   H. R. Schaefer, W. v. Drachenfels, W. Paul: Z. Phys. **A305**, 213 (1982)
6.6   P. R. S. Gomes, J. Byrne: J. Phys. **B13**, 3975 (1980)
6.7   S. Hultberg, B. Nagel, P. Olsen: Ark. Fys. **38**, 1 (1969)

6.8   S. J. Blakeway, W. Gelletly, H. R. Faust, K. Schreckenbach, J. Byrne: J. Phys. **B16**, 4565 (1983)
6.9   R. H. Pratt, A. Ron, H. K. Tseng: Rev. Mod. Phys. **45**, 273 (1973)
6.10   H. K. Tseng, R. H. Pratt: Phys. Rev. **A9**, 752 (1974)
6.11   H. A. Tolhoek: Rev. Mod. Phys. **28**, 277 (1956)
6.12   H. D. Bremer, H. C. Dehne, H. C. Lewin, H. Mais, R. Neumann, R. Rossmanith, R. Schmidt: In *High-Energy Physics with Polarized Beams and Polarized Targets*, ed. by C. Joseph, J. Soffer (Birkhäuser, Basel 1981) pp. 52, 469
6.13   A. A. Sokolov, I. M. Ternov: Sov. Phys. Dokl. **8**, 1203 (1964)
6.14   J. D. Jackson: Rev. Mod. Phys. **48**, 417 (1976)
6.15   L. I. Schiff: *Quantum Mechanics*, 3rd ed. (McGraw-Hill, New York 1968) Eq. (45.22)
6.16   J. G. Learned, L. K. Resvanis, C. M. Spencer: Phys. Rev. Lett. **35**, 1688 (1975)
6.17   R. D. Kohaupt, G.-A. Voss: Ann. Rev. Nucl. Part. Sci. **33**, 67 (1983)
6.18   H. D. Bremer, J. Kewenisch, H. C. Lewin, H. Mais, G. Ripken, R. Rossmanith, R. Schmidt, D. P. Barber: In *High-Energy Spin Physics – 1982*, ed. by G. M. Bunce, AIP Conference Proceedings No. 95 (AIP, New York 1982) p. 400
6.19   R. F. Schwitters, A. M. Boyarski, M. Breidenbach et al.: Phys. Rev. Lett. **35**, 1320 (1975)
6.20   B. W. Montague: Phys. Rep. **113**, 1 (1984)

Chapter 7

7.1   H. C. Siegmann: Phys. Rep. **17**, 37 (1975)
7.2   M. Campagna, D. T. Pierce, F. Meier, K. Sattler, H. C. Siegmann: Adv. Electron. Electron Phys. **41**, 113 (1976)
7.3   H. C. Siegmann, F. Meier, M. Erbudak, M. Landolt: Adv. Electron. Electron Phys. **62**, 1 (1984)
7.4   J. Kirschner: "Polarized Electrons at Surfaces", in *Springer Tracts Mod. Phys.*, Vol. 106 (Springer, Berlin, Heidelberg 1985)
7.5   R. Raue, H. Hopster, R. Clauberg: Phys. Rev. Lett. **50**, 1623 (1983)
7.6   G. Busch, M. Campagna, H. C. Siegmann: J. Appl. Phys. **41**, 1044 (1970)
7.7   K. Sattler, H. C. Siegmann: Phys. Rev. Lett. **29**, 1565 (1972)
7.8   K. Sattler, H. C. Siegmann: Z. Phys. **B20**, 289 (1975)
7.9   F. Meier, D. Pescia, M. Baumberger: Phys. Rev. Lett. **49**, 747 (1982)
7.10   F. Meier, G. L. Bona, S. Hüfner: Phys. Rev. Lett. **52**, 1152 (1984)
7.11   D. Mauri, M. Landolt: Phys. Rev. Lett. **47**, 1322 (1981)
7.12   G. Busch, M. Campagna, H. C. Siegmann: Phys. Rev. **B4**, 746 (1971)
7.13   W. Eib, B. Reihl: Phys. Rev. Lett. **40**, 1674 (1978)
7.14   R. Feder, W. Gudat, E. Kisker, A. Rodriguez, K. Schröder: Solid State Commun. **46**, 619 (1983)
7.15   M. Landolt, Ph. Niedermann, D. Mauri: Phys. Rev. Lett. **48**, 1632 (1982)
7.16   W. Eib, S. F. Alvarado: Phys. Rev. Lett. **37**, 444 (1976)
7.17   W. Gudat, E. Kisker, E. Kuhlmann, M. Campagna: Phys. Rev. **B22**, 3282 (1980)
7.18   E. Kisker, W. Gudat, E. Kuhlmann, R. Clauberg, M. Campagna: Phys. Rev. Lett. **45**, 2053 (1980)
7.19   D. T. Pierce, C. E. Kuyatt, R. J. Celotta: Rev. Sci. Instrum. **50**, 1467 (1979)
7.20   E. Kisker, R. Clauberg, W. Gudat: Rev. Sci. Instrum. **53**, 1137 (1982)
7.21   H. Hopster, R. Raue, G. Güntherodt, E. Kisker, R. Clauberg, M. Campagna: Phys. Rev. Lett. **51**, 829 (1983)
7.22   R. Clauberg, H. Hopster, R. Raue: Phys. Rev. **B29**, 4395 (1984)
7.23   E. Kisker, K. Schröder, M. Campagna, W. Gudat: Phys. Rev. Lett. **52**, 2285 (1984); R. Feder, A. Rodriguez, U. Baier, E. Kisker: Solid. State Commun. **52**, 57 (1984)

288   References

7.24   E. P. Wohlfahrt: Phys. Rev. Lett. **38**, 524 (1977)
7.25   V. Korenmann, R. E. Prange: Phys. Rev. Lett. **53**, 186 (1984)
7.26   D. T. Pierce, H. C. Siegmann: Phys. Rev. **B9**, 4035 (1974)
7.27   T. E. Feuchtwang, P. H. Cutler: Surf. Sci. **75**, 401 (1978);
         T. E. Feuchtwang, P. H. Cutler, D. Nagy, R. H. Good, Jr.: Surf. Sci. **75**, 490 (1978)
7.28   S. F. Alvarado: Z. Phys. **B33**, 51 (1979)
7.29   J. Unguris, A. Seiler, R. J. Celotta, D. T. Pierce, P. D. Johnson, N. V. Smith: Phys. Rev.
         Lett. **49**, 1047 (1982)
7.30   H. Scheidt, M. Glöbl, V. Dose, J. Kirschner: Phys. Rev. Lett. **51**, 1688 (1983);
         R. Feder, A. Rodriguez: Solid State Commun. **50**, 1033 (1984)
7.31   G. Obermair: Z. Phys. **217**, 91 (1968)
7.32   B. A. Politzer, P. H. Cutler: Phys. Rev. Lett. **28**, 1330 (1972);
         D. Nagy, P. H. Cutler, T. E. Feuchtwang: Phys. Rev. **B19**, 2964 (1979)
7.33   G. Chrobok, M. Hofmann, G. Regenfus: Phys. Lett. **26A**, 551 (1968)
7.34   N. Müller: Phys. Lett. **54A**, 415 (1975)
7.35   M. Landolt, M. Campagna: Phys. Rev. Lett. **38**, 663 (1977)
7.36   W. Eckstein, N. Müller: Appl. Phys. **6**, 71 (1975)
7.37   M. Landolt, Y. Yafet: Phys. Rev. Lett. **40**, 1401 (1978)
7.38   E. Kisker, G. Baum, A. H. Mahan, W. Raith, B. Reihl: Phys. Rev. **B18**, 2256 (1978)
7.39   W. Nolting, B. Reihl: J. Magn. Magn. Mat. **10**, 1 (1979)
7.40   D. M. Edwards: J. Phys. **C16**, L327 (1983)
7.41   G. Chrobok, M. Hofmann: Phys. Lett. **57A**, 257 (1976)
7.42   E. Kisker, W. Gudat, K. Schröder: Solid State Commun. **44**, 591 (1982)
7.43   H. Hopster, R. Raue, E. Kisker, G. Güntherodt, M. Campagna: Phys. Rev. Lett. **50**, 70
         (1983)
7.44   J. Unguris, D. T. Pierce, A. Galejs, R. J. Celotta: Phys. Rev. Lett. **49**, 72 (1982)
7.45   J. A. D. Matthew: Phys. Rev. **B25**, 3326 (1982)
7.46   D. R. Penn, S. P. Apell, S. M. Girvin: Phys. Rev. Lett. **55**, 518 (1985)
7.47   D. Mauri, R. Allenspach, M. Landolt: Phys. Rev. Lett. **52**, 152 (1984)
7.48   H. Hopster, R. Raue, R. Clauberg: Phys. Rev. Lett. **53**, 695 (1984)
7.49   J. Kirschner, D. Rebenstorff, H. Ibach: Phys. Rev. Lett. **53**, 698 (1984)
7.50   W. Eckstein: Z. Phys. **203**, 59 (1967)
7.51   R. Loth: Z. Phys. **203**, 66 (1967)
7.52   M. Erbudak, G. Ravano: Phys. Lett. **91A**, 367 (1982)
7.53   U. Heinzmann, K. Jost, J. Kessler, B. Ohnemus: Z. Phys. **251**, 354 (1972)
7.54   K. Koyama, H. Merz: Z. Phys. **B20**, 131 (1975)
7.55   C. Kittel: *Quantum Theory of Solids* (Wiley, New York 1963) Chap. X
7.56   J. R. Chelikowsky, M. L. Cohen: Phys. Rev. Lett. **32**, 674 (1974);
         Phys. Rev. **B14**, 556 (1976)
7.57   D. T. Pierce, F. Meier: Phys. Rev. **B13**, 5484 (1976)
7.58   R. Allenspach, F. Meier, D. Pescia: Appl. Phys. Lett. **44**, 1107 (1984)
7.59   M. Erbudak, B. Reihl: Appl. Phys. Lett. **33**, 584 (1978)
7.60   D. T. Pierce, G. C. Wang, R. J. Celotta: Appl. Phys. Lett. **35**, 220 (1979)
7.61   P. Zürcher, F. Meier, N. E. Christensen: Phys. Rev. Lett. **43**, 54 (1979)
7.62   J. Reyes, J. S. Helman: Phys. Rev. **B16**, 4283 (1977)
7.63   M. Wöhlecke, G. Borstel: Phys. Rev. **B24**, 2321, 2857 (1981)
7.64   R. Allenspach, F. Meier, D. Pescia: Phys. Rev. Lett. **51**, 2148 (1983)
7.65   D. Pescia, F. Meier: Surf. Sci. **117**, 302 (1982);
         G. Borstel, M. Wöhlecke: Phys. Rev. **B28**, 3153 (1983)
7.66   A. Eyers, F. Schäfers, G. Schönhense, U. Heinzmann, H. P. Oepen, K. Hünlich, J.
         Kirschner, G. Borstel: Phys. Rev. Lett. **52**, 1559 (1984)
7.67   J. Kirschner, R. Feder, J. F. Wendelken: Phys. Rev. Lett. **47**, 614 (1981)
7.68   R. Feder: J. Phys. **C14**, 2049 (1981)
7.69   R. Feder: Phys. Scr. **T4**, 47 (1983)

7.70   D. T. Pierce, R. J. Celotta: Adv. Electron. Electron Phys. **56**, 219 (1981)
7.71   E. G. McRae: Rev. Mod. Phys. **51**, 541 (1979)
7.72   M. R. O'Neill, M. Kalisvaart, F. B. Dunning, G. K. Walters: Phys. Rev. Lett. **34**, 1167 (1975)
7.73   C. J. Davisson, L. H. Germer: Phys. Rev. **33**, 760 (1929)
7.74   C. E. Kuyatt: Phys. Rev. **B12**, 4581 (1975)
7.75   R. Feder: Phys. Rev. **B15**, 1751 (1977)
7.76   P. J. Jennings, R. O. Jones: Surf. Sci. **71**, 101 (1978)
7.77   M. Kalisvaart, M. R. O'Neill, T. W. Riddle, F. B. Dunning, G. K. Walters: Phys. Rev. **B17**, 1570 (1978);
       J. K. Lang, K. D. Jamison, F. B. Dunning, G. K. Walters, M. A. Passler, A. Ignatiev, E. Tamura, R. Feder: Surf. Sci. **123**, 247 (1982)
7.78   G.-C. Wang, R. J. Celotta, D. T. Pierce: Phys. Rev. **B23**, 1761 (1981)
7.79   R. Feder, N. Müller, D. Wolf: Z. Phys. **B28**, 265 (1977)
7.80   P. Bauer, R. Feder, N. Müller: Solid State Commun. **36**, 249 (1980)
7.81   S. F. Alvarado, H. Hopster, R. Feder, H. Pleyer: Solid State Commun. **39**, 1319 (1981);
       S. F. Alvarado, M. Campagna, H. Hopster: Phys. Rev. Lett. **48**, 51, 1768 (1982)
7.82   J. Kirschner, R. Feder: Phys. Rev. Lett. **42**, 1008 (1979)
7.83   G.-C. Wang, B. I. Dunlap, R. J. Celotta, D. T. Pierce: Phys. Rev. Lett. **42**, 1349 (1979)
7.84   R. Feder, J. Kirschner: Surf. Sci. **103**, 75 (1981)
7.85   E. G. McRae, D. T. Pierce, G.-C. Wang, R. J. Celotta: Phys. Rev. **B24**, 4230 (1981)
7.86   G. Malmström, J. Rundgren: J. Phys. **C14**, 4937 (1981)
7.87   R. O. Jones, P. J. Jennings: Phys. Rev. **B27**, 4702 (1983)
7.88   R. Feder: Solid State Commun. **31**, 821 (1979)
7.89   J. C. Slater: *The Self-Consistent Field for Molecules and Solids*, Vol. 4 (McGraw-Hill, New York 1974)
7.90   P. W. Palmberg, R. E. DeWames, L. A. Vredevoe: Phys. Rev. Lett. **21**, 682 (1968)
7.91   T. Suzuki, N. Hirota, H. Tanaka, H. Watanabe: J. Phys. Soc. Jpn. **30**, 888 (1971)
7.92   R. J. Celotta, D. T. Pierce, G.-C. Wang, S. D. Bader, G. P. Felcher: Phys. Rev. Lett. **43**, 728 (1979)
7.93   D. T. Pierce, R. J. Celotta, J. Unguris, H. Siegmann: Phys. Rev. **B26**, 2566 (1982);
       J. Unguris, D. T. Pierce, R. J. Celotta: Phys. Rev. **B29**, 1381 (1984)
7.94   G. Waller, U. Gradmann: Phys. Rev. **B26**, 6330 (1982)
7.95   S. F. Alvarado, R. Feder, H. Hopster, F. Ciccacci, H. Pleyer: Z. Phys. **B49**, 129 (1982)
7.96   R. Feder, S. F. Alvarado, E. Tamura, E. Kisker: Surf. Sci. **127**, 83 (1983)
7.97   E. Tamura, B. Ackermann, R. Feder: J. Phys. **C17**, 5455 (1984)
7.98   S. F. Alvarado, D. Weller: Verhandl. Deutsche Physikal. Gesellschaft (VI) **19**, 260 (1984)
7.99   E. Tamura, R. Feder: Solid State Commun. **44**, 1101 (1982)
7.100 E. Tamura, R. Feder: Surf. Sci. **139**, L191 (1984)
7.101 J. Kirschner: Phys. Rev. **B30**, 415 (1984)
7.102 C. Rau: Comments Solid State Phys. **9**, 177 (1980)
7.103 D. W. Gidley, A. R. Köymen, T. W. Capehart: Phys. Rev. Lett. **49**, 1779 (1982)

Chapter 8

8.1   V. Bargmann, L. Michel, V. L. Telegdi: Phys. Rev. Lett. **2**, 435 (1959)
8.2   H. J. Meister: Z. Phys. **166**, 468 (1962)
8.3   P. S. Farago: Adv. Electron. Electron Phys. **21**, 1 (1965)
8.4   W. Wübker, R. Möllenkamp, J. Kessler: Phys. Rev. Lett. **49**, 272 (1982)
8.5   M. J. Alguard, J. E. Clendenin, R. D. Ehrlich, V. W. Hughes, J. S. Ladish, M. S. Lubell, K. P. Schüler, G. Baum, W. Raith, R. H. Miller, W. Lysenko: Nucl. Instrum. Meth. **163**, 29 (1979)

8.6   W. Gehenn, R. Haug, M. Wilmers, H. Deichsel: Z. Angew. Phys. **28**, 142 (1969)
8.7   K. Jost, F. Kaussen, J. Kessler: J. Phys. **E14**, 735 (1981)
8.8   K. Schackert: Z. Phys. **213**, 316 (1968)
8.9   T. Heindorff, J. Höfft, E. Reichert: J. Phys. **B6**, 477 (1973)
8.10   J. Kirschner: *Polarized Electrons at Surfaces*, in *Springer Tracts Mod. Phys.*, Vol. 106 (Springer, Berlin, Heidelberg 1985)
8.11   G.-C. Wang, R. J. Celotta, D. T. Pierce: Phys. Rev. **B23**, 1761 (1981)
8.12   J. van Klinken: Nucl. Phys. **75**, 161 (1966);
8.13   H. Boersch, R. Schliepe, K. E. Schriefl: Nucl. Phys. **A163**, 625 (1971)
8.14   R. Raue, H. Hopster, E. Kisker: Rev. Sci. Instrum. **55**, 383 (1984)
8.15   L. A. Hodge, T. J. Moravec, F. B. Dunning, G. K. Walters: Rev. Sci. Instrum. **50**, 5 (1979)
8.16   L. G. Gray, M. W. Hart, F. B. Dunning, G. K. Walters: Rev. Sci. Instrum. **55**, 88 (1984)
8.17   H. C. Siegmann, D. T. Pierce, R. J. Celotta: Phys. Rev. Lett. **46**, 452 (1981);
       R. J. Celotta, D. T. Pierce, H. C. Siegmann, J. Unguris: Appl. Phys. Lett. **38**, 577 (1981)
8.18   M. Erbudak, N. Müller: Appl. Phys. Lett. **38**, 575 (1981)
8.19   M. Erbudak, G. Ravano: J. Appl. Phys. **52**, 5032 (1981)
8.20   D. T. Pierce, S. M. Girvin, J. Unguris, R. J. Celotta: Rev. Sci. Instrum. **52**, 1437 (1981)
8.21   V. W. Hughes, R. L. Long, Jr., M. S. Lubell, M. Posner, W. Raith: Phys. Rev. **A5**, 195 (1972)
8.22   R. J. Celotta, D. T. Pierce: Adv. At. Mol. Phys. **16**, 101 (1980)
8.23   V. D. Ob'edkov, I. Kh. Mossallami: Vestn. Leningr. Univ. **22**, 43 (1971)
8.24   E. Karule: J. Phys. **B5**, 2051 (1972)
8.25   D. M. Campbell, H. M. Brash, P. S. Farago: Phys. Lett. **36A**, 449 (1971)
8.26   R. Krisciokaitis, W. Y. Tsai: Nucl. Instrum. Meth. **83**, 45 (1970);
       R. Krisciokaitis, W. K. Peterson: In *Electronic and Atomic Collisions, Abstracts of Papers, VIII ICPEAC, Belgrade 1973* (Institute of Physics, Belgrade 1973) p. 257
8.27   P. S. Farago: Rep. Prog. Phys. **34**, 1055 (1971)
8.28   G. Baum, U. Koch: Nucl. Instrum. Meth. **71**, 189 (1969)
8.29   P. F. Wainwright, M. J. Alguard, G. Baum, M.S. Lubell: Rev. Sci. Instrum. **49**, 571 (1978)
8.30   R. Möllenkamp, U. Heinzmann: J. Phys. **E15**, 692 (1982)
8.31   W. von Drachenfels, U. T. Koch, Th. M. Müller, W. Paul, H. R. Schaefer: Nucl. Instrum. Meth. **140**, 47 (1977)
8.32   P. J. Keliher, R. E. Gleason, G. K. Walters: Phys. Rev. **A11**, 1279 (1975)
8.33   L. A. Hodge, F. B. Dunning, G. K. Walters: Rev. Sci. Instrum. **50**, 1 (1979)
8.34   L. G. Gray, K. W. Giberson, Chu Cheng, R. S. Keiffer, F. B. Dunning, G. K. Walters: Rev. Sci. Instrum. **54**, 271 (1983)
8.35   P. J. Keliher, F. B. Dunning, M. R. O'Neill, R. D. Rundel, G. K. Walters: Phys. Rev. **A11**, 1271 (1975)
8.36   E. Kisker, G. Baum, A. H. Mahan, W. Raith, B. Reihl: Phys. Rev. **B18**, 2256 (1978)
8.37   E. Garwin, F. Meier, D. T. Pierce, K. Sattler, H. C. Siegmann: Nucl. Instrum. Meth. **120**, 483 (1974)
8.38   D. T. Pierce, R. J. Celotta, G.-C. Wang, W. N. Unertl, A. Galejs, C. E. Kuyatt, S. R. Mielczarek: Rev. Sci. Instrum. **51**, 478 (1980)
8.39   C. K. Sinclair: In *High-Energy Physics with Polarized Beams and Polarized Targets*, ed. by C. Joseph, J. Soffer (Birkhäuser, Basel 1981) p. 27
8.40   P. Souder, A. Barber, W. Bertozzi, G. Cates, G. Dodson, T. J. Gay, M. Goodman, V. W. Hughes, S. Kowalski, M. S. Lubell, A. Magnon, C. P. Sargent, R. Schaefer, W. Turchinetz, R. Wilson: In *High-Energy Spin Physics – 1982*, ed. by G. M. Bunce, AIP Conference Proceedings No. 95 (AIP, New York 1982) p. 574; E. Reichert, ibid. p. 580
8.41   W. Raith: In *Electronic and Atomic Collisions*, ed. by J. Eichler, I. V. Hertel, N. Stolterfoht (North-Holland, Amsterdam 1984) p. 107

8.42 D. Conrath, T. Heindorff, A. Hermanni, N. Ludwig, E. Reichert: Appl. Phys. **20**, 155 (1979)
8.43 E. Reichert, K. Zähringer: Appl. Phys. **A29**, 191 (1982)
8.44 B. Reihl, M. Erbudak, D. M. Campbell: Phys. Rev. **B19**, 6358 (1979)
8.45 J. Kirschner, H. P. Oepen, H. Ibach: Appl. Phys. **A30**, 177 (1983)
8.46 C. Y. Prescott, W. B. Atwood, R. L. A. Cottrell et al.: Phys. Lett. **77B**, 347 (1978); **84B**, 524 (1979)
8.47 C. S. Feigerle, D. T. Pierce, A. Seiler, R. J. Celotta: Appl. Phys. Lett. **44**, 866 (1984)
8.48 D. T. Pierce, R. J. Celotta: In *Optical Orientation*, ed. by F. Meier, B. P. Zakharchenya (North-Holland, Amsterdam 1984) p. 259
8.49 S. F. Alvarado, F. Ciccacci, M. Campagna: Appl. Phys. Lett. **39**, 615 (1981); M. Campagna, S. F. Alvarado: In *High-Energy Spin Physics – 1982*, ed. by G. M. Bunce, AIP Conference Proceedings No. 95 (AIP, New York 1982) p. 566
8.50 F. Ciccacci, S. F. Alvarado, S. Valeri: J. Appl. Phys. **53**, 4395 (1982)
8.51 A. Rich, J. C. Wesley: Rev. Mod. Phys. **44**, 250 (1972)
8.52 T. Kinoshita: Comments At. Mol. Phys. **12**, 227 (1983)
8.53 J. C. Wesley, A. Rich: Phys. Rev. **A4**, 1341 (1971), corrected by S. Granger and G. W. Ford: Phys. Rev. Lett. **28**, 1479 (1972)
8.54 F. H. Combley: Rep. Prog. Phys. **42**, 1889 (1979)
8.55 F. Bloch: Physica **19**, 821 (1953)
8.56 L. D. Landau, E. M. Lifshitz: *Quantum Mechanics* (Pergamon, London 1965)
8.57 G. Gräff, F. G. Major, R. W. H. Roeder, G. Werth: Phys. Rev. Lett. **21**, 340 (1968); G. Gräff, E. Klempt, G. Werth: Z. Phys. **222**, 201 (1969)
8.58 H. G. Dehmelt: Adv. At. Mol. Phys. **3**, 53 (1967); **5**, 109 (1969); F. L. Walls, T. S. Stein: Phys. Rev. Lett. **31**, 975 (1973); D. Wineland, P. Ekstrom, H. Dehmelt: Phys. Rev. Lett. **31**, 1279 (1973)
8.59 R. S. Van Dyck, Jr., P. B. Schwinberg, H. G. Dehmelt: Phys. Rev. Lett. **38**, 310 (1977); H. Dehmelt: In *Atomic Physics 7*, ed. by D. Kleppner, F. Pipkin (Plenum, New York 1981) p. 337; P. B. Schwinberg, R. S. Van Dyck, Jr., H. G. Dehmelt: Phys. Rev. Lett. **47**, 1679 (1981); D. J. Wineland, W. M. Itano: Adv. At. Mol. Phys. **19**, 135 (1983)
8.60 P. S. Cooper, M. J. Alguard, R. D. Ehrlich et al., Phys. Rev. Lett. **42**, 1386 (1979)
8.61 L. V. Knight: Thesis, Stanford University (1965)
8.62 N. S. Craigie, K. Hidaka, M. Jacob, F. M. Renard: Phys. Rep. **99**, 69 (1983)
8.63 J. D. Bjorken: Phys. Rev. **D1**, 1376 (1970); J. D. Bjorken: In *High-Energy Spin Physics – 1982*, ed. by G. M. Bunce, AIP Conference Proceedings No. 95 (AIP, New York 1982) p. 268
8.64 V. W. Hughes, J. Kurti: Ann. Rev. Nucl. Part. Sci. **33**, 611 (1983)
8.65 G. Baum, M. R. Bergström, P. R. Bolton et al., Phys. Rev. Lett. **51**, 1135 (1983)
8.66 M. J. Alguard, W. W. Ash, G. Baum et al., Phys. Rev. Lett. **37**, 1261 (1976); **41**, 70 (1978)
8.67 M. J. Alguard, W. W. Ash, G. Baum et al., Phys. Rev. Lett. **37**, 1258 (1976)
8.68 P. S. Cooper, M. J. Alguard, R. D. Ehrlich et al., Phys. Rev. Lett. **34**, 1589 (1975)
8.69 W. Eckstein: Z. Phys. **203**, 59 (1967)
8.70 R. Loth: Z. Phys. **203**, 66 (1967)
8.71 M. Erbudak, G. Ravano: Phys. Lett. **91A**, 367 (1982)
8.72 J. Unguris, D. T. Pierce, A. Galejs, R. J. Celotta: Phys. Rev. Lett. **49**, 72 (1982)
8.73 K. Koike, K. Hayakawa: Jpn. J. Appl. Phys. **23**, L187 (1984); Appl. Phys. Lett. **45**, 585 (1984)
8.74 R. J. Celotta, D. T. Pierce: In *Microbeam Analysis*, ed. by K. F. J. Heinrich (San Francisco Press, San Francisco 1982) p. 469
8.75 W. Hilgner, J. Kessler: Z. Phys. **221**, 305 (1969)
8.76 D. W. Walker: Phys. Rev. Lett. **20**, 827 (1968)

8.77   A. C. Yates: Phys. Rev. Lett. **20**, 829 (1968); Phys. Rev. **176**, 173 (1968)
8.78   W. Hilgner, J. Kessler, E. Steeb: Z. Phys. **221**, 324 (1969)
8.79   J. Kessler, J. Lorenz, H. Rempp, W. Bühring: Z. Phys. **246**, 348 (1971)
8.80   W. Thiemann: Naturwiss. **61**, 476 (1974)
8.81   L. Keszthelyi: Origins Life **11**, 9 (1981)
8.82   W. A. Bonner, M. A. van Dort, M. R. Yearian: Nature **258**, 419 (1975)
8.83   L. A. Hodge, F. B. Dunning, G. K. Walters, R. H. White, G. J. Schroepfer: Nature **280**, 250 (1979)
8.84   Ya. B. Zel'dovich, D. B. Saakyan: Sov. Phys. – JETP **51**, 1118 (1980)
8.85   A. Rich, J. Van House, R. A. Hegstrom: Phys. Rev. Lett. **48**, 1341 (1982);
       D. W. Gidley, A. Rich, J. Van House, P. W. Zitzewitz: Nature **297**, 639 (1982);
       R. A. Hegstrom: Nature **297**, 643 (1982)
8.86   P. S. Farago: J. Phys. **B13**, L 567 (1980)
8.87   M. J. M. Beerlage, P. S. Farago, M. J. Van der Wiel: J. Phys. **B14**, 3245 (1981)
8.88   P. S. Farago: J. Phys. **B14**, L743 (1981)
8.89   N. A. Cherepkov: J. Phys. **B16**, 1543 (1983)
8.90   W. A. Onishchuk: Sov. Phys. – JETP **55**, 412 (1982)
8.91   T. T. Gien: J. Phys. **B15**, 4617 (1982)
8.92   N. F. Mott, H. S. W. Massey: *The Theory of Atomic Collisions* (Clarendon, Oxford 1965) Chap. IX

Subject Index